G. Herbert Vogel
Process Development

G. Herbert Vogel

Process Development

From the Initial Idea to the
Chemical Production Plant

Prof. Dr. G. Herbert Vogel
TU Darmstadt
Ernst Berl Institute of
Chemical Engineering and
Macromolecular Science
Petersenstraße 20
64287 Darmstadt
Germany

■ All books published by Wiley-VCH are carefully produced. Nevertheless, authors, editors, and publisher do not warrant the information contained in these books, including this book, to be free of errors. Readers are advised to keep in mind that statements, data, illustrations, procedural details or other items may inadvertently be inaccurate.

Library of Congress Card No.: applied for

British Library Cataloguing-in-Publication Data:
A catalogue record for this book is available from the British Library.

Bibliographic information published by Die Deutsche Bibliothek
Die Deutsche Bibliothek lists this publication in the Deutsche Nationalbibliografie; detailed bibliographic data is available in the Internet at <http://dnb.ddb.de>

© 2005 Wiley-VCH Verlag GmbH & Co. KGaA, Weinheim

All rights reserved (including those of translation into other languages). No part of this book may be reproduced in any form – by photoprinting, microfilm, or any other means – nor transmitted or translated into a machine language without written permission from the publishers. Registered names, trademarks, etc. used in this book, even when not specifically marked as such, are not to be considered unprotected by law.

printed in the Federal Republic of Germany
printed on acid-free paper.

Composition   Mitterweger & Partner Kommunikationsgesellschaft mbH, Plankstadt
Printing   betz-druck GmbH, Darmstadt
Bookbinding   Litges & Dopf Buchbinderei GmbH, Heppenheim

ISBN-13: 978-3-527-31089-0
ISBN-10: 3-527-31089-4

For my children
Birke, Karl, Anke and *Till*

Preface

The idea behind this book is to facilitate the transition of academics from university to the chemical industry. It will enable not only all bachelor and doctoral graduates in natural sciences and engineering, but also economics students, who will shortly enter the chemical industry to make a smooth start to their careers and to hold competent discussions with experienced industrial chemists and engineers.

This book is intended both for experienced workers in the industry who are looking for a concise reference work as a guideline for problem solving, and for students studying technical chemistry as an aid to revision.

Chapters 3 to 6 describe the development and evaluation of production processes and the execution of projects from the chemist's viewpoint. The various aspects chemical, engineering, materials science, legal, economic, safety, etc. that must be taken into account prior to and during the planning, erection, and startup of a chemical plant are treated.

The necessary basic knowledge is provided in Chapter 2: The Chemical Production Plant and its Components. It deals with important subdisciplines of technical chemistry such as catalysis, chemical reaction engineering, separation processes, hydrodynamics, materials and energy logistics, measurement and control technology, plant safety, and materials selection. Thus, it acts as a concise textbook within the book that saves the reader from consulting other works when such information is required. A comprehensive appendix (mathematical formulas, conversion factors, thermodynamic data, material data, regulations, etc.) is also provided.

This book is based on the industrial experience that I gathered at BASF AG, Ludwigshafen from 1982 to 1993 in the development, planning, construction, and startup of petrochemical production plants. Therefore, the choice of topics and the approach are necessarily subjective. Here, I am especially grateful to my mentors in these subjects, Dr. Gerd Dümbgen and Dr. Fritz Thiessen.

This book could not have been realized without the help of Dr.-Ing. Gerd Kaibel (BASF AG, Ludwigshafen), who made available his extensive industrial experience and provided Chapter 2.3 on thermal and mechanical separation processes.

I thank Prof. Dr. Wilfried J. Petzny (formerly EC Erdölchemie, Cologne) for checking the manuscript, and for constructive criticism and comments, which have gratefully been incorporated.

Finally, I am grateful to Dieter Böttiger (TU Darmstadt) for preparing numerous figures, and my eldest daughter Birke Vogel for proofreading the manuscript. Any errors in the book are entirely my responsibility.

Darmstadt and Ludwigshafen, February 2005 G. Herbert Vogel

Contents

1	**Introduction** *1*	
1.1	The Goal of Industrial Research and Development *3*	
1.2	The Production Structure of the Chemical Industry *4*	
1.3	The Task of Process Development *11*	
1.4	Creative Thinking *12*	
2	**The Chemical Production Plant and its Components** *13*	
2.1	The Catalyst *16*	
2.1.1	Catalyst Performance *20*	
2.1.1.1	Selectivity *20*	
2.1.1.2	Activity *20*	
2.1.1.3	Lifetime *22*	
2.1.1.4	Mechanical Strength *25*	
2.1.1.5	Production Costs *25*	
2.1.2	Characterization of Catalysts *28*	
2.1.2.1	Chemical Composition *28*	
2.1.2.2	Nature of the Support Material *29*	
2.1.2.3	Promoters *29*	
2.1.2.4	Phase Composition *29*	
2.1.2.5	Particle Size *30*	
2.1.2.6	Pore Structure *30*	
2.1.2.7	Surface Structure *30*	
2.1.2.8	Byproducts in the Feed *31*	
2.1.3	Kinetics of Heterogeneous Catalysis *31*	
2.1.3.1	Film Diffusion *32*	
2.1.3.2	Pore Diffusion *35*	
2.1.3.3	Sorption *38*	
2.1.3.4	Surface Reactions *42*	
2.1.3.5	Pore Diffusion and Chemical Reaction *45*	
2.1.3.6	Film Diffusion and Chemical Reaction *50*	
2.2	The Reactor *51*	
2.2.1	Fundamentals of Chemical Reaction Technology *52*	
2.2.1.1	Ideal Reactors *56*	
2.2.1.2	Reactors with Real Behavior *60*	

Process Development. From the Initial Idea to the Chemical Production Plant. G. Herbert Vogel
Copyright © 2005 WILEY-VCH Verlag GmbH & Co. KGaA, Weinheim
ISBN: 3-527-31089-4

2.2.1.3	Nonisothermal reactors 68
2.2.1.4	Design of Reactors 74
2.3	Product Processing (Thermal and Mechanical Separation Processes) 80
2.3.1	Heat Transfer, Evaporation, and Condensation 80
2.3.1.1	Fundamentals 80
2.3.1.2	Dimensioning 86
2.3.2	Distillation, Rectification 94
2.3.2.1	Fundamentals of Gas–Liquid Equilibria 94
2.3.2.2	One-Stage Evaporation 99
2.3.2.3	Multistage Evaporation (Rectification) 102
2.3.2.4	Design of Distillation Plants 108
2.3.4.5	Special Distillation Processes 132
2.3.3	Absorption and Desorption, Stripping, Vapor-Entrainment Distillation 136
2.3.3.1	Fundamentals 136
2.3.3.2	Dimensioning 137
2.3.3.3	Desorption 142
2.3.3.4	Vapor-Entrainment Distillation 142
2.3.4	Extraction 143
2.3.4.1	Fundamentals 144
2.3.4.2	Dimensioning 146
2.3.4.3	Apparatus 154
2.3.5	Crystallization 155
2.3.5.1	Fundamentals 156
2.3.5.2	Solution Crystallization 159
2.3.5.3	Melt Crystallization 161
2.3.5.4	Dimensioning 164
2.3.6	Adsorption, Chemisorption 165
2.3.7	Ion Exchange 167
2.3.8	Drying 167
2.3.9	Special Processes for Fluid Phases 169
2.3.10	Mechanical Processes 170
2.4	Pipelines, Pumps, and Compressors 172
2.4.1	Fundamentals of Hydrodynamics 172
2.4.2	One-phase Flow in Pipelines 174
2.4.3	Pumps 179
2.4.4	Compressors 184
2.5	Energy Supply 186
2.5.1	Steam and Condensate System 186
2.5.2	Electrical Energy 188
2.5.3	Cooling Water 188
2.5.4	Refrigeration 189
2.5.5	Compressed Air 189
2.6	Product Supply and Storage 190

2.7	Waste Disposal [Rothert 1992]	*192*
2.7.1	Off-Gas Collection System and Flares	*192*
2.7.2	Combustion Plants for Gaseous and Liquid Residues	*192*
2.7.3	Special Processes for Off-Gas Purification	*194*
2.7.4	Wastewater Purification and Disposal	*197*
2.7.4.1	Clarification Plant	*197*
2.7.4.2	Special Processes for Wastewater Purification	*199*
2.7.5	Slop System	*201*
2.8	Measurement and Control Technology	*202*
2.8.1	Metrology	*202*
2.8.1.1	Temperature Measurement	*202*
2.8.1.2	Pressure Measurement	*205*
2.8.1.3	Measuring Level	*205*
2.8.1.4	Flow Measurement	*206*
2.8.2	Control Technology	*211*
2.8.3	Control Technology	*218*
2.9	Plant Safety	*220*
2.10	Materials Selection	*224*
2.10.1	Important Materials and their Properties	*226*
2.10.1.1	Mechanical Properties and Thermal Stability	*228*
2.10.1.2	Corrosion Behavior	*229*
2.10.2	Metallic Materials	*233*
2.10.3	Nonmetallic Materials	*234*
3	**Process Data**	*237*
3.1	Chemical Data	*239*
3.1.1	Heat of Reaction	*239*
3.1.2	Thermodynamic Equilibrium	*240*
3.1.3	Kinetics	*245*
3.1.4	Selectivity and Conversion as a Function of the Process Parameters	*257*
3.2	Mass Balance	*261*
3.3	Physicochemical Data	*263*
3.3.1	Physicochemical Data of Pure Substances	*264*
3.3.2	Data for Mixtures	*265*
3.4	Processing	*267*
3.5	Patenting and Licensing Situation	*267*
3.6	Development Costs	*270*
3.7	Location	*271*
3.8	Market Situation	*272*
3.9	Raw Materials	*273*
3.10	Plant Capacity	*276*
3.11	Waste-Disposal Situation	*277*
3.12	End Product	*278*

4	**Course of Process Development** 279	
4.1	Process Development as an Iterative Process 281	
4.2	Drawing up an Initial Version of the Process 284	
4.2.1.	Tools used in Drawing up the Initial Version of the Process 287	
4.2.1.1	Data Banks 287	
4.2.1.2	Simulation Programs 288	
4.2.1.3	Expert Systems 292	
4.3.	Checking the Individual Steps 294	
4.4	The Microplant: The Link between the Laboratory and the Pilot Plant 296	
4.5	Testing the Entire Process on a Small Scale 297	
4.5.1	Miniplant Technology 297	
4.5.1.1	Introduction 297	
4.5.1.2	Construction 298	
4.5.1.3	The Limits of Miniaturization 300	
4.5.1.4	Limitations of the Miniplant Technology 302	
4.5.2	Pilot Plant 302	

5	**Planning, Erection, and Start-Up of a Chemical Plant** 305	
5.1	General Course of Project Execution 307	
5.2	Important Aspects of Project Execution 314	
5.2.1	Licensing 314	
5.2.2	Safety Studies 317	
5.2.3	German Industrial Accident Regulation (Störfallverordnung) 320	
5.2.4	P&I Flow Sheets 321	
5.2.5	Function Plans 322	
5.2.6	Technical Data Sheets 323	
5.2.7	Construction of Models 323	
5.2.8	Preparation of Other Documents 325	
5.3	Commissioning 326	
5.4	Start-Up 327	

6	**Process Evaluation** 329	
6.1	Preparation of Study Reports 331	
6.1.1	Summary 332	
6.1.2	Basic Flow Diagram 332	
6.1.3	Process Description and Flow Diagram 332	
6.1.4.	Waste-Disposal Flow Diagram 334	
6.1.5.1	Introduction 335	
6.1.5.2	ISBL Investment Costs 336	
6.1.5.3	OSBL Investment Costs 339	
6.1.5.4	Infrastructure Costs 339	
6.1.6	Calculation of Production Costs 339	
6.1.6.1	Feedstock Costs 340	
6.1.6.2.	Energy Costs 341	
6.1.6.3	Waste-Disposal Costs 347	

6.1.6.4	Staff Costs	*348*
6.1.6.5	Maintenance Costs	*349*
6.1.6.6	Overheads	*349*
6.1.6.7	Capital-Dependent Costs (Depreciation)	*349*
6.1.7	Technology Evaluation	*350*
6.1.8	Measures for Improving Technical Reliability	*352*
6.1.9	Assessment of the Experimental Work	*357*
6.2	Return on Investment	*358*
6.2.1	Static Return on Investment	*358*
6.2.2	Dynamic Return on Investment	*360*
6.3	Economic Risk	*361*
6.3.1	Sensitivity Analysis	*361*
6.3.2	Amortization Time	*363*
6.3.3	Cash Flow	*363*
7	**Trends in Process Development**	*365*
8	**Appendix**	*371*
8.1	Mathematical formulas	*373*
8.2	Constants	*382*
8.3	List of elements with relative atomic masses bonding radius and melting and boiling points	*382*
8.4	Conversion of various units to SI units	*385*
8.5	Relationships between derived and base units	*390*
8.6	Conversion of concentrations for binary mixtures of dissolved component A in solvent B	*390*
8.7	van der Waals constants a and b and critical values for some gases	*391*
8.8	Heat capacities of some substances and their temperature dependance	*392*
8.9	Thermodynamic data of selected organic compounds	*393*
8.10	Order of magnitude of the reaction enthalpy $\Delta_R H$ for selected industrial reactions [Weissermel 1994]	*394*
8.11	Antoine parameters of selected organic compounds	*397*
8.12	Properties of water	*399*
8.12.1	Formulas for calculating the physicochemical properties of water between 0 or 150 °C (T in °C, P in bar)	*399*
8.12.2	Properties of water ρ = density, c_p = heat capacity, a = thermal expansion coefficient, λ = thermal conductivity, η = viscosity coefficient	*402*
8.12.3	Density ρ / kg m^{-3} of water at different temperatures and pressures	*404*
8.12.4	Specific heat capacity c_p/kJ kg^{-1} K^{-1} of water at different temperatures and pressures	*405*
8.12.5	Dynamic viscosity $\eta/10^{-6}$ kg m^{-1} s^{-1} of water at different temperatures and pressures	*406*
8.12.6	Self-diffusion coefficient D/m^2 s^{-1} of water at different temperatures and pressures.	*407*

8.12.7	Thermal expansion coefficient $\beta/10^{-3}$ K of water at different temperatures and pressures	408
8.12.8	Thermal conductivity $\lambda/10^{-3}$ W m^{-1} K^{-1} of water at different temperatures and pressures	409
8.12.9	Negative base ten logarithm of the ionic product of water pK_W/mol^2 kg^{-2} at different temperatures	410
8.12.10	Relative static dielectric constant ε_r of water as a function of pressure and temperature	410
8.13	Properties of dry air (molar mass: $M = 28.966$ g mol^{-1})	411
8.13.1	Real gas factor $r = pV/RT$ of dry air at different temperatures an pressures	412
8.13.2	Specific heat capacity c_p in kJ kg^{-1} K^{-1} of dry air at different temperatures and pressures	412
8.13.3	Dynamic viscosity $\eta/10^{-3}$ mPa s of dry air at different temperatures and pressures	413
8.13.4	Thermal conductivity λ/W m^{-1} K^{-1} of dry air at different temperatures and pressures	413
8.14	Dimensionless characteristic numbers	414
8.15	Important German regulations for handling of substances	416
8.16	Hazard and safety warnings	416
8.17	The 25 largest companies of the world in 2000	420
8.18	The 25 largest companies in Germany in 2000	421
8.19	Surface analysis methods	422
9	**References**	*425*
	Subject Index	*465*

1
Introduction

1.1
The Goal of Industrial Research and Development

In the chemical industry (Figure 1-1) about 7% of turnover is spent on research and development [Jarhbuch 1991, VCI 2000, VCI 2001] (Table 1.1 and Appendix 8.16). This sum is of the same order of magnitude as the company profit or capital investment. The goal of research management is to use these resources to achieve competitive advantages [Meyer-Galow 2000]. After all, the market has changed, from a national sellers' market (demand > supply) to a world market with ever-increasing competition. This in turn has affected the structure of the major chemical companies. In the 1990s integrated, highly diversified companies (e.g., Hoechst, ICI, Rhone-Poulenc) developed into specialists for bulk chemicals (Dow/UCC, Celanese), fine and specialty chemicals (Clariant, Ciba SC), and agrochemical and pharmaceutical formulations (Aventis, Novartis) [Felcht 2000, Perlitz 2000].

Unlike consumer goods such as cars and clothes, most commercial chemical products are "faceless" (e.g., hydrochloric acid, polyethylene), and as a rule the customer is therefore only interested in sales incentives such as price, quality, and availability. All the research activities of an industrial enterprise must therefore ultimately boil down to three basic competitive advantages, namely, being *cheaper* and/or *better* and/or *faster* than the competitor. The AND combination offers the greatest competive advantage and is thus known as the world-champion strategy. However, more often one must settle for the OR combination. The qualitive term *cheaper* can be quantified by means of a *production cost analysis*. Initially, it is sufficient to examine the coarse structure of the production costs. Thus, each item in Table 1-2 can be analysed individually and the

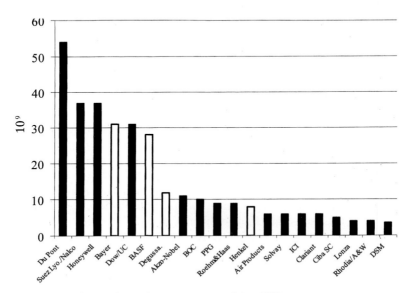

Fig. 1-1 Market capital of major chemical companies [Mayer-Galow 2000].

Process Development. From the Initial Idea to the Chemical Production Plant. G. Herbert Vogel
Copyright © 2005 WILEY-VCH Verlag GmbH & Co. KGaA, Weinheim
ISBN: 3-527-31089-4

Tab. 1-1 Growth of the German chemical industry [VCI 2001].

	1990	1995	1996	1997	1998	1999	2000
Turnover (10^9 EURO)	83.5	92.1	89.5	96.6	95.8	97.1	108.6
Employees in thousands	592	536	518	501	485	478	470
Investment in material goods	6.5	5.8	6.4	6.4	6.9	6.9	7.2
R&D experditure (10^9 EURO)	5.4	5.3	5.8	6.1	7.0	7.3	7.9

Tab. 1-2 Coarse structure of production costs.

Materials costs	
Energy costs	variable costs (production-dependent)
Disposol costs	
Personnel costs	
Workshop costs	
Depreciation	fixed costs (production-independent)
Other costs	
Σ Production costs	

total system optimized. The competive advantage *better* now refers not only to availability and product quality, but also to the environmental compatibility of the process [Gärtner 2000], and the quality assurance concept, delivery time, and exclusivity of the supplier, etc.

1.2
The Production Structure of the Chemical Industry

If the production structure of the chemical industry is examined [Petrochemie 1990, BASF 1999, Petzny 1999], it is seen that there are only a few hundred major basic products and intermediates that are produced on a scale of at least a few thousand to several million tonnes per annum worldwide. This relatively small group of key products, which are in turn produced from only about ten raw materials, are the stable foundation on which the many branches of refining chemistry (dyes, pharmaceuticals, etc.), with their many thousands of often only short-lived end products, are based [Amecke 1987]. This has resulted in the well-known chemical family tree (Figure 1-2), which can also be regarded as being synonymous with an intelligent integrated production system, with synergies that are often of critical importance for success.

A special characteristic of the major basic products and intermediates is their longevity. They are statistically so well protected by their large number of secondary products

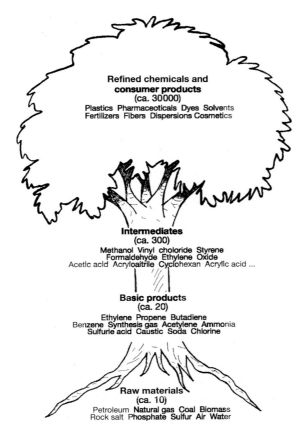

Fig. 1-2 Product family tree of the chemical industry: starting from raw materials and progressing through the basic products and intermediates, to the refined chemicals and final consumer products, as well as specialty chemicals and materials [Quadbeck 1990, Jentzsch 1990, Chemie Manager 1998, Raichle 2001].

and their wide range of possible uses that they are hardly affected by the continuous changes in the range of products on sale. Unlike many end products, which are replaced by better ones in the course of time, they do not themselves have a life cycle. However, the processes for producing them are subject to change. This is initiated by new technical possibilities and advances opened up by research, but is also dictated by the current raw material situation (Figure 1.3, Table 1.3).

In the longer term, an oil shortage can be expected in 40 to 50 years, and this will result in increased use of natural gas. The fossil fuel with the longest future is coal, with reserves for more than 500 years. The question whether natural gas reserves in the form of methane hydrate, in which more carbon is stored than in other fossil raw materials, will be recoverable in the future cannot be answered at present, since these lie in geographically unfavorable areas (permafrost regions, continental shelves of the oceans, deep sea).

1 Introduction

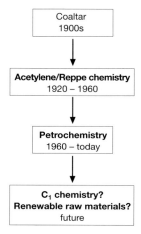

Fig. 1-3 How the raw material base of the chemical industry has changed with time [Graeser 1995, Petzny 1997, Plotkin 1999, Van Heek 1999].

In the case of *basic products* and *intermediates* it is not the individual chemical product but the production process or technology which has a life cycle. For example, Figure 1-4 shows the life cycles of the acrylic acid and ethylene oxide processes [Jentzsch 1990, Ozero 1984]. To remain competetive here the producer must be the price leader for his process. Therefore, strategic factors for success are [Felcht 2000]:

- Efficient process technology
- Exploiting economy of scale by means of world-scale plants

Tab. 1-3 World production (in 10^6 t/a) of the most important energy and raw materials sources.

	1994	1997
Fossil raw materials		
Coal	3568	3834
Oil	3200	3475
Lignite	950	914
Natural gas [10^9 m^3]	2162*)	2300*)
Renewable raw materials		
Cereals	1946	1983
Potatoes	275	295
Pulses	57	55
Meat	199	221
Sugar	111	124
Fats (animal and plant)	/	ca. 100

*) 1 t SKE (German coal unit) = 882 m^3 Natural gas = 0.7 t oil equivalent = 29.3×10^6 kJ

- Employing a flexible integrated system at the production site
- Professional logistics for large product streams.

The demands made on process development for *fine chemicals* differ considerably from those of basic products and intermediates (Figures 1-5 and 1-6). In addition to the

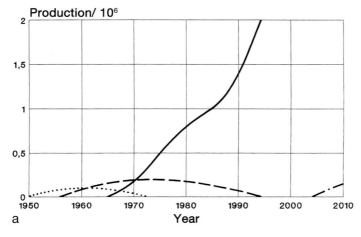

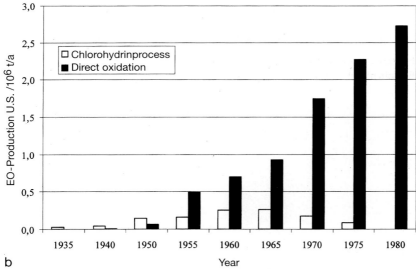

Fig. 1-4 Life cycles of production processes
a) Acrylic acid processes
.... Cyanohydrin and propiolactone processes
Reppe process
– – – – – Heterogeneously catalyzed propylene oxidation
(2000: 3.456×10^6 t, 2003 (estimated): 4.8×10^6 t [Vogel 2001])
–·–·–·– New process ?
b) Ethylene oxide process.

above-mentione boundary conditions of better and/or cheaper, *time to market* (production of the product at the right time for a limited period) and *focused R & D effort* are of importance here. Only a few fine chemicals, such as vanillin, menthol, and ibuprofen, reach the scale of production and lifetime of bulk chemicals. Futher strategic factors for success are [Felcht 2000]:

- Strategic development partnerships with important customers
- The potential to develop complex multistep organic syntheses
- A broad technology portfolio for the decisive synthetic methods
- Certified pilot and production plants
- Repuatation as a competent and reliable supplier.

Specialty chemicals are complex mixtures whose value lies in the synergistic action of their ingredients. Here the application technology is decisive for market success. The manufacturer can no longer produce all ingredients, which can lead to a certain state of dependence. Strategic factors for successful manufacturers are [Felcht 2000, Willers 2000]:

- Good market knowledge of customer requirements.
- A portfolio containing numerous *magic ingredients*
- Good technical understanding of the customer systems
- Technological breadth and flexibility.

Active substances such as pharmaceuticals and agrochemicals can only be economically marketed while they are under patent protection, before suppliers of generic products enter the market. Therefore, producers of such products cannot simply concentrate on costly research. As soon as possible after clinical trials and marketing approval, worldwide sales of the product must begin so that the remaining patent time can be used for

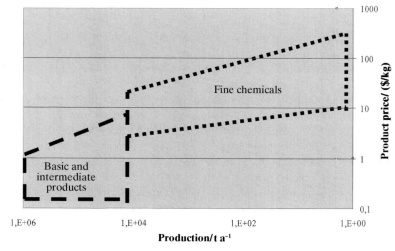

Fig. 1-5 Order of magnitude of product prices as a function of production volume for basic products and intermediates and for fine chemicals [Metivier 2000].

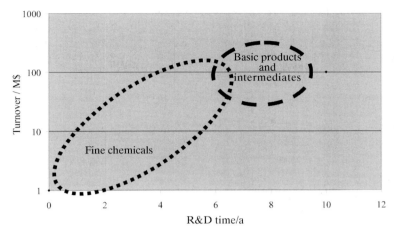

Fig. 1-6 Comparison between bulk and fine chemicals with regard to turnover and the development time of the corresponding process [Metivier 2000].

gaining customers. In contrast, the actual chemical production of the active substance is of only background importance. The required precursors can be purchased from suppliers, and the production of the active substance can be farmed out to other companies. Strategic factors for success of active substance manufacturers are [Felcht 2000]:

- Research into the biomolecular causes of disease and search for targets for pharmacological activities
- Efficient development of active substances (high-throughput screening, searching for and optimizing lead structures, clinical development)
- Patent protection
- High-performance market organization.

Enterprises which already have competitive advantages must take account of the *technology S curve* [Marchetti, 1982, Marquadt 1999] in their research and development strategy (Figure 1-7). The curve shows that as the research and development expenditure on a given technology increases, the productivity of this expenditure decreases with time [Krubasik 1984]. If enterprises are approaching the limits of a given technology, they must accept disproportionately high research and development expenditure, with the result that the contribution made by these efforts to the research objectives of *cheaper* and/or *better* becomes increasingly small, thereby always giving the competitor the opportunity of catching up on the technical advantage. On the other hand, it is difficult for a newcomer to penetrate an established market. But, as Japanese and Korean companies have shown in the past, it is not impossible. Figure 1-8 shows the so-called learning curve for a particular chemical process. It is a double-logarithmic plot (power law $y = x^n$) of the production cost as a function of the cumulative production quantity, which can be regarded as a measure for experience with the process. With increasing experience the production costs of a particular product drop. How-

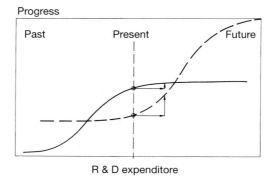

Fig. 1-7 The technology S curve [Specht 1988, Blumenberg 1994]: The productivity of the research and development expenditure increases considerably on switching from basic technology (——) to a new trend-setting technology (– – –).

ever, an overseas competitor who can manufacture the same product in a new plant with considerably lower initial cost can catch up with the inland producer who has produced 10×10^6 t after only about 100 000 t of production experience (Figure 1-8), and can then produce more cheaply.

Once an enterprise has reached the upper region of the product or technology S curve, the question arises whether it is necessary to switch from the standard technology to a new *pace-setting technology* in order to gain a new and sufficient competitive advantage [Perlitz 1985, Bönecke 2000]. Figure 1-7 depicts this switch to a new technology schematically and shows that on switching from a basic technology to a new pace-setting technology, the productivity of the research and development sector increases appreciably, and substantial competitive advantages can thus be achieved [Miller 1987, Wagemann 1997].

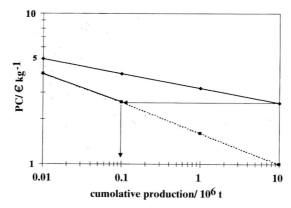

Fig. 1-8 Learning curve: production costs (PC) as a function of cumulative production, which can be regarded as measure of experience with the process, in a double-logarithmic plot [Semel 1997]: diamonds: inland producer, squares: overseas competitor (for discussion, see text).

The potential of old technologies for the development of cheaper and/or better is only small, whereas new technologies have major potential for achieving competitive advantages. It is precisely on this innovative activity that the prosperity of highly developed countries with limited raw material sources such as Germany and Japan is based, since research represents an investment in the future with calculable risks [Mittelstraß 1994], whereas capital investments in the present are based on existing technology.

To assess whether a research and development strategy of better and/or cheaper is still acceptable in the long term for a given product or production process, the R & D management must develop an *early warning system* [Collin 1986, Jahrbuch 1991, Steinbach 1999, Fild 2001] that determines the optimum time for switching to a new product or a new technology [Porter 1980, Porter 1985]. Here it is decisive to have as much up-to-date infomation on competitors as possible. This information can be obtained not only from the patent literature but also from external lectures, conferences, company publications, and publicly accessible documents submitted to the authorities by competitors (Section 3.5). Since industrial research is very expensive, instruments for controlling the research budget are required [Christ 2000, Börnecke 2000, Kraus 2001], for example:

- A cost/benefit analysis for a particular product area, whereby the benefit is determined by the corresponding user company sector.
- A portfolio analysis (Section 3.8) to answer the questions:
  - Where are we now?
  - Where do we want to be in 5 or 10 years?
  - What do we have to do to now to get there?
- An ABC analysis for controlling the R&D resources, based on the rule of thumb that
  - 20% of all products account for 80% of turnover, or
  - 20% of all new developments acccount for 80% of the development costs.

It is therefore important to recognize which 20% these are in order to set the appropriate priorities (A = important, profitable, high chance of success; B = low profitability; C = less important tasks with low profitability).

The way in which chemical companies organize their research varies and depends on the product portfolio [Harrer 1999, Eidt 1997]. Mostly it involves a mixture of the two extremes: pure centralized research on the one hand, and decentralized research (research exclusivelyin the company sectors) on the other [Hänny 1984].

1.3
The Task of Process Development

The task of process development is to extrapolate a chemical reaction discovered and researched in the laboratory to an industrial scale, taking into consideration the economic, safety, ecological, and juristic boundary conditions [Harnisch 1984, Semel 1997, Kussi 2000]. The starting point is the laboratory apparatus, and the outcome of development is the production plant; in between, process development is required. The following account shows how this task is generally handled. Although the sequence of steps in the development process described is typical, it is by no means obligatory, and it is only possible to outline the basic framework.

1.4
Creative Thinking

Numerous methods for creative thinking are described in the literature [Schlicksupp 1977, Börnecke 2000]. In the daily routine of work there simply is no time for important things such as coming up with ideas for new processes and products. Therefore, every year plans shoud be made in advance for:

- Visiting conferences, including ones that are outside of one's own specialist area
- Visiting research establishments (institutes, universities, etc.)
- Excursions to companies
- Regular discussions with planners and heads of department.

Here intensive discussions can lead to new ideas that can later be evaluated. Regular browsing in the literature can also be a source of inspiration.

Tab. 1-4 Creative methods for generating ideas [Schlicksupp 1977, Bornecke 2000, Kraus 2001].

Method group	Characteristics	Important representatives
Brainstorming its variations	Uninhibited discussion in which criticism is not permitted; fantastic ideas and spontaneous associations should be expressed	• Brainstorming • Discussion 66
Brainwriting Methods	Spontaneous writing down of ideas on forms or sheets; circulation of forms	• Features 635 • Brainwriting-pool • Ideas Delphi
Methods creative orientation	Following certain principles in the search for a solution	• Heuristic principles • Bionics
Methods of creative confrontation	Simulation of solution finding by confrontation with meaning contents that apparently have no correction to the problem	• Synectics • BBB methods • Semantic intuition
Methods of systematic structuring	Splitting the problem into partial problems; solving partial problems and combining to give a total solution; systematization of possible solutions	• Morphological box • Morphological Taublau? • Sequential morphology • Problem solving tree
Methods of systematic problem specification	Revealing the core questions of a problem by systematically and hierarchically structured approach	• Progressive abstraction • K-J-methods • Hypothese s matrix • Relevance tree

2
The Chemical Production Plant and its Components

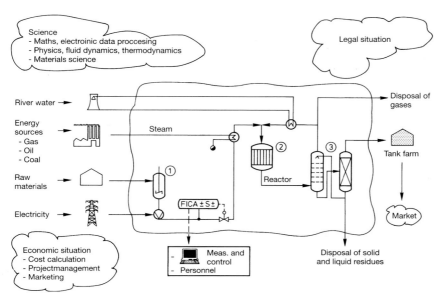

Fig. 2-1 Principle structure of a chemical plant. Around the production plant itself with starting material preparation, the reactor, and product workup are a series of other auxilliary units without which operation would not be possible.

Like a living organism, the total chemical plant is more than the sum of its individual units (organs in the former, units in the latter) [GVC VDI 1997]. A properly functioning chemical plant requires harmonic cooperation of all units [Sapre 1995]. Figure 2-1 shows the most important units of a chemical plant. Since more than 85% of all reactions carried out industrially today require a catalyst [Romanov 1999], the *catalyst* can be regarded as the true *core of the plant* [Misono 1999]. The development of the chemical industry is largely detetined by the development and introduction of new cataytic processes. In 1995 the market value of all catalyts worldwide was $6 \cdot 10^9$ \$ (polymerization 36%, production of chemicals 26%, oil processing 22%, emission control 16%) [Quadbeck 1997, Felcht 2001, Senkan 2001].

The chemical reaction that takes place at the active site of the catalyst determines the *design of the reactor* [Bartholomew 1994]. The reactor in turn determines the pretreatment of the starting material (size reduction, dissolution, mixing, filtration, sieving, etc.) and the *product workup* (rectification, extraction, crystallization, filtration, drying, etc.). From this follows the required *infrastructure*, including waste disposal, tank farms, energy supply, safety devices, and so on. Because of the pyramid structure shown in Figure 2-2, planning errors vary in their severity. If a catalyst behaves only slightly differently in the plant than in the R&D phase (e.g., activity, selectivity, lifetime, mechanical stability), then this will have dramatic effects on the entire plant, which may even have to be scrapped. Planning errors made further up in

Process Development. From the Initial Idea to the Chemical Production Plant. G. Herbert Vogel
Copyright © 2005 WILEY-VCH Verlag GmbH & Co. KGaA, Weinheim
ISBN: 3-527-31089-4

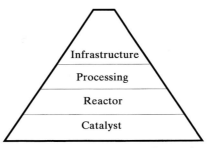

Fig. 2-2 Pyramid structure of process development. Without a functional catalyst as basis, process development is pointless [Sapre 1995].

the pyramid can mostly be eliminated by retrofitting of apparatus. An integrated process design, as is described in Chapter 4, only makes sense if the performance of the catalyst is essentially fixed. Because of the importance of catalysis for process development, the process engineer must have sufficient knowledge in this area to evaluate the state of catalyst development [Bisio 1997, Armor 1996]. Therefore, the important fundamentals of catalysis [Ertl 1997] are discussed in the following section.

2.1
The Catalyst

Not only must a catalyst accelerate a chemical reaction (*activity*), it must also direct it towards the desired product (*selectivity*).

Johann W. Döbereiner (1780–1849) was the first to discover the catalytic action of the noble metal platinum on a hydrogen/oxygen mixture and to exploit it commercially (Döbereiner's lighter), without being aware of the term "catalysis". Instead, he referred to the "effect of contact" or "contact processes". Ten years later Jacob J. Berzelius was the first to use the term "catalyst" and stated [Schwenk 2000]:

"*The catalytic force seems to lie in the fact that certain bodies through their mere presence bring to life relationships that would otherwise only slumber at these temperatures...*"
"*We have ground to believe that in living plants and animals thousands of catalytic processes take place between the fluids and tissues.*"

According to the definition by Wilhelm Ostwald (1853–1932), which is still valid today, a catalyst is a substance that changes the rate of a chemical reaction without appearing in the end product [Ertl 1994, Fehlings 1999]. This does not necessarily mean that the catayst remains unchanged. The generally desired increase in the rate of the reaction is usually associated with a lowering of the activation energy of the rate-determining step (Figure 2.1-1).

Catalysis (from the greek kata = down and lyein = loosen) can be divided into areas such as homogeneous, heterogeneous, bio-, photo-, and electrocatalysis (Figure 2.1-2, Table 2.1-1).

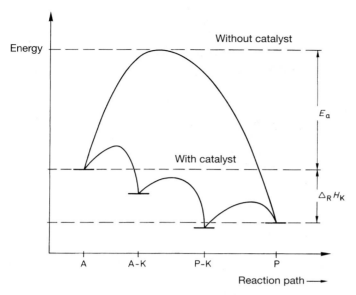

Fig. 2.1-1 Principle of catalysis for the example of conversion of the starting material A to product P with the catalytically active species K:
A + K → A − K
A − K → P − K
P − K → P + K.
$\Delta_R H$ is the reaction enthalpy and E_a the activation energy of the uncatalyzed process.

In *homogeneous catalysis* the catalyst (e.g., an acid, a base, or a transition metal complex) is dissolved in a liquid phase. An example from the chemical industry is hydroformylation, in which olefins react with synthesis gas (CO/H_2 mixture) on cobalt or rhodium catalysts to give aldehydes [Weissermel 2003]. The practical applications of homogeneous catalysis date back to the eighth century. At this time mineral acids were used to prepare ether by dehydration of ethanol. [Thomas 1994].

In *heterogeneous catalysis* the catalyst is present in solid form, and the reaction takes place at the fluid/solid phase boundary. The well-known Döbereiner lighter [Thomas 1994] was the first example of commercial exploitation of heterogeneous catalysis. Today attempts are made to combine the advantages of both types by fixing homogeneous catalysts on solid support materials (*catalyst immobilization*).

Of the around 7000 naturally occurring enzymes more than 3000 are currently known, which catalyze an enormous variety of chemical reactions. From this practically inexhaustible natural resource, currently only about 75 enzymes are used industrially. The world market for industrial enzymes is estimated at about $ 10^9. However, wider application of these *biocatalysts* in chemical synthesis is often hampered by inherent disadvantages: high catalytic activity of conventional enzymes can generally only be achieved in a narrow window of pH and temperature in aqueous medium,

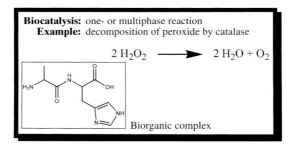

Fig. 2.1-2 Classification of catalysis into the three important classes of heterogeneous, homogeneous, and biocatalysis.

so that economical exploitation, which is further disfavored by large reactor volumes, is not promising.

Catalysis is one of the *key technologies* [Felcht 2000, p. 97], both today and in the future, and can be ranked with other fields in which advances lead to a chain of innovations [Pasquon 1994, VCI 1995], for example, biotechnology and information techgnology, even though advances in catalysis are often regarded by the general public as less spectacular. In the past many approaches to research on catalysis were proposed [Schlögl 1998]. However, the methods used today for the development of new catalysts differ from those used at the time of Carl Bosch (1874–1949), Alwin Mitasch (1869–1953), and Matthias Peer (1882–1965).

Tab. 2.1-1 Advantages and disadvantages of heterogeneous and homogeneous catalysis [Cavani 1997].

	Homogeneous catalysis	Heterogeneous catalysis
Advantages	• no Mass-transfer limitation • high selectivity • mild reaction conditions (50–200 °C)	• no catalyst separation • high thermal stability
Disadvantages	• low stability of catalyst complexes • catalyst separation • corrosion problems • toxic wastewater after catalyst recycling • contamination of product with catalyst • High costs of catalyst losses (noble metal complexes)	• lower selectivity • temperature control of highly exothermic reactions • Mass-transfer limitations • high mechanical stability required • harsh reaction conditions (> 250 °C)

In earlier days mass screening of catalysts, that is, testing a large number of solid-state samples in an even larger number of experiments, was customary. For example, during the development of ammonia synthesis from atmospheric nitrogen and hydrogen (Haber–Bosch process), 3000 differently prepared and doped iron oxides were tested in about 20 000 experiments in order to find the optimal catalyst. Production of ammonia by this process, which began in 1910 at the Oppau plant with a rate of 30 t/d, was a major achievement in cataysis research and engineering.

The modern approach is characterized by two keywords: *interdisciplinary cooperation* and *rational catalyst design*. In the course of time cataysis research has split into different areas, each of which has developed its own arsenal of methods. The four main foundations are solid-state science, chemical reaction technology, microscopic modeling, and surface science. The interdisciplinary exchange of knowledge between these areas allows many pieces of the puzzle to be fitted together in a short time to give a total picture that brings us a step closer to understanding catalysis. In an iterative process the results of the individual research groups lead to a proposal for a catalyst modification that is based on an understanding of the physicochemical processes. This cycle is repeated until the catalyst has been improved and the catalytic mechanism has been understood (rational catalyst design).

Combinatorial chemistry with the tool of *high-throughput screening* is also used for catalyst screening [Maier 1999, Maier 2000]. With the aid of largely automated laboratory apparatus it is possible to prepare and screen many thousands of catalysts per year for a given reaction [Senkan 1998, Senkan 1999, Senkan 2001].

2.1.1
Catalyst Performance

For the process developer knowledge of the following catalyst properties is important:

- Selectivity
- Activity
- Lifetime
- Mechanical stability (heterogeneous catalysts)
- Production costs.

2.1.1.1
Selectivity

Selectivity (see Section 3.1.4) influences all items in the production costs table (Table 1-1), especially those for starting materials and waste disposal. Drawing up a mass balance, which is the basis of every pilot plant, without knowing the selectivty is pointless.

2.1.1.2
Activity

The catalyst activity largely determines the size of the reactor. The absolute measure of the activity is the factor A_{cat} by which the reaction rate increases (Eq. 2.1-1) or the difference in activation energy ($E_A - E_{A,cat}$, Eq. 2.1-2) under otherwise constant reaction conditions:

$$r_{\text{with cat}} = A_{cat} \cdot r_{\text{without cat}} \qquad (2.1-1)$$

where

$$A_{\text{Kat}} \propto \exp\left(E_A - E_{A,\text{Kat}}\right)/RT. \qquad (2.1-2)$$

r = conversion rate in moles time^{-1}

In the early stages of development, exact data on the reaction rate as a function of the processs parameters is often not available, because their determination is very laborious. Therefore the activity is characterized by the following "soft" quantities:

Conversion as a measure of activity
Under otherwise constant reaction conditions (T, p, c_i, etc.) a more active catalyst results in higher conversion.

Reaction temperature as a measure of activity
To perform a given reaction at constant conversion, a more active catalyst requires a lower temperature than a less active catalyst (Figure 2.1-3).

Space-time yield as a measure of activity

The space-time yield (STY) is the amount of product that is produced by 1 kg of catalyst in unit time (Eq. 2.1-3).

$$STY/h^{-1} = \frac{product\ mass/time}{catalyst\ mass}. \qquad (2.1-3)$$

Similar quantities that are often used in the literature and which express the load on the catalyst are the liquid hourly space velocity (LHSV, Eq. 2.1-4), the gas hourly space velocity (GHSV, Eq. 2.1-5), and the weight hourly space velocity (WHSV, Eq. 2.1-6).

$$LHSV/h^{-1} = \frac{liquid\ volume/time}{catalyst\ volume} \qquad (2.1-4)$$

$$GHSV/h^{-1} = \frac{gas volume/time}{catalyst/volume} \qquad (2.1-5)$$

$$WHSV/h^{-1} = \frac{mass/time}{catalyst/masse}. \qquad (2.1-6)$$

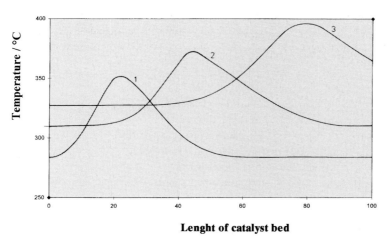

Fig. 2.1-3 Temperature profile in a catalyst tube as a measure of activity (schematic).

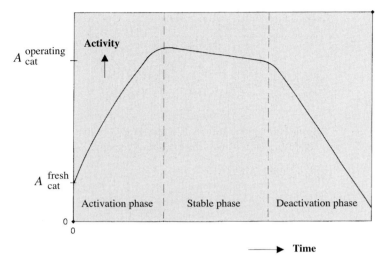

Fig. 2.1-4 Example of how the activity of a freshly introduced catalyst changes with time. The activity/time behavior shown here is often found for mixed-oxide catalysts.

2.1.1.3
Lifetime

The activity and/or selectivity of a catalyst can change during its use in the laboratory or in a production process. After a short initial phase, in which the catalyst performance often increases, the performance of the catalyst decreases with time (Figure 2.1-4). More than 90% of expenditure in industrial catalysis is related to problems of catalyst deactivation [Ostrovskii 1997, Forzatti 1999].

One reason for this is that the catalyst structure (surface, subsurface, bulk) is dependent on the chemical environment. In the initial phase the true catalytically active species are formed and the activity increases (activation phase). At the same time this process is accompamied by a deactivatation process, which can have different causes (Figure 2.1-5).

- *Deposits (fouling, coking)*.
  The catalyst surface becomes blocked by deposits, which can be formed by side reactions, for example, coke in cracking of high-boiling petroleum fractions, formation of polymers on the surface, and so on. Possible countermeasures are burning off the coke or adding substances that decrease the tendency for deposition (e.g., steam).

- *Poisoning*.
  The activity of the fresh catalyst is decreased by impurities in the feed gas which block active sites or coat the entire catalyst. In ammonia synthesis, *reversible poisoning* by oxygen, argon, and methane is known. This can be alleviated by flushing with a pure gas that is free of these components. *Irreversible poisoning* is caused by

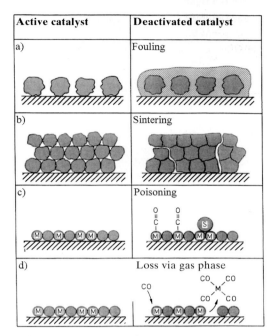

Fig. 2.1-5 Basic types of catalyst deactivation mechanisms.

components that form stable chemical compounds with the active sites. May metal catalysts can relatively easily poisoned. The main poisons are elements of Groups 15–17 (P, As, Sb, O, S, Se, Te, Cl, Br) in elemental form or as compounds. Certain metals (Hg, metal ions) and molecules with π bonds (e.g., CO, C_2H_4, C_2H_2, C_6H_6) also act as deactivators. Other widely occurring catalyst poisons include PH_3, AsH_3, H_2S, COS, SO_2, and thiophene [Foratti 1999]. A suitable countermeasure is to install a bed of absorbent or a sacrificial catalyst bed upstream of the reactor.

- *Aging and sintering.*
  Here the crystal structure of the catalyst changes. This can be caused by diffusion of active centers due to excessive temperature, or the pore structure may change. This can happen relatively quickly (e.g., in the production of phthalic anhydride or in ammonia synthesis) or slowly over a longer time period. A possible countermeasure is to artificially pre-age the catalyst (e.g., by tempering), so that the catalyst retains a constant activity over a longer time period.

- *Loss via the gas phase.*
  Here the active component in the catalyst has a finite vapor pressure and is carried out of the reactor. Examples are the loss of MoO_3 from Keggin-type heteropolyacids (vanadomolybdatophosphates) and the loss of $HgCl_2$ as the catalytically active species in the production of vinyl chloride from ethylene and HCl. A possible countermeasure is saturation of the feed with the active component.

Tab. 2.1-2 Examples of deactivation functions for the boundary conditions $D_{cat}(t=0) = D_{cat,0}$ and a deactivation rate constant k_D.

Linear	$\dot{D}_{cat} = -k_D$	$D_{cat}(t) = D_{cat,0} - k_D \cdot t$
Exponential	$\dot{D}_{cat} = -k_D \cdot D_{cat}$	$D_{cat}(t) = D_{cat,0} \cdot \exp(-k_D \cdot t)$
Hyperbolic	$\dot{D}_{cat} = -k_D \cdot D_{cat}^2$	$D_{cat}(t) = \frac{D_{cat,0}}{(D_{cat,0} \cdot k_D \cdot t - 1)}$

The aging process can be quantified by superimposing the chemical kinetics $r_{m,0}$ with a deactivation function $D_{cat}(t)$ (Table 2.1-2, Equation 2.1-7).

$$r_m = r_{m,0}(T, c_i, etc.) \cdot D_{cat}(t) \qquad (2.1-7a)$$

where

$$D_{cat}(t) = f(t; T, c_i, Re, etc.). \qquad (2.1-7b)$$

It is important for the process developer to know the lifetime of the catalyst, since this determines the *on-stream time* of the plant (theoretically 24 h d^{-1} × 365 d a^{-1} = 8760 h a^{-1}) and thus the investment costs. If a fixed-bed catalyst has to be changed once a year because the activity and or selectivty no longer allows economically viable operation, then this means a major reduction in on-stream time.

The best, but also the most laborious, method to determine the lifetime is a long-term test in an integrated reactor. Here the catalyst is operated under intensified conditions (high conversion, high temperature, high concentration, etc.) in so-called stress tests that allow statements about the lifetime to be rapidly made. This method is especially suitable for comparing different variants of a catalyst.

The deactivation mechanism must be determined, so that targeted countermeasures, some examples of which are given in Table 2.1-3, can be implemented.

The process developer must also consider what will become of the used catalyst after removal from the plant. Today processes for catalyst recycling (e.g., chemical regeneration) are preferred to disposal at waste dumps, which was practiced earlier.

Tab. 2.1-3 Measures to reduce catalyst deactivation.

Type of deactivation	Countermeasure
Coking	Addition of steam
Loss via gas phase	Presaturation
Catalyst poison in feed	Upstream absorber

2.1.1.4
Mechanical Strength

Catalysts must not only resist thermal and mechanical stress due to structural changes, for example, during reduction and regeneration, it is also subject to mechanical stress during transport and insertion in the reactor. The mechanical strength of heterogeneous catalysts primarily determines the pressure drop of the reaction gas across the reactor and hence the energy costs for the compressor or pump, because during installation of the shaped catalyst particles and as a result of vibrations and thermal expansion, powder is formed which can increase the pressure drop. Since the pressure drop of a catalyst bed also depends on the shape of the pellets, hollow shapes are preferable to solid spheres.

An initial qualitative assessment of the mechanical strength can made with a simple fingernail test (an acceptable catalyst should not be powderable between the thumbnails) or a simple drop test from a certain height. A quantitative measure of the mechanical strength is the lateral strength of the catalyst particles, which should exceed 15 N cm^{-2}.

2.1.1.5
Production Costs

The production costs of the catayst must be borne by the product and are therefore included in the raw materials calculation (Eq. 2.1-8)

$$RC = \frac{m_{cat} \cdot PC}{P \cdot t} \qquad (2.1-8)$$

RC = share of the catalyst in the raw materials costs of the product in € kg^{-1}
PC = production cost of the catalyst in € kg^{-1}
m_{cat} = mass of catalyst in the reactor in kg
P = production in kg a^{-1}
t = lifetime of catalyst in years.

Many catalysts are prepared by one of the following two methods [Perego 1997]:

a) Precipitation
The active components are dissolved and then precipitated under certain conditions (*T*, pH, stirring speed, etc.). The resulting precipitate is dried and subjected to a series of mechanical unit operations (filtration, drying, shaping, calcination, etc.; Figure 2.1-6). The catalyst mass thus obtained can be shaped in pure form by pressing or extrusion or applied as a thin layer on steatites (magnesium silicate) spheres (shell catalysts).

b) Impregnation
A suitable porous catalyst support (Table 2.1-4) is impregnated with a solution of the active components and then dried and calcined (Figure 2.1-7). This method is pre-

ferred for expensive active components such as noble metals because high degrees of dispersion (mass of active components on the surface relative to the total mass of active components) can be achieved.

Information on specific production methods can be found in the literature [Pinna 1998]. Impregnated catalysts are mainly produced batchwise with discontinuous process steps. Therefore, continuous quality control of the individual catalyst batches is vital (e.g., testing of mechanical strength, performance tests in screening reactors). The process developer must pay special attention to transferring the laboratory recipe to industrial catalyst production. Test production should be carried out relatively early

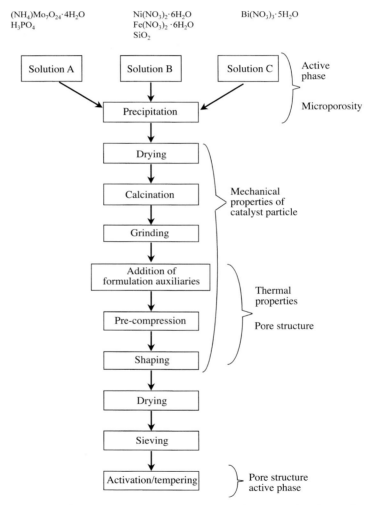

Fig. 2.1-6 Production of a Bi–Mo mixed-oxide catalyst, as used for the partial oxidation of propylene [Engelbach 1979].

Tab. 2.1-4 Support materials and the order of magnitude of their specific surface areas and mean pore diameters [Hagen 1996, Despeyroux 1993].

Support material	Specific surface area/m^2 g^{-1}	|d|/nm
Magnesium oxide	5–10	/
Silica gel	200–800	1–10
a-Aluminum oxide (corundum)	5–10	/
γ-Aluminum oxide	160–250	15
Activated carbon	500–1800	1–2

to minimize the risk of scaling up from the laboratory (e.g., 100 g of catalyst) to the industrial scale (up to 100 t of catalyst) [Pernicone 1997]. In the production of industrial heterogeneous catalysts a compromise has to be found between pellet size, reactor volume (important for high-pressure reactors), and pressure drop in fixed-bed.; 1/8- or 1/16-inch pellets are generally optimum.

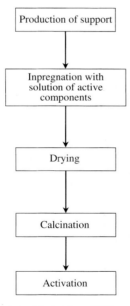

Fig. 2.1-7 Preparation of impregnated catalysts [Pinna 1998].

2.1.2
Characterization of Catalysts [Leofanti 1997 a, b]

The above-mentioned variables that characterize the performance of a catalyst depend not only on the external process parameters (T, p, c_i, etc.) but also in a complex manner on a series of other quantities:

$$\text{Catalyst performance} = f \left\{ \begin{array}{c} \text{chemical composition} \\ \text{support} \\ \text{promoters} \\ \text{phase composition} \\ \text{particle size} \\ \text{pore structure and size distribution} \\ \text{surface structure} \\ \text{Feed impurities} \\ \text{etc.} \end{array} \right\}$$

Some of these are dicussed in the following.

2.1.2.1
Chemical Composition

The chemical composition is usually expressed as the empirical formula of the active components, for example, $H_{3.1}Mo_4VP_3O_x$. Determination of the chemical composition by wet chemical analysis, atomic absorption spectroscopy, or X-ray fluorescence can reveal errors in the production process (e.g., weighing of starting materials), but it gives no information on catalyst performance.

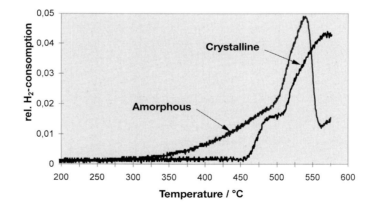

Fig. 2.1-8 TP reduction spectrum of an Mo–V mixed-oxide catalyst with hydrogen as probe molecule [Böhling 1997]. In spite of equal BET surface areas (ca. 16 m^2 g^{-1}) the amorphous catalyst is considerably more active.

2.1.2.2
Nature of the Support Material

The support material used to prepare shell or impregnated catalysts cannot be regarded a priori as inert. It often has a major influence on catalyst performance. This influence can be investigated by instationary (temperature-programmed measurements [Drochner 1999], Figure 2.1-8) or stationary kinetic experiments. Widely used "inert" support materials are listed in Table 2.1-4.

2.1.2.3
Promoters

Promoters are "secret" ingredients that are often added to catalysts for reasons of patent law. They have no intrinsic catalytic activity but, by interacting with the actual catalyst mass, they can considerably influence the performance of the catalyst, which has led to their reputation as being "black magic". They have a direct or indirect effect on the catalytic cycle, for example, by stabilizing certain structures or modifying sorption properties (Table 2.1-5). Only with the advent of modern methods of surface science has it become possible to understand their role in the catalysis mechanism.

2.1.2.4
Phase Composition

The catalytically active phase is often formed from the fresh catalyst under reaction conditions. This can be recognized by a change in crystal structure from the new to the used catalyst. However, only in situ methods (e.g., XRD, EXAFS, XANES, DRIFTS [Krauß 1999, Dochner 1999a]; see also Appendix 8.18) can identify the catalytically active phase, but only when it exceeds a ceratin size (> 3 nm).

Tab. 2.1-5 Examples of promoters for heterogeneous catalysis [Hagen 1996].

Catalyst	Promoter	Effect
Ethylene oxide catalyst, silver-based	calcium	increases selectivity
Ammonia catalyst, iron-based	K_2O	lowers binding energy between Fe and N_2
	Al_2O_3	lowes rate of sintering of metallic iron
Acrylic acid catalyst, based on Mo/V mixed oxide	copper	lowers reaction temperature

2.1.2.5
Particle Size

The particle size (Figure 2.1.9) can have a decisive influenec on the performance of a catalyst. The electronic properties change with decreasing particle size due to shifting of the valence and conduction bands, especially in the mesoscopic range (2–50 nm).

2.1.2.6
Pore Structure

Attaining an economic space–time yield (Equation 2.1-3) requires a sufficient number of active centers. This is achieved by having a large internal surface area in the form of pores. Table 2.1-6 lists the standard classification of pores according to their diameter. Methods for determining the pore size distribution and the mean pore diameter are mercury porosimetry and the BET method (see Section 2.1.3.3) [Kast 1988, Wijngaarden 1998].

2.1.2.7
Surface Structure

A smooth surface generally has lower catalytic activity then a rough one, because active atoms on corners, edges, and terraces have higher energy. However, characterization of such catalytically active sufaces remains an unsolved problem. Many of the methods for investigating surface structure [Niemensverdriet 1993] involve the impact of a beam of particles such as electrons or ions on the catalyst and measuring changes

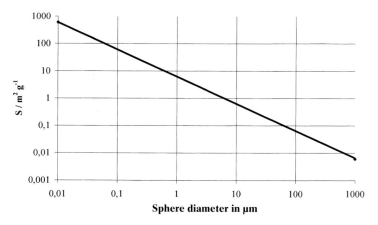

Fig. 2.1-9 Specific external surface area S of a sphere of density ρ as a function of diameter d.

$$S/(m^2 g^{-1}) = \frac{6}{\rho/(g\ cm^{-3}) \cdot d/\mu m}.$$

Tab. 2.1-6 Classification of pores.

Pore type	Pore diameter/nm
Macropores	> 50
Mesopores	2–50
Micropores	< 2
Submicropores	< 1

in, for example, their energy or impulse, which give information on the structure of the solid surface. However, these methods can only be used in vacuum (ex situ). To investigate catalysts under in situ conditions, methods using electromagnetic radiation (NMR, IR, UV, VIS, X ray), which can penetrate a reactive atmosphere (see Appendix 8.18), must be used. An established method for such investigations is DRIFT spectroscopy [Krauß 1999, Drochner 1999a].

2.1.2.8
Byproducts in the Feed

Impurities in the starting material stream can have positive effects (e.g., addition of halogen-containing compounds in silver-catalyzed ethylene oxide synthesis, but mostly their effects on catalyst performance are negative. In each case the catalyst must be tested with the original feed (e.g., from an integrated pilot plant).

2.1.3
Kinetics of Heterogeneous Catalysis [Santacesaria 1997, Santacesaria 1997a]

The kinetics of a heterogeneously catalyzed reaction consist of a series of sequential steps (Figure 2.1-10):

- A1: Convection
- A2: Film diffusion   } Mars transport of starting materials
- A3: Poren diffusion

- A4: Adsorption of starting materials
- A*5: Surfacereaction   } *(Mikrokinetics)*
- P6: Desorption of products

- P7: Poren diffusion
- P8: Film diffusion   } Mass transport of products
- P9: Convection

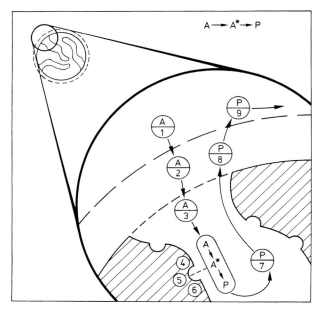

Fig. 2.1-10 Overall course of a heterogeneously catalyzed surface reaction (schematic; see text for discussion).

which in their totality are known as macrokinetics (chemical reaction with external mass transfer) [Santacesaria 1997]. The actual chemical reaction, including the processes of sorption of the reactants on the catalyst surface, is referred to as microkinetics (chemical reaction without external mass transfer) [Emig 1997].

The art of the catalyst specialist is to determine which is the rate-determining step. To this end the effective reaction rate must be measured (macrokinetics), and it must be determine, whether the actual chemical reaction rate (microkinetics) is inhibited by the external material transport. With the aid of this knowledge, the catalyst can be modified such that this limiting step is accelerated, which leads to an increase in the space-time yield and directly affects the reactor costs. The mechanisms of the individual transport steps are discussed in the following.

2.1.3.1
Film Diffusion

The transport of starting material A through the laminar boundary layer (Figure 2.1-11) that forms on the outer surface of the catalyst is described by Fick's law via the mass-transfer coefficient k_G (Equation 2.1-9).

$$r_{\text{eff}} = k_G \cdot a \cdot (A_G - A_s) \qquad (2.1-9)$$

k_G = mass-transfer coefficient in m s^{-1}
a = specific outer surface area in m^2 m^{-3}
A_G = concentration of starting material A in the gas phase in mol L^{-1}
A_s = concentration of starting material A on the surface in mol L^{-1}.

For the sake of simplicity, the chemical reaction is treated as a first-order reaction (Equation 2.1-10).

$$r_{\text{eff}} = k_{\text{eff}} \cdot A_G = k \cdot A_s \qquad (2.1-10)$$

k_{eff} = effective reaction rate constant
k = true reaction rate constant without the influence of film diffusion.

In the stationary state Equations (2.1-9) and (2.1-10) are equal and can be solved for the unknown A_s (Equation 2.1-11).

$$A_s = \frac{A_G}{1 + \dfrac{k}{k_G \cdot a}}. \qquad (2.1-11)$$

Together with Equation (2.1-10) one obtains Equation (2.1-12)

$$r_{\text{eff}} = k_{\text{eff}} \cdot A_G = \frac{1}{\dfrac{1}{k} + \dfrac{1}{k_G \cdot a}} \cdot A_G. \qquad (2.1-12)$$

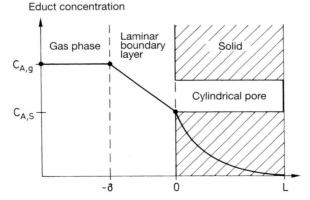

Fig. 2.1-11 Concentration profile of the educt A during mass transfer through a laminar boundary layer of thickness δ (external mass-transfer resistance) and a cylindrical catalyst pore of length L (internal mass-transfer resistance).

and the cases of Equations (2.1-13) and (2.1-14) can be distinguished.

a) $k \gg k_G \cdot a$ (strong external mass-transfer resistance)
$$r_{\text{eff}} = k_G \cdot a \cdot A_G \qquad (2.1-13)$$

b) $k \ll k_G \cdot a$ (no external mass-transfer resistance)
$$r_{\text{eff}} = k \cdot A_G. \qquad (2.1-14)$$

The magnitude of the inhibition by external mass transfer is described by the so-called external catalyst efficiency (ratio of the reaction rate with mass-transfer limitation to that without mass-transfer limitation; Equation 2.1-15).

$$\eta_{\text{ext}} = \frac{k_{\text{eff}}}{k} = \frac{1}{1 + \dfrac{k}{k_G \cdot a}}. \qquad (2.1-15)$$

From this results Equation (2.1-16) for the measured reaction rate.

$$r_{\text{eff}} = \eta_{\text{ext}} \cdot k \cdot A_G. \qquad (2.1-16)$$

Equations for other reaction orders are given in the literature [Baerns].

To determine whether the measured reaction rate is influenced by external mass transfer, the Sherwood number (Appendix 8-14) of the catalyst is calculated (Equation 2.1-17)

$$Sh = \frac{k_G \cdot \delta}{D_A} = \frac{\text{total mass transfer}}{\text{mass transfer by diffusion}}, \qquad (2.1-17)$$

which in turn is a function of the Schmidt number Sc and Reynolds number Re (Equation 2.1-18a)

$$Sh = f\left(Sc = \frac{\nu}{D_A}; Re = \frac{u \cdot \delta}{\nu}\right) \qquad (2.1-18a)$$

ν = kinematic viscosity
D_A = self-diffusion coefficient of component A
u = flow velocity
δ = thickness of the laminar boundary layer.

The quantities D_i and δ must be estimated (Section 2.1.3.2). In analogy to heat transport Equation (2.1-18b) is obtained with the Nusselt number Nu (see Section 2.3.1).

$$Sh = Nu \cdot \left(\frac{Sc}{Pr}\right)^{1/3} \qquad (2.1-18b)$$

2.1.3.2
Pore Diffusion [Keil 1999]

The overall reaction rate cannot be faster than the transport of the starting materials into the interior of the pores (Figure 2.1-11) where the catalytically active sites are located. The most important types of molecular transport are:

- Diffusion in free space
- Pore diffusion
- Knudsen diffusion
- Surface diffusion
- Poiseuille diffusion.

Fick's laws are the mathematical basis for the description of diffusion.

Fick's *first law*[2] [Fick 1855]
The particle flow density is proportional to the concentration gradient (Equation 2.1-19):

$$\frac{1}{A} \cdot \frac{dn_A}{dt} = -D_A \cdot grad\ (A(x,t)) \qquad (2.1-19)$$

A   = area
A   = concentration of A in mol L^{-1} at location x and time t
D_A = diffusion coefficient for spatially isotropic and isothermal diffusion of A.

Fick's *second law*
This describes the change in concentration with space and time due to diffusion (Equation 2.1-20):

$$\frac{\partial A(x,t)}{\partial t} = div\ [D_A(A(x,t)) \cdot grad(A)]. \qquad (2.1-20)$$

If D_A is independent of concentration, Equation (2.1-20) simplifies to Equation (2.1-21):

$$\frac{\partial A(x,t)}{\partial t} = D_A \frac{\partial^2 A(x,t)}{\partial x^2}. \qquad (2.1-21)$$

Next we will discuss the individual diffusion mechanisms.

Diffusion in *free space*
This type of diffusion takes place when the pore diameter is large compared to the mean free path of the gas molecules. In this case collisions between the molecules represent the main barrier to diffusion. In contrast to micropore diffusion, macropore diffusion, which is largely determined by this mechanism, is not an activated process. The order of magnitude of the diffusion coefficient for a gas A can be estimated by means of the simple kinetic gas theory (Equation 2.1-22)

---

2) Adolf Fick, physiologist (1829–1901).

$$D_{A,G} \approx \frac{2}{3} \cdot \bar{u}_A \cdot \Lambda_{AA} \propto \frac{T^{1,5}}{P}, \qquad (2.1-22)$$

where the mean molecular velocity is given by Equation (2.1-23):

$$\bar{u}_A = \sqrt{\frac{8 \cdot RT}{\pi \cdot M_A}} \qquad (2.1-23)$$

M_A = molecular mass

and the mean free path by Equation (2.1-24):

$$\Lambda_{AA} = \frac{RT}{\sqrt{2} \cdot N_L \cdot \pi \cdot \sigma_{AA}^2 \cdot P_A} \qquad (2.1-24)$$

σ_{AA} = collision diameter of molecule A (Table 2.1-7)
P_A = partial pressure of gas A.

Equations that allow a more accurate calculation can be found in the literature [Baerns 1999]. For the description and estimation of the diffusion coefficients of liquids, the reader is referred to the specialist literature [Fei 1998]. The following orders of magnitude can be given for the self-diffusion coefficients:

- Gases at standard pressure $10^{-5}\,\text{m}^2\,\text{s}^{-1}$
- "Normal" liquids at ambient temperature $10^{-9}\,\text{m}^2\,\text{s}^{-1}$
- Viscous liquids. $10^{-10}\,\text{m}^2\,\text{s}^{-1}$

Tab. 2.1-7 Collision diameters for some industrially important gases [Baerns 1999].

Gas	collision diameter/nm
Hydrogen	0.283
Oxygen	0.347
Nitrogen	0.380
Water	0.283
Carbon monoxide	0.369
Carbon dioxide	0.394
Methane	0.376
Ethane	0.444
Propane	0.512
Isobutene	0.528

Table 2.1-8 lists selected binary diffusuion coefficients of gases at standard temperature.

Pore diffusion
Diffusion in large pores can be desribed by the laws of diffusion in free space, provided two effects are taken into account:

a) Porosity
The surface area of the pore openings makes up only a fraction ε_P of the outer surface area of a porous solid (Equation 2.1-25).

$$\varepsilon_P = \frac{\sum \text{area of pore openings}}{\text{outside catalyst surface area}}. \qquad (2.1-25)$$

Typical values lie in the range $0.2 < \varepsilon_P < 0.7$.

b) Labyrinth factor
The deviation of the pore geometry from the ideal cylindrical form is taken into account by the labyrinth factor χ, which often lies on the order of magnitude of 1/2 to 1/6. Thus, the effective diffusion coefficient in pores is is obtained from the binary diffusion coefficient according to Equation (2.1-26).

$$D_{AB,\ eff}^{Pore} = \varepsilon_P \cdot \chi \cdot D_{AB}. \qquad (2.1-26)$$

Knudsen[3] diffusion
In small pores or at low pressure the mean free path Λ_{AA} (Equation 2.1-24) of the diffusing molcules can become comparable to the pore diameter d. In this case collisions of the molecules with the pore walls are the main barrier to diffusion. When $\Lambda_{AA}/d < 0.2$, the mean the mean free path Λ_{AA} in Equation (2.1-22) must be replaced by the pore diameter d (Equation 2.1-27).

$$D_A^{Knud} \approx \frac{1}{3} \cdot d \cdot \bar{u}_A \quad \propto T^{0.5}. \qquad (2.1-27)$$

Tab. 2.1-8 Experimentally determined diffusion coefficients of gases at standard temperature.

Gas in gas	Diffusion coefficient $D_{AB}/10^{-5} m^2\ s^{-1}$
H_2/CH_4	8.1
H_2O/CH_4	2.9
CO_2/CH_4	1.5
N_2/O_2	2.3

---

3) Martin Knudsen, Danish physicist (1871–1949).

Thus the Knudsen diffusion coefficient is less strongly temperature sensitive than the free-space diffusion coefficient and, in particular, is independent of pressure and gas composition.

Surface diffusion
Molecules that are physisorbed on the surface of a solid or are located in micropores ($d < 2$ nm) move by a series of jumps between different adsorption sites, whereby steric effects are of major importance. This type of diffusion is an activated process which, similar to diffusion in liquids [Vogel 1981, 1982], can be described by an Arrhenius expression (Equation 2.1-28):

$$D_A^{Surface} = D_0 \cdot exp\left(-\frac{E_A^{Diff}}{RT}\right). \qquad (2.1-28)$$

The surface diffusion coefficient for small molecules at room temperature is on the order of magnitude of 10^{-11} m² s⁻¹.

Poiseuille flow
When the pressure drop along a macropore is large, a pressure-induced laminar flow can arise. This additional contribution to mass transfer is know as Poiseuille flow.

2.1.3.3
Sorption

Sorption (adsorption/desorption) [Kast 1988] is the interaction of the reactants with the surface of the catalyst (adsorbent + adsorptive → adsorbate). Depending on the nature and strength of binding (Table 2.1.9), a distinction is made between physisorption (geometrical structure and electronic properties of the free particles and of the free surface are largely retained, adsorption enthalpy of ca. 20–50 kJ mol⁻¹, that is, of the same order of magnitude as the enthalpy of condensation) and chemisorption (dramatic change in the electronic structure of the free particles and the catalyst surface, adsorption enthalpy greater than 50 kJ mol⁻¹).

The sorption of a given species A is characterized by its sorption isotherm, which is obtained by measuring the loading Θ_A (= loaded area/total area) of the solid (adsorbent) with component A (adsorptive) as a function of partial pressure or of gas concentration and temperature (Equation 2.1-29, Figure 2.1-12).

$$\Theta_A(P_A, T) = \frac{m_A}{m_{solid}} = \frac{adsorbed\ mass\ of\ absorptive}{dry\ mass\ of\ adsorbent}. \qquad (2.1-29)$$

For kinetic modeling, the loading can also be defined in terms of the number of particles (Equation 2.1-30a) or the number of active sites (*) of the catalyst (Equation 2.1-30b).

$$\Theta_A(P_A, T) = \frac{n_A}{n_{A,mono}} = \frac{\text{moles of adsorbed gas A on the surface}}{\text{moles of adsorbed gas A in a monolayer}} \quad (2.1-30a)$$

$$\Theta_A(P_A, T) = \frac{(A)}{(A)+(*)} = \frac{\text{sites occupied by A}}{\text{total number of all active sites}}. \quad (2.1-30b)$$

Various models for the quantitative description of adsorption isotherms can be found in the literature, the most important of which are the Langmuir[4] and BET models, since their parameters can be interpreted in physical terms.

Langmuir isotherm for two species
Requirements of the model

- Monomolecular coverage of the active sites (*) with the species A and B
- No interaction of the adsorbed species with one another

Chemical model

$$A(\text{gas}) + (*) \xrightarrow{k_A} (A)$$

$$(A) \xrightarrow{k_{-A}} A(\text{gas}) + (*)$$

Tab. 2.1-9 Strength ranges of different types of binding and bond strengths.

Type of binding	Binding energy/kJ mol^{-1}
Van der Waals interaction	1–5
Hydrogen bonding	10–30
Ionic bonding	50–100
Metallic bonding	100–300
Covalent bonding	200–400
O-H	465
C-H	415
C-Cl	331
C-N	306
C=N	615
C-O	360
C=O	737
C-C	348
C=C	511

---

4) Irving Langmuir, American physicist (1881–1957).

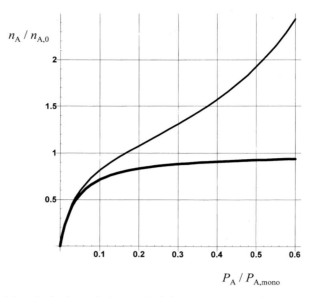

Fig. 2.1-12 Adsorption isotherm of substance A. Thick curve: Langmuir isotherm ($K_A = 25$, $P_A^0 =$ saturation vapor pressure = 1, Equation 2.1-33); Thin curve: BET model (c = 25, Equation 2.1-36).

$$B(gas) + (*) \xrightarrow{k_B} (B)$$

$$(B) \xrightarrow{k_{-B}} B(gas) + (*)$$

Mathematical model (Equation 2.1-30b)

$$\frac{d\Theta_A}{dt} = k_A \cdot P_A \cdot (1 - \Theta_A - \Theta_B) - k_{-A} \cdot \Theta_A = 0 \qquad (2.1-31)$$

$$\frac{d\Theta_B}{dt} = k_B \cdot P_B \cdot (1 - \Theta_A - \Theta_B) - k_{-B} \cdot \Theta_B = 0 \qquad (2.1-32)$$

where $1 = \Theta_A + \Theta_B + \Theta_{(*)}$.

Solving this system of equations gives Equation (2.1-33):

$$\Theta_A = \frac{K_A \cdot P_A}{1 + K_A \cdot P_A + K_B \cdot P_B} \quad \text{with } K_A = \frac{k_A}{k_{-A}} \text{ and } K_B = \frac{k_B}{k_{-B}}. \qquad (2.1-33)$$

An analogous equation applies for B. After linearization of the above equation (Equation 2.1-34), the adsorption constants K_A and K_B can be obtained by plotting P_A/Θ_A versus P_A or P_B and, after gas-kinetic derivation of the model [Jakubith 1998], interpreted in the form of Equation (2.1-35).

$$\frac{P_A}{\Theta_A} = P_A + \frac{1}{K_A} + \frac{K_B}{K_A} \cdot P_B \tag{2.1-34}$$

$$K_A = \frac{1}{k_{-A} \cdot \sqrt{2 \cdot RT \cdot \pi \cdot M}} \cdot exp\left(\frac{\Delta_{des}H_A - \Delta_{ad}H_A}{RT}\right). \tag{2.1-35}$$

$\Delta_{des}H_A$ = heat of desorption
$\Delta_{ad}H_A$ = heat of adsorption
k_{-A} = rate constant of desorption

Brunauer–Emmet–Teller isotherm [Brunauer 1938]
Requirements of the model
In an extension of the langmuir isotherm, it is assumed that additional adsorption layers can form on part of the surface before a complete monolayer has formed. In the first adsorbed layer the adsorption energy (condensation and binding energies) is released, while for the second and subsequent layers, only the condensation energy is liberated.

Mathematical solution

$$\Theta_A = \frac{n_A}{n_{A,mono}} = \frac{C \cdot (P_A/P_A^0)}{[1 - (P_A/P_A^0)] \cdot [1 + (C-1) \cdot (P_A/P_A^0)]} \tag{2.1-36}$$

The dimensionless constant $C = exp\{-(\Delta_{ad}H_A + \Delta_V H_A)/RT\}$ depends on the condensation energy of A [Kast 1988]. Rearrangement of this equation gives the linearized form (Equation 2.1-37).

$$\frac{(P_A/P_A^0)}{n_A \cdot (1 - (P_A/P_A^0))} = \frac{(C-1)}{n_{A,mono} \cdot C} \cdot (P_A/P_A^0) + \frac{1}{n_{A,mono} \cdot C}\bigg|_T. \tag{2.1-37}$$

Hence, from a plot of $(P_A/P_A^0)/[n_A(1-P_A/P_A^0)]$ versus P_A/P_A^0 the amount of A absorbed in the monolayer $n_{A,mono}$ and the constant C can be obtained directly from the ordinate intersection and the slope. This equation does not apply to the phenomenon of capillary condensation, the description of which requires the introduction of a further parameter [Brunauer 1940]. Plotting n_A versus P_A/P_A^0 gives adsorption isotherms whose form gives information about the adsorbate. Brunauer, Emmet, and Teller classified them into six types according to the porosity of the adsorbent and its interaction with the adsorptive (Figure 2.1-13) [Kast 1988, IUPAC 1985].

With the aid of these models the surface area of a catalyst can be determined by experimentally measuring the amount of absorbed gas A in a monolayer. The surface area A is obtained from the quantity $n_{A,mono}$ and the area S_A occupied by a probe molecule according to Equation (2.1-38).

$$S = n_{A,mono} \cdot N_L \cdot S_A. \qquad (2.1-38)$$

Generally dinitrogen is used as probe gas ($S_{N2} = 1.62 \times 10^{-19}\,m^2$), and evaluation is performed with Equation (2.1-37) at 77 K. The thus-obtained value is the BET surface area S_{BET}. Dividing this by the mass of the solid gives the specific BET surface area s_{BET} in $m^2\,g^{-1}$.

2.1.3.4
Surface Reactions

The heterogeneously catalyzed reaction of starting materials A and B can follow different mechanisms [Claus 1996, Ertl 1990]. The models of Langmuir-Hinshelwood, Eley-Rideal, and Mars-van Krevelen are widely used in practice.

Langmuir – Hinshelwood[5] kinetics
Requirements of the model

- The starting materials A and B are adsorbed on the surface and react in the adsorbed state to give product P.

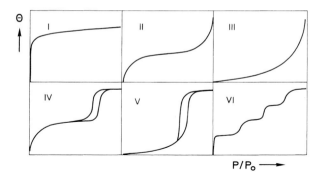

Fig. 2.1-13 The six types of adsorption isotherms.
Type I: Microporous materials and chemisorption. In micropores the adsorption potentials of the two walls overlap, and this leads to amplification of the adsorption potential [Everett 1976]. This results in increased adsorption energy at very small relative pressures.
Type II: Nonporous, finely divided solids.
Type III: for example, water on hydrophobic substances.
Types IV and V: Capillary condensation in mesopores. Initially, a multilayer is adsorbed on the capillary walls. When the pressure is further increased, liquids droplets form preferentially at sites where the curvature fulfills the Kelvin equation. When two opposing droplets touch, the pore is filled. On desorption, pores whose radius is smaller than the Kelvin radius are emptied. The adsorption branch indicates the extent of the pores, and the desorption branch the size of the pore openings [Evertt 1976].
Type VI: Stepped isotherms, for example, nitrogen on some activated carbons.

---

5) C. N. Hinshelwood (1897–1967) [Laidler 1987].

Chemical model:

$$\begin{aligned} A(\text{gas}) + (*) &\xrightleftharpoons{K_A} (A), \text{ fast} \\ B(\text{gas}) + (*) &\xrightleftharpoons{K_B} (B), \text{ fast} \\ (A) + (B) &\xrightarrow{k_{LH}} (P), \text{ slow} \\ (P) &\leftrightarrow P + (*), \text{ fast} \end{aligned} \qquad (2.1-39)$$

Mathematical model:

$$r_{\text{tot}} = k_{LH} \cdot \Theta_A \cdot \Theta_B.$$

With Equation (2.1.33) this gives the classical Langmuir-Hinshelwood expression (Equation 2.1-40).

$$r_{\text{tot}} = k_{LH} \cdot \frac{K_A \cdot K_B \cdot P_A \cdot P_B}{(1 + K_A \cdot P_A + K_B \cdot P_B)^2}. \qquad (2.1-40)$$

Eley – Rideal[6] kinetics
Requirements of the model

- The starting material A is adsorbed on the surface and reacts with starting material B from the gas phase to give product P.

Chemical model:

$$\begin{aligned} A(\text{gas}) + (*) &\xrightleftharpoons{K_A} (A), \text{ fast} \\ (A) + B(\text{gas}) &\xrightarrow{k_{ER}} (P), \text{ slow} \\ (P) &\leftrightarrow P + (*), \text{ fast} \end{aligned} \qquad (2.1-41)$$

Mathematical model:

$$r_{\text{tot}} = k_{ER} \cdot \Theta_A \cdot P_B. \qquad (2.1-42)$$

With Equation (2.1.33) this gives the classical Eley–Rideal expression (Equation 2.1-42):

$$r_{\text{tot}} = k_{ER} \cdot \frac{K_A \cdot P_A \cdot P_B}{(1 + K_A \cdot P_A)^1}. \qquad (2.1-43)$$

The two mechanisms can be distinguished by determining the microkinetics or by means of concentration-pulse experiments (Figure 2.1-14).

---

6) E. K. Rideal (1890–1974) [Laidler 1987].

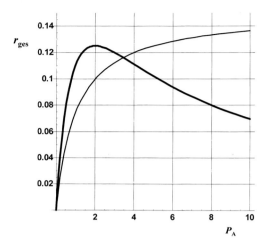

Fig. 2.1-14 Comparison of the total reaction rate as a function of the partial pressure of the starting material under the assumption of a Langmuir–Hinshelwood (thick curve, Equation 2.1-40) and an Eley–Rideal mechanism (thin curve, Equation 2.1-43).

Mars-van Krevelen[7] kinetics [Mars 1954]
Requirements of the model

- The oxidation of starting materials with oxygen on transition metal oxides often follws this two-step mechanism:
  - Oxidation of the starting material by the catalyst with simultaneous reduction of the oxide
  - Re-oxidation of the catalyst by molecular oxygen.

Chemical model:

$$A(gas) + (O) \xrightarrow{k_1} P(gas) + (*) \qquad (2.1-44)$$

$$(*) + \frac{1}{2} O_2 \xrightarrow{k_2} (O)$$

Mathematical model:

$$r_A = k_1 \cdot \Theta_O \cdot P_A$$
$$r_{O2} = k_2 \cdot (1 - \Theta_O) \cdot P_{O2}^n \qquad (2.1-45)$$

where Θ_O is the degree of oxygen coverage.

If ν_A molecules of oxygen are required for oxidation of starting material A, Equation (2.1-46) applies:

---

7) Dirk Willem van Krevelen, industrial chemist, honorary doctor of the TU Darmstadt (1915–2001).

$$r_A = \frac{1}{v_A} \cdot r_{O2} = k_1 \cdot P_A \cdot \Theta_O = \frac{1}{v_A} \cdot k_2 \cdot P_{O2}^n \cdot (1 - \Theta_O). \tag{2.1-46}$$

And with Equations (2.1-47) and (2.1-45), Equation (2.1-48) is obtained.

$$\Theta_O = \frac{k_2 \cdot P_{O2}^n}{v_A \cdot k_1 \cdot P_A + k_2 \cdot P_{O2}^n} \tag{2.1-47}$$

$$r_A = \frac{1}{\dfrac{1}{k_1 \cdot P_A} + \dfrac{v_A}{k_2 \cdot P_{O2}^n}}. \tag{2.1-48}$$

Figure 2.1-15 shows the reaction rate as a function of the concentration.

2.1.3.5
Pore Diffusion and Chemical Reaction [Emig 1993, Forni 1999, Keil 1999]

The interplay between the diffusion of the starting material in the pores and the reaction of the starting material on the catalytically active walls of the pores can be derived from the general mass balance (Equation 2.1-49).

$$D_{\text{eff}} \cdot \frac{d^2 A}{dx^2} - k \cdot A^n = 0. \tag{2.1-49}$$

If the concentration A is taken relative to the contration of A at the entrance to the pore A_0, and the pore length relative to the mean pore length L, then after rearrangement, Equation (2.1-50) is obtained:

$$\frac{d^2(A/A_0)}{d(x/L)^2} = \varphi^2 \cdot (A/A_0)^n \tag{2.1-50}$$

with the dimensionless quantity ϕ, first introduced by Thiele and known as the Thiele modulus (Equation 2.1-51) [Thiele 1939, Weisz 1954]:

$$\varphi = L \cdot \sqrt{\frac{k \cdot A_0^{n-1}}{D_{\text{eff}}}}. \tag{2.1-51}$$

This gives the ratio of the reaction rate without the influence of pore diffusion to the diffusive mass transfer in the pore. The above differential equation can be solved completely for simple catalyst geometries and simple kinetics.

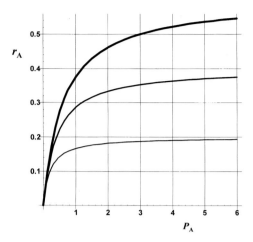

Fig. 2.1-15 Rate of consumption of A as a function of P_A for different oxygen partial pressures ($P_{O2} = 0.2$, 0.4, 0.6 from top to bottom, $k_A = 1$, $k_{O2} = 1$, $n = 1$) according to Equation (2.1-48). With increasing P_A, r_A ceases to increase because re-oxidation becomes rate-determining.

Example 2.1-1

For $n = 1$, $A_0 = 1$, and $L = 1$ one obtains from Equation (2.1-50):

$$\frac{d^2(A(x))}{dx^2} = \varphi^2 \cdot A(x). \tag{2.1-52}$$

From the boundary conditions $A(x = 0) = 1$ and $A(L = 1)$ the following solution results:

$$A(x) = \frac{A_0 \cdot exp(-\varphi \cdot x)}{1 - exp(2 \cdot L \cdot \varphi)} \cdot [exp(2 \cdot \varphi \cdot x) - exp(2 \cdot L \cdot \varphi)]. \tag{2.1-53}$$

Figure 2.1-16 shows the concentration curves of A for different Thiele moduli.

Cylindrical Single Pore (Length L, Diameter d_p) and nth-Order Reaction

$$\frac{A(x)}{A_0} = \frac{\cosh[\varphi_z \cdot (1 - x/L)]}{\cosh \varphi_z} \tag{2.1-54}$$

where

$$\varphi_z = L \cdot \sqrt{\frac{4 \cdot k \cdot A_0^{n-1}}{d_p \cdot D_{eff}}}, \tag{2.1-55}$$

is the Thiele modulus for a cylindrical pore.

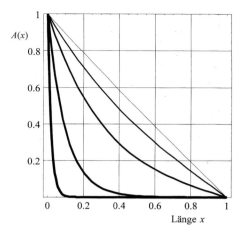

Fig. 2.1-16 Concentration of component A as a function of length x at different Thiele moduli according to Equation (2.1-53). ϕ = 0.5, 1.5, 3, 10, 50 from top to bottom.

Catalyst sphere (radius r) and nth-order reaction

$$\frac{A(x)}{A_0} = \frac{\sinh\left[\varphi_K \cdot (x/R)\right]}{(x/R) \cdot \sinh \varphi_K} \qquad (2.1-56)$$

where

$$\varphi_K = \sqrt{\frac{n+1}{2}} \cdot R \cdot \sqrt{\frac{k \cdot A_0^{n-1}}{D_{\text{eff}}}}, \qquad (2.1-57)$$

is the Thiele modulus for a sphere.

With increasing temperature the Thiele modulus increases due to the Arrhenius[8] dependence of k. The starting material A reacts faster and faster at the boundary zones of the catalyst, that is valuable catalyst material (e.g., noble metal) in the interior of the pores is now longer used. In this case the use of a shell catalyst is more economical.

To better judge this situation, a degree of pore use is introduced (also known as degree of catalyst use η_{cat}; Figure 2.1-17, Equation 2.1-58), which expresses the ratio of the mean reaction rate to the maximum possible reaction rate, that is, without the influence of diffusion. Inserting the course of the concentration in the cylindrical pore (Equation 2.1-54) into this equation gives, after integration, Equation (2.1-59):

---

8) Svante Arrhenius, Swedish physicist and chemist (1859–1927).

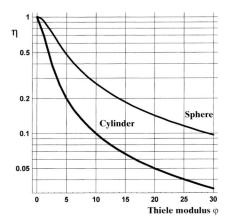

Fig. 2.1-17 Degree of catalyst exploitation as a function of Thiele number for an individual pore (Equation 2.1-59) and a catalyst sphere (Equation 2.1-60).

$$\eta_{\text{cat}} = \frac{\int_0^L k(T) \cdot A(x)^n \cdot dx}{k(T) \cdot A_0^n}, \qquad (2.1-58)$$

$$\eta_{\text{cat}}^{\text{cyl}} = \frac{\tanh \varphi_z}{\varphi_z} \begin{cases} \approx 1 \text{ for } \varphi_z < 0.3 \\ 1/\varphi_z \text{ for } \varphi_z > 3 \end{cases} \qquad (2.1-59)$$

while for a catalyst sphere (Equation 2.1-56) Equation (2.1-60) is obtained:

$$\eta_{\text{cat}}^{\text{sphere}} = \frac{3}{\varphi_K}\left\{\coth \varphi_K - \frac{1}{\varphi_K}\right\} \begin{cases} \approx 1 \text{ for } \varphi_K < 0.3 \\ 3/\varphi_K \text{ for } \varphi_K > 3 \end{cases}. \qquad (2.1-60)$$

The influence of mass-transfer resistance can be investigated by determining the temperature dependence of the total process. Since Equation (2.1-61) applies for the measured effective reaction rate $r_{m,\text{meas}}$ of a first-order reaction:

$$r_{m,\text{meas}} = (\eta_{\text{Kat}} \cdot k) \cdot A_0 = k_{\text{meas}} \cdot A_0, \qquad (2.1-61)$$

then for the kinetically controlled range ($\eta \approx 1$) with an Arrhenius expression:

$$\ln r_{m,\text{meas}} = \left(-\frac{E_a}{R}\right) \cdot \frac{1}{T} + \ln(A_0 \cdot k_0). \qquad (2.1-62)$$

Together with Equation (2.1-60) and for pure diffusion control in the pores, Equation (2.1-63) results:

$$r_{m,meas} = \frac{3}{\varphi_K} \cdot k \cdot A_0. \qquad (2.1-63)$$

Inserting Equation (2.1-57) for ϕ_K and an Arrhenius expression for k gives Equation (2.1-64).

$$\ln r_{m,meas} = \left(-\frac{E_a}{2 \cdot R}\right) \cdot \frac{1}{T} + \ln\left(\frac{3 \cdot \sqrt{k_0} \cdot A_0 \cdot \sqrt{D}}{R}\right). \qquad (2.1-64)$$

Plotting $\ln r_{m,\,meas}$ or $\ln k_{meas}$ versus $1/T$ gives a curve with two inflexion points, which indicate a change in the nature of the reaction limitation. A prerequisite is that the mechanism, and hence the underlying activation energy, of the reaction does not change.

The difficulty in the direct determination of η_{cat}, that is, the extent of mass-transfer limitation, $r_{m,meas}$ can be directly measured, but k in the following equation (Equations 2.1-61, 2.1-60, and 2.1-57)

$$r_{m,meas} = \frac{3 \cdot \sqrt{D_{eff}}}{R} \cdot \left(\coth\left[\frac{R \cdot \sqrt{k}}{\sqrt{D_{eff}}}\right] - \frac{\sqrt{D_{eff}}}{R \cdot \sqrt{k}}\right) \cdot \sqrt{k} \cdot A_0 \qquad (2.1-65)$$

is not directly accessible. If k is regarded as an adjustable parameter and is fitted to the to the above nonlinear equation, then from the measurement of $r_{m,meas}$ for a given A_0 the true reaction rate k results as a fitting parameter and hence the Thiele modulus or degree of catalyst use according to Equation (2.1-60).

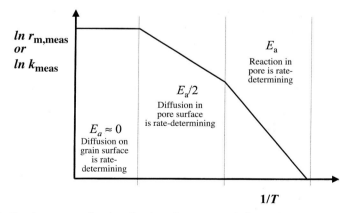

Fig. 2.1-18 Reaction rate as a function of reciprocal temperature (schematic).

Decisive for the scaleup of reactors is knowing whether mass-transfer limitation is present. Criteria for this can be found in the literature [Baerns 1999, Emig 1997]. All require the knowledge of:

- Catalyst geometry (e.g., grain diameter k_{grain})
- Fluid-phase diffusion coefficient D_A
- Pore diffusion coefficient D_A^{Pore}
- Physicochemical properties of the gases
- Experimentally determined total reaction rate r_{eff} [Wijngaarden 1998].

For example, if for a first-order reaction ($n = 1$) Equation (2.1-66) applies, then the effectivity is greater than 0.95.

$$\frac{r_{V,eff} \cdot d_{grain}^2}{4 \cdot (1 - \varepsilon) \cdot D_A \cdot A_{gas}} \leq 0.6 \tag{2.1-66}$$

2.1.3.6
Film Diffusion and Chemical Reaction

The interaction between diffusion of the starting material in the laminar gas film (1) and its reaction in the liquid film (2) can be described analogously to the treatment in Section 2.1.3.5. Assuming that A is converted in a first-order reaction (A + B(excess) → P), it follows from Equation (2.1-49):

$$D_{eff} \cdot \frac{d^2 A}{dx^2} - k \cdot A = 0.$$

or

$$\frac{d^2(A/A_0)}{d(x/\delta)^2} - Ha^2 \cdot (A/A_0) = 0 \tag{2.1-67}$$

δ = thickness of the laminar boundary layer

$$Ha = \delta \cdot \sqrt{\frac{k}{D_{eff}}} = \text{Hatta number.} \tag{2.1-68}$$

Hence, under the assumption of stationary nonequilibrium conditions, the particle flux of A that is consumed by chemical reaction is equal to that transported through the phase boudary at $x = 0$. Under the boundary conditions:

$$A(x = 0) = A^*, \quad A(x = \delta) = 0$$

$$\frac{dB}{dx} = 0, \quad B(x = \delta) = B_1,$$

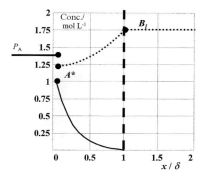

Fig. 2.1-19 Concentration profile across the phase boundary area (Hatt number = 4). See text for discussion.

the solution of the differential Equation (2.1-67) (see Figure 2.1-19) is:

$$A(x) = \frac{\sinh\left[Ha \cdot (1 - x/\delta)\right]}{\sinh\left[Ha\right]} \cdot A^* \text{ for } x \in [0, \delta]. \qquad (2.1-69)$$

2.2
The Reactor

The reactor is the heart of every chemical plant, even though in terms of space requirements and investment it is often the smallest component of the plant. Because of its central function it fundamentally influences all other plant components; consider, for example, the measures necessary to recycle unconverted starting material. Therefore, besides the choice of catalyst, the selection of the reactor type is the most important step in process development [Donati 1997].

With the aid of chemical reaction technology [Levenspiel 1980, Hofmann 1983], the theoretical basis of which was only developed in the mid-1900s, the reactor can be designed. To this end, the questions of:

- Reaction conditions (process parameters)
- Size of the reactor
- Shape of the reactor
- Operating mode,

must be answered for a given reaction, which can be carried out without problems in laboratory apparatus, and required production performance while taking into account the optimal operating point [Damköhler 1936, Platzer 1996]. In simple terms, the optimal operating point is that at which the return on investment is highest. This depends on all parameters that affect the production costs and the required investment. For simple reactions (no problems with selectivity such as in ammonia synthesis) optimization of the reaction technology generally aims at maximization of the turnover. For complex chemical reactions this is generally supplemented by yield and performance optimization.

2.2.1
Fundamentals of Chemical Reaction Technology

The reactor types in use today can be classified according to different criteria:

- The number of phases involved (homogeneous or heterogeneous)
- Operating mode (limiting cases: batch and continuous)
- Temperature control (limiting cases: isothermal and adiabatic)
- Residence-time behavior (limiting cases: plug flow and complete backmixing)
- Mixing of the starting materials (macro- and micromixing with the limiting cases of ideal micromixing and complete segregation).

In practice a combination of the criteria is used (e.g., homogeneous, continuous, adiabatic, plug flow).

This variety of reactor types reflects the complex interaction between chemical reaction and mass and heat transfer. In spite of this complexity, in principle every reactor can be described by the fundamental balance equations for the preservation of mass, energy, and impetus. These form a system of five partial differential equations (PDEs), coupled through temperature, concentration, and three rate vectors [Damköhler 1936, Platzer 1996, Adler 2000].

PDE 1: The law of the conservation of energy in the form of the extended continuity equation describes the change in concentration in space and time of component A that reacts in the reactor according to $v_A A \rightarrow$ products (A in mol L^{-1}; Equation 2.2-1):

$$\frac{\partial A(x, y, z, t)}{\partial t} = -div(A \cdot \vec{u}) + div(D_A \cdot grad\ A) + v_A \cdot r_V \pm \beta_A \cdot a_S \cdot \Delta A. \quad (2.2-1)$$

The first term on the right-hand side is the so-called convective term (flow term of forced convection), where $\vec{u}$ is the vector field of the flow velocity. The second term is the so-called conductive term (effective diffusion, Section 2.1.3). The third term is the reaction term (Section 3.1.3), with the reaction rate per unit volume r_V and the stoichiometric coefficient v_A. The final term is the mass-transfer term with the mass transfer coefficient β_A (Section 2.1.3), the specific mass-transfer area a_S, and the concentration difference ΔA between boundary surface and bulk phase.

PDE 2: The law of conservation of enthalpy (mechanical energy and nuclear energy are usually not considered) describes the temperature distribution in the reactor (Equation 2.2-2):

$$\frac{\partial T(x,y,z,t)}{\partial t} = - \operatorname{div}(T \cdot \vec{u}) + \operatorname{div}\left(\frac{\lambda}{c_P \cdot \rho} \cdot \operatorname{grad} T\right)$$
$$+ \frac{(-\Delta_R H) \cdot r_V}{c_P \cdot \rho} \pm \frac{a \cdot A_w \cdot (T_w - T)}{c_P \cdot \rho \cdot V_R} \quad (2.2-2)$$

λ = thermal conductivity coefficient in J m^{-1} K^{-1} s^{-1}
c_P = heat capacity at constant pressure in J kg^{-1} K^{-1}
ρ = density in kg m^{-3}
A_W = heat-exchange area in m^2
$\Delta_R H$ = reaction enthalpy in J mol^{-1}
a = heat-transfer coefficient in J m^{-2} K^{-1} s^{-1}
T_W = wall temperature in K
V_R = reactor volume in m^3
r_V = reaction rate per unit volume in mol m^{-3} s^{-1}.

As in PDE 1, the individual terms on the right-hand side correspond to heat transport by convection, effective heat conduction, generation of heat by a chemical reaction, and heat exchange across a surface (from left to right).

PDEs 3–5: The law of the conservation of impetus descibes the variation with time of the velocity distribution of the fluid in the three directions of space (Equation 2.2-3):

$$\frac{\partial \vec{u}(x,y,z,t)}{\partial t} = \vec{a} - \frac{1}{\rho} \cdot \operatorname{grad}(P) + \operatorname{grad}\left(\frac{\eta}{\rho} \cdot \operatorname{div}(\vec{u})\right) \quad (2.2-3)$$

a = acceleration vector in m s^{-2}
η = viscosity coefficient in Pa s
ρ = density in kg m^{-3}
P = pressure in Pa.

In an initial approximation, this impetus balance, which relates vectors that contain no reaction term, is generally treated separately ($\partial u/\partial z \approx 0$). Therefore, the equations for the conservation of mass and enthalpy are most relevant for reactor design.

The formulation used up to now has the advantage that it is not bound to a coordinate system. In Cartesian coordinates Equations (2.2-1a) and (2.2-2a) are obtained for energy and mass balances:

$$\frac{\partial A(x,y,z,t)}{\partial t} = -\left[\frac{\partial(u_x \cdot A)}{\partial x} + \frac{\partial(u_y \cdot A)}{\partial y} + \frac{\partial(u_z \cdot A)}{\partial z}\right]$$
$$+ \left[\frac{\partial}{\partial x}\left(D_x \frac{\partial A}{\partial x}\right) + \frac{\partial}{\partial y}\left(D_y \frac{\partial A}{\partial y}\right) + \frac{\partial}{\partial z}\left(D_z \frac{\partial A}{\partial z}\right)\right] \quad (2.2-1a)$$
$$+ \nu_A \cdot r_V \pm \beta_A \cdot a_S \cdot \Delta A$$

$$\frac{\partial T(x,y,z,t)}{\partial t} = -\left[\frac{\partial(u_x \cdot T)}{\partial x} + \frac{\partial(u_y \cdot T)}{\partial y} + \frac{\partial(u_z \cdot T)}{\partial z}\right]$$
$$+ \left[\frac{\partial}{\partial x}\left(\frac{\lambda_x}{\rho \cdot c_P} \frac{\partial T}{\partial x}\right) + \frac{\partial}{\partial y}\left(\frac{\lambda_y}{\rho \cdot c_P} \frac{\partial T}{\partial y}\right) + \frac{\partial}{\partial z}\left(\frac{\lambda_z}{\rho \cdot c_P} \frac{\partial T}{\partial z}\right)\right]$$
$$+ \frac{r_V \cdot (-\Delta_R H)}{\rho \cdot c_P} \pm \frac{a \cdot A_W \cdot (T_W - T)}{\rho \cdot c_P \cdot V_R} \qquad (2.2-2a)$$

while for the mass balance in cylindrical coordinates (Appendix 8.1) Equation (2.2-1b) applies

$$\frac{\partial A(r,\theta,z,t)}{\partial t} = -\left[\frac{\partial(u_r \cdot A)}{\partial r} + \frac{1}{r}\frac{\partial(u_\theta \cdot A)}{\partial \theta} + \frac{\partial(u_z \cdot A)}{\partial z}\right]$$
$$+ \frac{1}{r}\frac{\partial}{\partial r}\left(D_r \cdot r \cdot \frac{\partial A}{\partial r}\right) + \frac{1}{r^2}\frac{\partial}{\partial \theta}\left(D_\theta \frac{\partial A}{\partial \theta}\right) + \frac{\partial}{\partial z}\left(D_z \frac{\partial A}{\partial z}\right)$$
$$+ \nu_A \cdot r_V \pm \beta_A \cdot a_S \cdot \Delta A, \qquad (2.2-1b)$$

or, when the angle Θ is constant and D is isotropic, Equation (2.2-1c).

$$\frac{\partial A(r,z,t)}{\partial t} = -\left[\frac{\partial(u_r \cdot A)}{\partial r} + \frac{\partial(u_z \cdot A)}{\partial z}\right] + D_r \cdot \left(\frac{1}{r} \cdot \frac{\partial A}{\partial r} + \frac{\partial^2 A}{\partial r^2}\right)$$
$$+ D_z \frac{\partial^2 A}{\partial z^2} + \nu_A \cdot r_V \pm \beta_A \cdot a_S \cdot \Delta A. \qquad (2.2-1c)$$

When all variables are symmetrically distributed about the cylindrical axis, the energy balance in cylindrical coordinates is given by Equation (2.2-2b):

$$\frac{\partial T(r,z,t)}{\partial t} = -\left[\frac{\partial(u_r \cdot T)}{\partial r} + \frac{\partial(u_z \cdot T)}{\partial z}\right] + \frac{\lambda_r}{\rho \cdot c_P}\left[\frac{\partial^2 T}{\partial r^2} + \frac{1}{r}\frac{\partial T}{\partial r}\right]$$
$$+ \frac{\lambda_z}{\rho \cdot c_P}\frac{\partial^2 T}{\partial z^2} + \frac{r_V \cdot (-\Delta_R H)}{\rho \cdot c_P} \pm \frac{a \cdot A_W \cdot (T_W - T)}{\rho \cdot c_P \cdot V_R}. \qquad (2.2-2b)$$

With further simplifications, such as spatial independence of the rate and the transport coefficients, in one-dimensional Cartesian coordinates (positional coordinate x), they read as Equations (2.2-4) and (2.2-5):

$$\frac{\partial A(x,t)}{\partial t} = -u \cdot \frac{\partial A}{\partial x} + D_A \cdot \frac{\partial^2 A}{\partial x^2} + \nu_A \cdot r_V \pm \beta_A \cdot a_S \cdot \Delta A \qquad (2.2-4)$$

$$\frac{\partial T(x,t)}{\partial t} = -u \cdot \frac{\partial T}{\partial x} + \left(\frac{\lambda}{c_P \cdot \rho}\right) \cdot \frac{\partial^2 T}{\partial x^2} + \left(\frac{(-\Delta_R H)}{c_P \cdot \rho}\right) \cdot r_V$$
$$\pm \left(\frac{a \cdot A_W}{V_R \cdot c_P \cdot \rho}\right) \cdot (T_W - T). \qquad (2.2-5)$$

Other equations that are required for the description and design of reactors follow [Bartholomew 1994]:

- Thermal state equations of the form $P(\rho,T)$ such as the ideal gas law (Equation 2.2-6):

$$P(\rho, T) = \left(\frac{R}{M}\right) \cdot \rho \cdot T. \qquad (2.2-6)$$

M = molar mass
R = 8.313 J mol^{-1} K^{-1}

- Phase equilibrium relationships of the form $y_i(x_i)$ such as Raoult's law (Equation 2.2-7; see Section 2.3.2):

$$y_A = \frac{x_A \cdot P_A^0(T)}{x_A \cdot P_A^0(T) + (1 - x_A) \cdot P_B^0(T)} \qquad (2.2-7)$$

y_A = mole fraction of component A in the gas phase
x_A = mole fraction of component A in the liquid phase
$P_i^0(T)$ = vapor pressure of the pure components i (i = A, B).

- Reaction rate equations of the form $r(c_i, T, P)$

  Microkinetics (Section 2.1.3), for example, first-order rate equations or Langmuir–Hinshelwood kinetics (Equation 2.1-40).

  Macrokinetics of the form $r_{m,meas} = \eta_{cat} r_m$, where η_{cat} is the degree of catalyst use (Equation 2.1-58).

The problem of chemical reaction engineering is that – similar to quantum mechanics with the Schrödinger equation – complete solutions for this system of differential equations, which are coupled through the quantities c, T, and $\vec{u}$ can only be found for simple limiting cases, the so-called ideal reactors. In these simplified cases it is assumed that the hydrodynamic behavior can be described by the limiting cases of complete backmixing (continuous stirred tank) and plug flow (ideal tubular reactor), and that the reactors are operated isothermally and isobarically, so that the enthalpy and impetus balances can be dispensed with. Furthermore, only homogeneous opertion (one phase) is considered, and hence there are no exchange terms between individual phases. The design equations are thus reduced to solving the mass balance.

2.2.1.1
Ideal Reactors

Continuous stirred tank reactor (CSTR)
This ideal reactor is the easiest to solve mathematically. The balance space (Figure 2.2-1) can be extended over the entire reactor since, because of the boundary condition of complete backmixing, this reactor type is gradient-free.

For the reaction A → P differential Equation (2.2-4) can be replaced by algebraic Equation (2.2-8):

$$\dot{n}_{A,0} - \dot{n}_A = \dot{V} \cdot (A_0 - A) = V_R \cdot r_V \rightarrow r_V = \frac{A_0 - A}{\tau}, \qquad (2.2-8)$$

$\dot{V}$ = volume flow (no change due to reaction)
V_R = effective reactor volume
A = concentration of A in mol L^{-1}.

Hence the change in the molar flux $\Delta \dot{n}_A$ of starting material A is equal to the amount of substance that has reacted ($V_R r_V$).

For a reaction that is nth order in starting material A: $\mathring{A} = -kA^n$ and the mean residence time $\tau = V_R/\dot{V}$ is given by Equation (2.2-9a)

$$\frac{A_0 - A}{\tau} = k \cdot A^n \qquad (2.2-9a)$$

which rearranges to Equation (2.2-9b).

$$A^n = \frac{A_0}{k\tau} \cdot \left(1 - \frac{A}{A_0}\right). \qquad (2.2-9b)$$

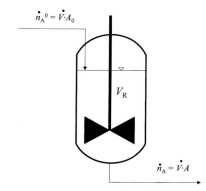

Fig. 2.2-1 Ideal continuous stirred tank (CSTR).

Equations (2.2-10)–(2.2-11) give solutions for certain values of n:

$$n = 0 : \quad \frac{A}{A_0} = 1 - \frac{k\tau}{A_0} = 1 - Da \qquad (2.2-10)$$

$$n = 1 : \quad \frac{A}{A_0} = \frac{1}{1 + k\tau} = \frac{1}{1 + Da} \qquad (2.2-11)$$

$$n = 2 : \quad \frac{A}{A_0} = \frac{1}{2 \cdot k\tau \cdot A_0} \cdot \left[\sqrt{(4 \cdot k\tau \cdot A_0 + 1)} - 1 \right]$$

$$= \frac{1}{2 \cdot Da} \cdot \left[\sqrt{(4 \cdot Da + 1)} - 1 \right] \qquad (2.2-12)$$

where $Da = k\tau A_0^{n-1}$ is the dimensionless Damköhler number.

If, in a bimolecular reaction of A with B, the starting material B is present in excess E, then $B = A + E$ and Equation (2.2-8) then reads:

$$\frac{A_0 - A}{\tau} = k \cdot A \cdot (A + E) \qquad (2.2-13)$$

with the solution:

$$\frac{A}{A_0} = \frac{1}{A_0} \cdot \left[\sqrt{\left(\frac{1}{2 \cdot k\tau} + \frac{E}{2}\right)^2 + \frac{A_0}{k\tau}} - \left(\frac{1}{2 \cdot k\tau} + \frac{E}{2}\right) \right]. \qquad (2.2-14)$$

With known kinetics [here $k(T)$] the reactor (here the volume of the reactor V_R) can be designed for given specifications [here the conversion $U = (A_0-A)/A_0$)] and volumetric flow $\dot{V}$.

Ideal tubular reactor or batch stirred tank (PFR: plug flow reactor or STR: stirred tank reactor)

With the aid of Equation (2.2-4) and the boundary conditions for an ideal tubular reactor:

- $\vec{u} = |u| = \text{constant}$
- $\frac{\partial A}{\partial t} = 0$ (stationary) $\qquad (2.2\text{-}15)$

and the assumption of nth order kinetics, Equation 2.2-16 results:

$$0 = u \cdot \frac{dA}{dx} - k \cdot A^n, \qquad (2.2-16)$$

with the solution (Equation 2.2-17):

$$A = A_0 \cdot exp\,(-k\tau) \text{ for } n = 1 \qquad (2.2-17a)$$

$$A = \left\{(n-1)\cdot k\tau + A_0^{(1-n)}\right\}^{1/(1-n)} \text{ for } n \neq 1, \qquad (2.2-17b)$$

where $\tau = L/u$ is the mean residence time and L is the length of the reactor.

If in a bimolecular reaction of A with B the starting material B is present in excess E then $B = A + E$ and Equation (2.2-18) applies:

$$A = \frac{A_0 \cdot E}{(A_0 + E)\cdot exp\,(E\cdot k\tau) - A_0}. \qquad (2.2-18)$$

The same relationship is obtained for the boundary conditions of an ideal batch stirred tank:

- $\dfrac{\partial A}{\partial x} = 0$ (ideal backmixing, i.e., no concentration gradient).

Inserting these boundary conditions in general Equation (2.2-4) gives Equation (2.2-19)

$$\frac{\partial A}{\partial t} = 0 + 0 + v_A \cdot r_V. \qquad (2.2-19)$$

For nth order kinetics one obtains the same solution as above (Equations 2.2-17 and 2.2-18), except that the variable τ is replaced by the absolute reaction time.

Evidently the ideal tubular reactor and the batch stirred tank have the same behavior (Equation 2.2-20):

$$\text{Kessel: } \frac{\partial A}{\partial t} = v_A \cdot r_V \xrightarrow{u=dx/dt} \text{Rohr: } \frac{\partial A}{\partial x} = \frac{1}{u}\cdot v_A \cdot r_V, \qquad (2.2-20)$$

hence, there is no distinction between a reacting volume element in a batch stirred tank and one in an ideal tubular reactor (prerequisite: no volume change).

The most important reactor equations for simple kinetics are summarized in Table 2.2-1.

Plotting the concentration ratio A_{tube}/A_{tank} for the three cases $n = -1, 0, +1$ (Figure 2.2-2) reveals that for the normal case ($n > 0$) a tubular reactor has advantages over a continuous stirred tank.

Tab. 2.2-1 Equations for ideal reactors for n^{th} order kinetics.

n	Ideal tubular reactor or batch stirred tank	continuous stirred tank
-1	$A = A_0 \cdot \sqrt{1 - \dfrac{2 \cdot k\tau}{A_0^2}}$	$A = \dfrac{A_0}{2}\left(1 + \sqrt{1 - \dfrac{k\tau}{A_0^2}}\right)$
0	$A = A_0 - k\tau$	$A = A_0 - k\tau$
$0{,}5$	$A = \left(\sqrt{A_0} - \dfrac{k\tau}{2}\right)^2$ für $\tau \prec 2\sqrt{A_0}/k$	$A = \dfrac{1}{2}\left(2 \cdot A_0 + k^2\tau^2 - k\tau\sqrt{4 \cdot A_0 + k^2\tau^2}\right)$
1	$A = A_0 \cdot \exp(-k\tau)$	$A = \dfrac{A_0}{1 + k\tau}$
2	$A = \dfrac{A_0}{1 + A_0 \cdot k\tau}$	$A = \dfrac{1}{2 \cdot k\tau} \cdot \left(\sqrt{4 \cdot k\tau \cdot A_0 + 1} - 1\right)$
n	$A = \left\{(n-1) \cdot k\tau + A_0^{(1-n)}\right\}^{1/(1-n)}$ für $n \neq 1$	$A^n = \dfrac{A_0}{k\tau} \cdot \left(1 - \dfrac{A}{A_0}\right)$
2 With excess E of $B = A + E$	$A = \dfrac{A_0 \cdot E}{(A_0 + E) \cdot \exp(E \cdot k\tau) - A_0}$	$\dfrac{A}{A_0} = \dfrac{1}{A_0} \cdot \left(\sqrt{\left(\dfrac{1}{2 \cdot k\tau} + \dfrac{E}{2}\right)^2 + \dfrac{A_0}{k\tau}} - \left(\dfrac{1}{2 \cdot k\tau} + \dfrac{E}{2}\right)\right)$

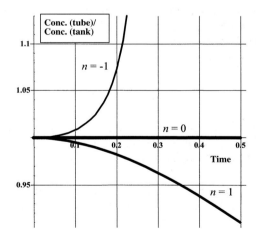

Fig. 2.2-2 Ratio of the concentration of A in a tubular reactor to that in a continuous tank ($k = 1$). For a reaction order $n = 1$, the ratio is always smaller than zero, that is, the starting material is more rapidly consumed in a tubular reactor than in a continuous tank. For $n > 1$ this situation is reversed. Only in the case of a zeroth order reaction are the two reactors equal.

Cascade of ideal stirred tanks

This can be regarded as the link between the ideal continuous stirred-tank reactor (tank number $N = 1$) and the ideal tubular reactor ($N \to \infty$). For a first-order reaction and a cascade of N identical tanks, Equation (2.2-21) is obtained:

$$A_N = A_0 \cdot \frac{1}{(1 + k\tau)^N}. \tag{2.2 – 21}$$

2.2.1.2
Reactors with Real Behavior

In practice, smaller or larger deviations from the above-described ideal behavior occur. The magnitude of the deviation and thus the first step in evaluating the true situation in a given reactor is an analysis of the residence-time behavior.

Residence-time behavior

The selectivity and conversion of a chemical reactor depend not only on the kinetics of the reaction but also on the time for which the reaction partners are available for the reaction, that is, the hydrodynamic behavior. By determining the residence-time behavior of a reactor, one can determine its deviation from ideal hydrodynamic behavior (boundary cases: plug flow and complete backmixing) and hence decide with which reactor model the real reactor can best be modeled.

The residence-time behavior can be determined experimentally with the aid of injection or displacement labeling, in which a nonreacting labeling material M is introduced at the reactor inlet. In Figure 2.2-3 the concentration of M in the volume flow to the reactor at $t = 0$ is suddenly increased from $M = 0$ to M_0 at $t > 0$ and the course of the concentration at the reactor outlet is measured, for example, by switching the feed from container B1 (containing fully deionized water) to B2 (containing 1%

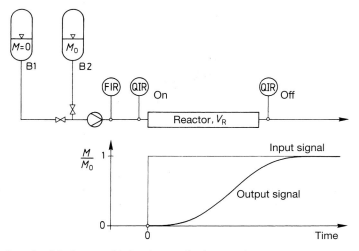

Fig. 2.2-3 Example of displacement labeling (see text for discussion).

Na$_2$SO$_4$) by means of two ball valves and measuring the electrical conductivity at the reactor outlet (conductivity meter Q). In practice, radioactive labels are used, since even traces which do not influence the reaction are readily detectable.

The result of the measurement is the cumulative residence-time curve $F(t)$ (Equation 2.2-22):

$$F(t) = \frac{M(t)}{M^0} \qquad (2.2-22)$$

$F(t) = 0$ for $t = 0$
$F(t) = 1$ for $t \to \infty$.

Differentiation of $F(t)$ (dimensionless) gives the residence-time distribution function (dimensions s^{-1}; Equation 2.2-23):

$$w(t) = \frac{dF(t)}{dt} \quad \text{bzw.} \quad F(t) = \int_0^t w(t')dt', \qquad (2.2-23)$$

which in the literature is often symbolized by $E(t)$ [Baerns 1999]. This function can be directly measured by means of injection labeling, in which the entire labeling substance is introduced in a very short time.

From the resulting residence-time function [$F(t)$ from displacement labeling or $w(t)$ from injection labeling], the mean residence time τ can be calculated (Equation 2.2-24).

$$\tau = \int_0^\infty t \cdot w(t)dt = \int_0^1 t \cdot dF(t) = \int_0^\infty (1 - F(t))dt. \qquad (2.2-24)$$

For practical evaluation, the integral is determined numerically, for example, from the cumulative distribution function (Equation 2.2-26):

$$\tau \approx \sum_i t_i \cdot \Delta F_i = \sum_i (1 - F_i) \cdot \Delta t_i. \qquad (2.2-25)$$

Analogously, Equation (2.2-26) is obtained for the scatter (variance) about the mean value:

$$\sigma^2 = \int_0^\infty (t - \tau)^2 \cdot w(t)dt = 2 \cdot \int_0^\infty (1 - F(t)) \cdot (t - \frac{\tau}{2})dt, \qquad (2.2-26)$$

or, after the introduction of discrete measured values, Equation (2.2-27) [Baerns 1999].

$$\sigma^2 \approx 2 \cdot \sum_i (2 - F_i) \cdot (t - \frac{\tau}{2}) \cdot \Delta t_i. \qquad (2.2-27)$$

2.2.1.2.2
Residence-time behavior of ideal reactors (Figures 2.2-4 and 2.2-5)

Ideal tubular reactor with plug flow

$$F(t) = 0 \text{ for } t < \tau \text{ und } 1 \text{ for } t \geq \tau \qquad (2.2-28)$$

$$w(t) = \infty \text{ for } t = \tau \text{ und } 0 \text{ otherwise} \qquad (2.2-29)$$

Laminar flow in a tubular reactor

$$F(t) = 1 - \frac{1}{4 \cdot (t/\tau)^2} \text{ for } (t/\tau) \geq 0.5 \qquad (2.2-30)$$

$$w(t) = \frac{1}{2 \cdot \tau} \cdot \frac{1}{(t/\tau)^3}. \qquad (2.2-31)$$

Ideal stirred tank

$$F(t) = 1 - \exp(-t/\tau) = F(t/\tau) \qquad (2.2-32)$$

$$w(t) = \frac{1}{\tau} \cdot \exp(-t/\tau) = \frac{1}{\tau} \cdot w(t/\tau). \qquad (2.2-33)$$

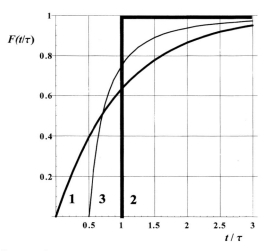

Fig. 2.2-4 Cumulative residence-time function $F(t/\tau)$ for an ideal continuous stirred tank (1), an ideal tubular reactor with plug flow (2), and laminar flow in a tubular reactor (3).

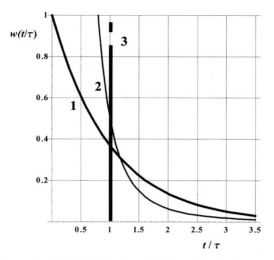

Fig. 2.2-5 Residence-time distribution function $w(t/\tau)$ for an ideal continuous stirred tank (1), an ideal tubular reactor with plug flow (2), and laminar flow in a tubular reactor (3). $\tau = 1$.

Residence-time behavior of real reactors

If the deviations are small then they can be described by the dispersion model (additional dispersive flow is is superimposed on the plug flow) or cell model (cascade of ideal stirred tanks). For larger deviations the calculation of nonideal reactors is generally difficult. A more simply treated special case occurs when the volume elements flowing through the reactor are macroscopically but not microscopically mixed (segregated flow). This case can be solved by the Hofmann–Schoenemann[9] method (see below).

Dispersion model [Qi 2001]

The dispersion model assumes that the residence-time distribution of a real tubular reactor can be regarded as the superimposition of the plug flow that is characteristic of the ideal tubular reactor and diffusionlike axial mixing, characterized by an axial dispersion coefficient D_{ax} which has the same dimensions as, but can be much larger than, the molecular diffusion coefficient. The following effects can contribute to the axial mixing:

- Convective mixing in the direction of flow due to eddy formation or turbulence.
- Different residence-time behaviors of particles moving along different flow lines owing to nonuniform distribution of the flow velocity over the cross section of the tube.
- The ubiquitous molecular diffusion.

In most practical cases the molecular diffusion is negligible in comparison to the other effects. The first two effects differ in the fact that the first can lead to backmixing, that

---

9) Karl Schoenemann, from 1948–1966 professor at the TU Darmstadt.

is, mass transfer against the flow under the influence of a concentration gradient; backmixing as a result of the second effect is impossible.

The cumulative residence-time function for a real tubular reactor can be derived from the general mass balance by taking into account the dispersion term (Equation 2.2-34):

$$\frac{\partial M}{\partial t} = -u_x \cdot \frac{\partial M}{\partial x} + D_{ax} \cdot \frac{\partial^2 M}{\partial x^2} \qquad (2.2-34)$$

M = concentration of the nonreacting labeling substance.

By introducing the dimensionless variables t/τ and x/L and by using the definition of the dimensionless Bodenstein[10] number Bo (Equation 2.2-35):

$$Bo = \frac{u_x \cdot L}{D_{ax}}, \qquad (2.2-35)$$

leads to differential Equation (2.2-36):

$$\frac{\partial M}{\partial (t/\tau)} = -\frac{\partial M}{\partial (x/L)} + \frac{1}{Bo} \cdot \frac{\partial^2 M}{\partial (x/L)^2}, \qquad (2.2-36)$$

the solution of which for an infinitely long reactor ($L \to \infty$) is given by Equation (2.2-37) [Fitzer 1989]:

$$\frac{M(t/\tau)}{M^0} = F(t/\tau) = \frac{1}{2} \cdot \left[1 - erf \left(\frac{\sqrt{Bo}}{2} \cdot \frac{1-(t/\tau)}{\sqrt{t/\tau}} \right) \right] \qquad (2.2-37)$$

with error function (Equation 2.2-38):

$$erf(x) = \frac{2}{\sqrt{\pi}} \cdot \int_0^x \exp(-t^2) \cdot dt. \qquad (2.2-38)$$

Differentiation of the cumulative residence-time function $F(t/\tau)$ (Equations 2.2-23 and 2.2-37) gives the corresponding residence-time distribution function (Equation 2.2-39):

$$w(t/\tau) = \frac{1+(t/\tau)}{4 \cdot (t/\tau)} \cdot \sqrt{\frac{Bo}{\pi \cdot (t/\tau)}} \cdot \exp\left(-\frac{Bo \cdot [1-(t/\tau)]^2}{4 \cdot (t/\tau)}\right). \qquad (2.2-39)$$

This equation, which is valid for $Bo > 0$, does not apply to laminar flow behavior. Figures 2.2-6 and 2.2-7 plot these functions for different Bo values.

---

10) Max E. A. Bodenstein, German physicist (1871–1942).

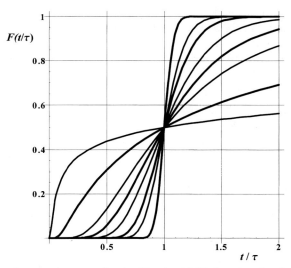

Fig. 2.2-6 Cumulative residence-time function (Equation 2.2-37) for various Bo values ($Bo = 0.1, 1, 5, 10, 20, 50, 100, 200$ from left to right).

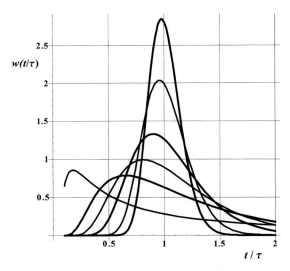

Fig. 2.2-7 Residence-time distribution function (Equation 2.2-39) for various Bo values ($Bo = 1, 5, 10, 20, 50, 100$ from left to right).

For the boundary cases, the following applies:

- $Bo \to \infty$: there is no axial mixing, that is the system fulfills the prerequisites of an ideal tubular reactor.
- $Bo = 0$: complete backmixing occurs, as is characteristic of an ideal stirred tank.

In order to describe the flow in a real tube, from the group of calculated curves that which best corresponds to the experimental curve is located. From the Bo value corresponding to this curve, the behavior of the system as a reactor can be estimated, and since u and L are known, D_{ax} can be calculated from Bo.

Cell model (Figures 2.2-8 and 2.2-9)
The true residence-time behavior of a reactor can be described with the aid of a reactor cascade. The characteristic quantity for the behavior is the number of reactors N, and the distribution functions are given by Equations (2.2-40) and (2.2-41):

$$F(t/\tau) = 1 - \left[\sum_{i=1}^{N} \frac{(N \cdot (t/\tau))^{i-1}}{(i-1)!} \right] \cdot \exp(-N \cdot t/\tau) \qquad (2.2-40)$$

$$w(t/\tau) = \frac{N \cdot (N \cdot t/\tau)^{N-1}}{(N-1)!} \cdot \exp(-N \cdot t/\tau). \qquad (2.2-41)$$

When the residence-time behavior differs only slightly from that of the ideal tubular reactor ($Bo > 40$), the dispersion and cell models are similar and Equation (2.2-42) applies:

$$Bo \approx 2 \cdot N. \qquad (2.2-42)$$

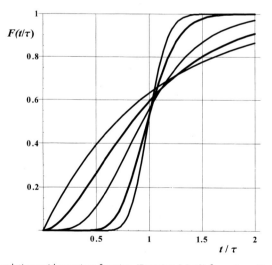

Fig. 2.2-8 Cumulative residence-time function (Equation 2.2-40) for various N values (N = 1, 2, 5, 20, 50 from left to right).

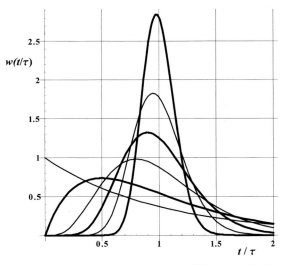

Fig. 2.2-9 Residence-time distribution function (Equation 2.2-41) for various N values ($N = 1, 2, 5, 20, 50$ from left to right).

For larger deviations from ideal behavior (dead zones, short circuits), there are the following possibilities for solving the problem:

a) Historically, engineers solved the problem with the aid of the similarity theory and the characteristic quantities derived therefrom (Buckinghham Π theory [Zlokarnik 2000]).
b) Nowadays one attempts to solve the system of differential equations with the help of high-performance computers (see Section 2.4.1) [Platzer 1996]. Nevertheless, this approach can easily become limited by the complexity of the real system. The reason for this is, as was mentioned above, difficuties in solving the system of nonlinear partial differential equations (Equations 2.2-1 to 2.2-3) and the large number of required data, which for real systems are often not readily accessible or are associated with large errors.

Hofmann[11]-Schoenemann[12] method

In a segregated flow each volume element can be regarded as a small ideal tank in which the reaction occurs for as along as the volume element resides in the reactor. The resulting concentration is calculated from Equation (2.2-43):

$$A_{ende} = \int_0^\infty A(t) \cdot w(t) \cdot dt \quad \text{bzw.} \quad U = 1 - \int_0^\infty \frac{A(t)}{A_0} \cdot w(t) \cdot dt, \qquad (2.2-43)$$

---

11) Hans Hofmann, professor emeritus at the TU Karlsruhe.
12) Karl Schoenemann, professor at the TU Darmstadt (1900–1984).

where $A(t)$ is the concentration of the starting material in a batch tank operated for time t, and $w(t)$ the residence-time spectrum of the volume element. $A(t)$ can be measured in a laboratory reactor; $w(t)$ must be measured in a technical reactor. Equation (2.2-43) can be solved numerically or graphically (Hofmann–Schoenemann method).

Example 2.2-1

For an ideal continuous stirred tank the corresponding equations are:

$$A(t) = A_0 \cdot \exp(-k \cdot t)$$

$$w(t) = \frac{1}{\tau} \cdot \exp\left(-\frac{t}{\tau}\right).$$

Insertion in Equation (2.2-43) and rearrangement gives:

$$A_{end} = \frac{A_0}{\tau} \int_0^\infty \exp\left\{-\left(k + \frac{1}{\tau}\right) \cdot t\right\} dt$$

and integration:

$$A_{end} = \frac{A_0}{1 + k\tau}$$

the known Equation (2.2-11) for a stirred-tank.

2.2.1.3
Nonisothermal reactors

The assumed isothermal operating mode of the ideal reactors is practically unrealizable for highly exothermic reactions in technical reactors, so that in practice, in addition to the mass balance (Equation 2.2-44), an energy balance (Equation 2.2-45) must be considered:

$$\frac{\partial A(x,t)}{\partial t} = -\bar{u} \cdot \frac{\partial A}{\partial x} + D_A \cdot \frac{\partial^2 A}{\partial x^2} + v_A \cdot r_V(T) \qquad (2.2-44)$$

$$\frac{\partial T(x,t)}{\partial t} = -\bar{u} \cdot \frac{\partial T}{\partial x} + \left(\frac{\lambda}{c_P \cdot \rho}\right) \cdot \frac{\partial^2 T}{\partial x^2} + \left(\frac{(-\Delta_R H)}{c_P \cdot \rho}\right) \cdot r_V(T)$$
$$+ \left(\frac{k_W \cdot A_W}{V_R \cdot c_P \cdot \rho}\right) \cdot (T_K - T) \quad (2.2-45)$$

k_W = heat-transfer coefficient in W m^{-2} K^{-1}
T_K = coolant temperature in K
for other symbols, see Equation (2.2-2).

The two equations are coupled through the temperature dependence of the reaction rate $r_V(T)$. Since this dependence is usually expressed by an Arrhenius term, this system of differential equations must be solved numerically, since the Arrhenius term can not be completely integrated. In the following, descriptions of the most important models are given with consideration of a nonisothermal mode of operation.

Example 2.2-2 [Wang 1999]

Chemistry: *Oxidation of an aqueous glucose solution with pure oxygen under supercritical conditions [Franck 1999], that is, homogeneous fluid phase.*

- $C_6(H_2O)_6 + 6\ H_2O \rightarrow 6\ CO_2 + 6\ H_2O$, $\Delta_R H$ = const. = -2802 kJ mol^{-1}

Kinetics: *pseudo-first order reaction with respect to glucose (A)*

- $A = A_0 \cdot \exp(-kt)$
- $k(380\,°C) = 0.0536$ s^{-1}
- $E_A = 91.3$ kJ mol^{-1}
- $k(T) = 1.08 \times 10^6 \exp(-91300/RT)$

Boundary conditions:

- Initial temperature $T_0 = 380\,°C = 653$ K
- Pressure $P = 300$ bar
- Initial concentration $A_0 = 1\%$ (g g^{-1}) = 11.11 mol m^{-3} (average density 200 kg m^{-3})

Material data:

- Mean density $\rho = 200$ kg m^{-3}
- Mean heat capacity $c_P = 6000$ J kg^{-1} K^{-1}
- Cooling conditions T_K = const. = $380\,°C = 653$ K
- Heat-transfer coefficient k_W = const. = 500 W m^{-2} K^{-1}

Continuous ideal tubular reactor with adiabatic reaction control

Taking into account the boundary conditions for an ideal tubular reactor and assuming a first-order reaction leads to Equations (2.2-46) and (2.2-47):

$$0 = -\bar{u} \cdot \frac{dA}{dx} - k(T) \cdot A \qquad (2.2-46)$$

$$0 = -\bar{u} \cdot \frac{dT}{dx} + \left(\frac{(-\Delta_R H)}{c_p \cdot \rho}\right) \cdot k(T) \cdot A. \qquad (2.2-47)$$

Example 2.2-2a

Inserting the above example conditions and assuming a flow rate of $u = 2 \text{ m s}^{-1}$:

$$\frac{dA}{dx} = -\frac{k(T) \cdot A}{\bar{u}} = -\frac{1{,}08 \cdot 10^6 \cdot \exp(-91300/RT) \cdot A}{2}$$

$$\frac{dT}{dx} = \frac{(-\Delta_R H)}{\rho \cdot c_p \cdot \bar{u}} \cdot k(T) \cdot A = \left(\frac{2802 \cdot 10^3}{6000 \cdot 200 \cdot 2}\right) \cdot 1.08 \cdot 10^6 \cdot \exp(-91300/RT) \cdot A.$$

The numerical solution gives the temperature profile along the tubular reactor (Figure 2.2-10).

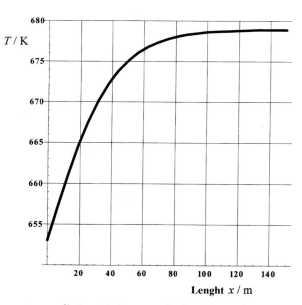

Fig. 2.2-10 Temperature profile in a tubular reactor (for boundary conditions, see text). The adiabatic temperture rise is 26 °C (Equation 2.2-50).

By dispensing with the temperature/length profile $T(x)$, the temperature rise as a function of conversion can be directly calculated by integration of Equation (2.2-44):

$$0 = -\bar{u} \cdot \frac{dT}{dx} + \left(\frac{(-\Delta_R H)}{c_p \cdot \rho}\right) \cdot \frac{-dA}{dt} \quad \text{where } \bar{u} = \frac{dx}{dt} \qquad (2.2-48)$$

$$T = T_0 + \frac{(-\Delta_R H) \cdot A_0}{c_p \cdot \rho} \cdot U. \qquad (2.2-49)$$

For complete conversion Equation (2.2-50) follows for the adiabatic temperature rise:

$$\Delta T_{ad} = \frac{(-\Delta_R H) \cdot A_0}{c_p \cdot \rho}. \qquad (2.2-50)$$

Ideal batch stirred tank with adiabatic reaction control
This is mathematically identical to the above case, since with the aid of the mean flow velocity $\bar{u}$ one can transform the length x into the time t ($dx = \bar{u}\, dt$).

Ideal batch stirred tank with polytropic reaction control
For the boundary conditions of an ideal batch stirred-tank reactor and a first-order reaction Equations (2.2-51) and (2.2-52) result from Equations (2.2-44) and (2.2-45):

$$\frac{\partial A(x,t)}{\partial t} = -k(T) \cdot A \qquad (2.2-51)$$

$$\frac{dT(x,t)}{dt} = \left(\frac{(-\Delta_R H)}{c_p \cdot \rho}\right) \cdot k(T) \cdot A + \left(\frac{k_W \cdot A_W}{V_R \cdot c_p \cdot \rho}\right) \cdot (T_W - T). \qquad (2.2-52)$$

As can be seen from Equation (2.2-44), the amount of heat removed increases linearly with the temperature in the reactor (for a constant coolant temperature), but the amount of heat increases exponentially according to the Arrhenius equation.

Example 2.2-2b

For a $1\,m^3$ spherical reactor (heat-exchange area $4.84\,m^2$), inserting the example conditions gives:

$$\frac{dA(x,t)}{dt} = -1.08 \cdot 10^6 \cdot \exp(-91300/RT) \cdot A$$

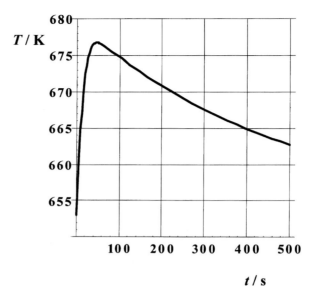

Fig. 2.2-11 Course of temperature with time in a discontinuous stirred tank (for boundary conditions, see text).

$$\frac{dT(x,t)}{dt} = \left(\frac{2802 \cdot 10^3}{6000 \cdot 200}\right) \cdot 1.08 \cdot 10^6 \, exp\,(-91300/RT) \cdot A$$
$$+ \left(\frac{500 \cdot 4.84}{1 \cdot 6000 \cdot 200}\right) \cdot (653 - T).$$

The numerical solution gives the temperature profile T(t) as a function of reaction time t (Figure 2.2-11).

Continuous ideal tubular reactor with polytropic reaction control

This is mathematically identical to the above case, since with the aid of the mean flow velocity $\bar{u}$ one can transform the length x into the time $x(dt = \bar{u}^{-1}dx)$.

Ideal continuous stirred tank with polytropic reaction control

Mathematically, the treatment of this case is particularly simple because, owing to the gradient-free operation, no system of coupled differential equations need be solved. For the mass balance of a continuous stirred tank and a simple irreversible reaction A → P, Equations (2.2-53)–(2.2-54) result from Equation (2.2-9a) for orders of reaction n:

$$n = 0: \quad U_A(T, \tau) = \frac{k(T) \cdot \tau}{A_0} \tag{2.2 – 53}$$

$$n = 1: \quad U_A(T, \tau) = \frac{k(T) \cdot \tau}{1 + k(T) \cdot \tau} \tag{2.2 – 54}$$

$$n = 2: \quad U_A(T, \tau) = 1 - \frac{1}{2k\tau \cdot A_0} \cdot \left[\sqrt{(4 \cdot k\tau \cdot A_0 + 1)} - 1\right]. \tag{2.2 – 55}$$

The amount of heat $\dot{Q}_R$ produced (consumed) by reaction is equal to the heat exchanged with the surroundings $\dot{Q}_W$ and the amount of heat transported by convection $\dot{Q}_K$ (Equation 2.2-56):

$$\dot{Q}_R = \dot{Q}_W + \dot{Q}_K. \tag{2.2 – 56}$$

The amount of heat produced (consumed) by reaction is given by Equation 2.2-57:

$$\dot{Q}_R(T, \tau) = (-\Delta_R H) \cdot (\dot{n}_A^0 - \dot{n}_A) = (-\Delta_R H) \cdot \dot{V} \cdot A_0 \cdot U_A(T, \tau). \tag{2.2 – 57}$$

The amount of heat exchanged with the surroundings is given by Equation 2.2-58:

$$\dot{Q}_W(T, \tau) = k_W \cdot A_W \cdot (T - T_K) \tag{2.2 – 58}$$

k_W = heat-transfer coefficient
A_W = heat-exchange area
T_K = coolant temperature (constant).

The amount of heat introduced into or removed from the reactor by flow is given by Equation (2.2-59).

$$\dot{Q}_K(T, \tau) = \dot{V} \cdot \rho \cdot c_P \cdot (T - T_0). \tag{2.2 – 59}$$

T_0 = inlet temperature

Equation (2.2-56) is most easily solved graphically by determining the mutual intersection (= operating point) of the S-shaped $\dot{Q}_R$ curve and the straight line ($\dot{Q}_W + \dot{Q}_K$), as shown in Figure 2.2-13. A stationary operating point is stable (unstable) if the slope of the $\dot{Q}_R$ curve is smaller (larger) than the slope of the ($\dot{Q}_W + \dot{Q}_K$) straight line [Müller-Erwin 1998, Ferino 1999].

Example 2.2-2c (Figure 2.2-12) _____

For a 1 m³ spherical reactor (heat-exchange area 4.84 m²) and a volumetric flow of 36 m³ h⁻¹ (corresponding to a residence time of 100 s), inserting the above example conditions gives Equation (2.2-60).

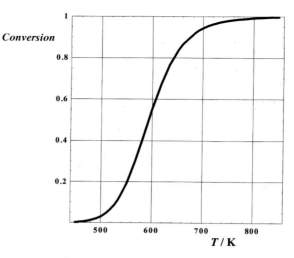

Fig. 2.2-12 Conversion as function of reactor temperature according to Equation (2.2-60).

$$U_A(T, \tau) = \frac{k(T) \cdot \tau}{1 + k(T) \cdot \tau} \quad \text{mit } k(T) = 1.08 \cdot 10^6 \cdot \exp(-91300/RT). \quad (2.2-60)$$

The following individual terms (in kJ s^{-1}) result:

$$\dot{Q}_R(T, \tau) = (-\Delta_R H) \cdot \dot{V} \cdot A_0 \cdot U_A(T, \tau) = 2\,802\,000 \cdot 0.01 \cdot 11.11 \cdot U_A(T, \tau)$$

$$\dot{Q}_W(T, \tau) = k_W \cdot A_W \cdot (T - T_K) = 500 \cdot 4.84 \cdot (T - 653)$$

$$\dot{Q}_K(T, \tau) = \dot{V} \cdot \rho \cdot c_p \cdot (T - T_0) = 0{,}01 \cdot 200 \cdot 6000 \cdot (T - 653)$$

As shown in Figure 2.2-13, the intersection of the heat-production and heat-removal curves gives a stationary reactor temperature of 672 K, which corresponds to a conversion of 90 %.

2.2.1.4
Design of Reactors

The wide variety of reactor designs results from the possible combinations of:

- Batch or continuous operation
- Heat of reaction (stongly exothermic to strongly endothermic)
- Temperature control (adiabatic to isothermal)
- Space–time yield (low to high)
- Reaction conditions (pressure, temperature, concentration)

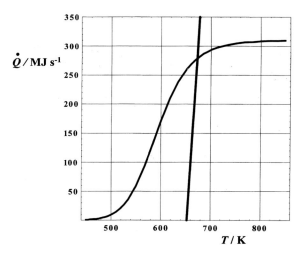

Fig. 2.2-13 Heat–temperature diagram (see text for discussion).

- Catalyst (heterogeneous, homogeneous, biocatalyst)
- Phases (homogeneous to multiphase).

However, in industrial practice a smaller number of reactor types are found, which can be classified according to the states of aggregation that they handle:

- Gas-phase reactors
- Liquid-phase reactors
- Gas-liquid reactors
- Gas-solid reactors, and so on.

A good overview can be found in [Ullmann2]. In the following only the three most important reactor types – the stirred-tank, tubular, and fluidized-bed reactors – are discussed in detail, together with a more recent development, namely the microreactor.

Stirred-Tank Reactor

The classical apparatus for homogeneous liquid-phase reaction systems is the stirred tank, which is preferentially operated in batch mode, since for normal reaction kinetics (order > 0) its continuous operation is disadvantageous; an exception here is its use in a cascade.

For smaller production quantities (rule of thumb: 10 000 t/a) and/or frequent product changes, it has advantages over the tubular reactor. Very long residence times can readily be achieved, and reaction conditions such as temperature, pH value, and catalyst concentration can be changed and optimized during the reaction time. In batch mode, product quality is subject to certain variations, and hence continuous process control is necessary.

Fig. 2.2-14 DIN standard stirred tank.

Volume/m^3	d_B^*/m	Heat-exchange area/m^2	Surface area to volume ratio/m^2 m^{-3}
0,1	0,508	0,80	8,0
0,25	0,700	1,48	5,9
1,0	1,20	3,87	3,9
2,50	1,60	7,90	3,2
6,30	2,00	13,1	2,1
10,0	2,40	18,7	1,9
25,0	3,00	34,6	1,4

* Diameter

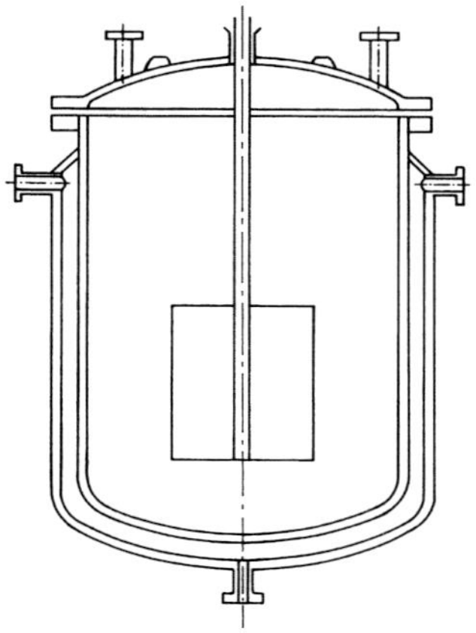

Stirred tanks are available in a large range of standardized sizes (Figure 2.2-14). For reasons of cost, wherever possible, standard dimensions and materials are preferred to custom designs.

One of the main tasks of the stirred tank is homogenization of the inflowing starting material and the reaction mixture with the aid of a suitable stirrer. To maintain a homogeneous reaction mixture in the vessel, the mixing time should be at most 10% of the time constant of the reaction (initial concentration divided by the reaction rate). Apart from the use of stirrers, it is also possible to mix the reactor contents by using jet mixers (injection of the circulating liquid) and loop reactors.

Heat exchange through the vessel wall and built-in pipe coil, if present (especially in larger tanks) also play a considerable role in reactor design. Figure 2.2-15 shows a widely used design in which the stirrer is replaced by a pump/jet system and heat exchange takes place in an external heat exchanger.

Tubular reactor

Compared to a stirred tank, the tubular reactor has a much larger ratio of heat-removing wall area to reactor contents and this therefore suitable for highly exothermic or endothermic reactions.

Tubular reactors are generally operated in the turbulent flow regime ($Re > 10^4$). The length to diameter ratio should be greater than 50, so that influence of backmixing processes is negligible relative convective transport due to forced flow in the reactor.

For slow reactions, for example, many liquid-phase reactions, the use of tubular reactors is limited by the required long residence times and the resulting low flow rates. A controlled reaction is no longer possible for $Re < 1000$, especially with low-molecular media. Heat and mass transfer can lead to secondary streams that can result in chaotic reactor behavior. In highly viscous media (e.g., in polmerization reactions), a laminar velocity profile becomes established and leads to a broad residence-time spectrum (Figure 2.2-5). This can be alleviated to some extent by internals

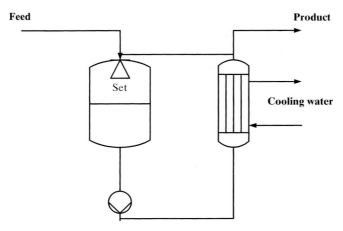

Fig. 2.2-15 Example of an intelligent stirred-tank design.

such as static mixers or packing elements, which lead to better evening out of radial concentration and temperature.

In heterogeneously catalyzed gas-phase reactions (e.g., partial oxidations, dehydrogenations) tubular reactors are often used as tube bundles (many catalyst-filled tubes in parallel; Figure 2.2-16) [Hofmann 1979, Adrigo 1999, Adler 2000].

Fluidized-bed reactor
In a fluidized-bed reactor a fluid passes from below through a layer of particles without entrainig them. The fluidization (mixing) is based on two parallel fundamental processes : mixing of the flowing fluid in the interparticle spaces, and mixing in the entire bed by the moving particles. In narrow reactors the residence-time behavior of the flowing fluid phase corresponds to that of a tubular reactor. For wide reactors and lower fluid velocities the behavior corresonds to that of a stirred tank.

Fluidized-bed reactors are preferably used for highly exothermic or endothermic reactions (e.g., synthesis of acrylonitrile by ammoxidation of propylene, $\Delta H_R = -502$ kJ mol^{-1}; synthesis of melamine from urea, $\Delta H_R = +472$ kJ mol^{-1}) to avoid local overheating (hot spots). The heat transfer is 5–10 times higher than in a fixed bed. In spite of the use of very small solid particles (< 300 μm, generally 50–10 μm) with a large specific surface area, the pressure drop is very low compared to a fixed bed. Figure 2.2-17 shows the pressure drop as a function of fluid velocity. Initially the pressure drop increases with increasing velocity (fixed- bed reactor, $\varepsilon = 0.4$–0.5) up to the minimum fluidization velocity (0.001–0.1 m s^{-1} relative to the empty tube), at which point the solid layer is loosened and passes into the fluidized state (up to 0.5 m s^{-1}). With further increases in velocity the pressure drop remains approximately constant, while the fluidized layer expands more and more and the movement of the particles becomes incresingly intense (fluidized-bed reactor, $\varepsilon = 0.7$–0.8).With a further increase in velocity, solid particles are carried out of the reactor with the fluid

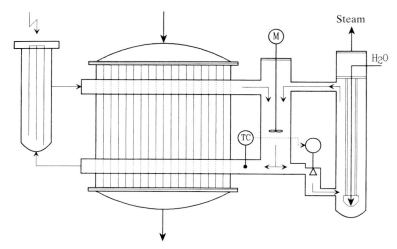

Fig. 2.2-16 Salt-bath tube-bundle reactor for partial oxidation on a heterogeneous catalyst. The heat of reaction is removed by a circulating salt bath, which is cooled by evaporation of condensate.

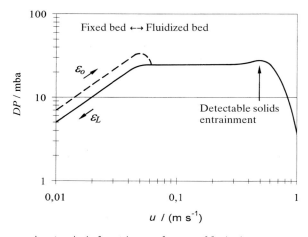

Fig. 2.2-17 Pressure drop in a bed of particles as a function of fluid velocity u.

Laminar Flow ($Re_{dP} = 10$ to 50): $\Delta P/L \propto \dfrac{(1-\varepsilon)^2}{\varepsilon^3} \cdot \dfrac{v \cdot u}{d_P^2}$

Turbulent flow ($Re_{dP} > 300$): $\Delta P/L \propto (1-\varepsilon) \cdot \dfrac{\rho \cdot u^2}{d_P}$

Fluidized bed: $\Delta P/L \propto (1-\varepsilon) \cdot (\gamma_{solid} - \gamma_{gas})$

$\varepsilon = V_{empty}/V_{tot}$
$v =$ kinematic viscosity
$d_P =$ particle diameter
$\rho =$ density
$\gamma =$ specific weight
where $Re_{dp} = \dfrac{u \cdot d_p \cdot \rho}{\eta}$.

For further discussion, see text

stream, and the amount of solids in the reactor can only be maintained by recycling the entrained solids (circulating fluidized bed [Contractor 1999], $\varepsilon = 0.9$–0.98). Beyond a certain fluid velocity, the system enters the realm of pneumatic conveying, in which the solid particles are carried out of the reactor without residence time, and the pressure drop increases noticeably due to acceleration of the particles.

In industrial fluidized-bed reactors inhomogeneities often occur (channeling, bubble formation, pulsating fluidized bed), which arise from the interaction of the hydrodynamic pressure with that due to the weight of the particles, and the size and state of the surface of the partivles. Heavy, irregularly shaped, and sharp-edged particles in particular tend to form such inhomogeneities.

Chemical fluidized-bed reactors can have diameters of up to 10 m. Their scale-up is associated with major risks (erosion of reactor walls and especially cooling tubes, bubble formation, solids entrainment, formation of fines), so that today the main application is syntheses with large heats of reaction (e.g., acrylonitrile, maleic anhydride, fluid cat cracking) and fluidized combustion.

Microreactors

Microreactor technology is concerned with chemical reactions and unit operations in components and systems whose characteristic dimensions range from the submillimeter down to the submicrometer region. Since, for given temperature and concentration differences, a decrease in system size leads to corresponding increase in the gradients of these quantities of state and hence to much higher driving forces for mass and heat transfer.

Currently, microreactors [Wörz 2000, Wörz2000a, GIT 1999, Ehrfeld 2000] are under discussion both as production reactors and as tools for reactor development. These reactors are characterized by reaction channels with diameters on the order of 10–100 μm and very high surface area to volume ratios (see also Figure 2.1-9). Therefore, microreactors are suitable for carrying out very fast reactions with large heats of reaction. Undesirable hot spots can thus be avoided. However, if fouling is expected, then the use of microreactore is questionable.

2.3
Product Processing (Thermal and Mechanical Separation Processes)

Separation tecnology, a cornerstone of chemical technology, has a diversity that virtually no other area can match. Separation processes require about 43% of energy consumption and 40–70% of the investment costs [Eissen 2002]. Thermal separation processes [Sattler 2001, Schönbucher 2002] such as rectification, mechanical processes such as filtration and size reduction, and chemical reactions (e.g., ion exchange) are used. Accordingly, the range of apparatus and auxiliary materials is huge, and so only a small selection can be presented here.

2.3.1
Heat Transfer, Evaporation, and Condensation

Besides pumps, heat exchangers are, in terms of numbers, the most common apparatus in chemical plants. They play a role in pretreatment of starting materials, achieving and maintaining reaction conditions, and above all in thermal separation processes. A thermally well designed plant contributes considerably to the economic viability of a process. Thus, following the second oil crisis, in the 1980s the chemical industry made major efforts to thermally optimize chemical plants (e.g., by means of pinch technology, see Section 6.1.6.2).

2.3.1.1
Fundamentals

Heat transfer is an irreversible process that always takes place from a point of higher to one of lower temperature. The nonstationary cooling of an apparatus can be approximated by a simple Exponential law (Equation 2.3.1-1)

$$\Delta T(t) = \Delta T_0 \cdot exp\left(-\frac{t}{\tau}\right), \tag{2.3.1 – 1}$$

with time constant τ, which depends on the available area, the heat-transfer coefficient, and the heat capacity. Thermal energy can be transported by the following three mechanisms:

Thermal radiation
Energy transfer by thermal radiation depends on the temperature T, the area A_W, and the structure of the surface ε, according to the Stefan[1]–Boltzmann[2] law (Equation 2.3.1-2)

---
1) Josef Stefan, Austrian physicist (1835–1893).
2) Ludwig Boltzmann, Austrian physicist (1844–1906).

$$\dot{Q}_{Strahlung} = \sigma_s \cdot \varepsilon \cdot A_W \cdot T^4 \qquad (2.3.1-2)$$

σ_S = 5.670 × 10^{-8} W m^{-2} K^{-4} (Stefan–Boltzmann constant)
ε = emissivity (e.g., polished Cu (100 °C) ca. 0.02, oxidized Cu (400 °C) ca. 0.7, polished Fe (100 °C) ca. 0.2, rusty iron (20 °C) ca. 0.65, concrete (20 °C) ca. 0.94, glass (20 °C) ca. 0.9, water (20 °C) ca. 0.95).

Polished metal surfaces have low emissivity because they absorb only little radiation. In contrast, glasss apparatus loses much heat by radiation. Therefore, rectification columns in miniplants, in spite of vacuum jacketing, require mirrored surfaces if adiabatic conditions are to be attained.

For practical calculation of the net radiative heat flux from body 2 (area A_{W2}, higher temperature T_2) to body 1 (area A_{W1}, lower temperature T_1), Equation (2.3.1-3) is used:

$$\dot{Q}_{21} = \sigma_s \cdot C_{21} \cdot A_{W2} \cdot (T_2^4 - T_1^4). \qquad (2.3.1-3)$$

The radiative exchange coefficient C_{21} depends on the geometrical arrangement of the two surfaces (e.g., Equations 2.3.1-4 and 2.1.3-5):

Parallel surfaces:
$$C_{21} = \frac{1}{\frac{1}{\varepsilon_1} + \frac{1}{\varepsilon_2} - 1} \qquad (2.3.1-4)$$

Cylinder in cylinder:
$$C_{21} = \frac{1}{\frac{1}{\varepsilon_2} + \frac{A_{W2}}{A_{W1}}\left(\frac{1}{\varepsilon_1} - 1\right)}. \qquad (2.3.1-5)$$

Thermal conduction

The heat flux due to thermal conduction depends on the temperature gradient and the area according to Fourier's[3] first law (Equation 2.3.1-6a):

$$\dot{Q}_{cond.} = \lambda(T) \cdot A_W \cdot \frac{dT}{dx} \qquad (2.3.1-6a)$$

λ = coefficient of thermal conductivity in W m^{-1} K^{-1}
(e.g., λ at 20 °C: Cu 392, Al 221, Fe 67, steel 46, glass 0.8, water 0.6, air 0.026)

Integration over a flat wall of thickness Δx_1 gives Equation (2.3.1-6b):

$$\dot{Q}_{cond., wall} \cdot \int_0^{\Delta x_1} dx = \lambda \cdot A_W \int_{T_0}^{T_1} dT \qquad (2.3.1-6b)$$

---

3) Jean-Baptiste Joseph Fourier, French mathematician and physicist (1768–1830).

and the temperature gradient is given by Equation (2.3.1-6c):

$$(T_1 - T_0) = \frac{\dot{Q}_{cond., wall}}{\lambda \cdot A_W} \cdot \Delta x_1. \quad (2.3.1-6c)$$

For stationary heat conduction through a tube wall Equation (2.3.1-7a) applies:

$$\dot{Q}_{cond., tube} = -\lambda(T) \cdot (2\pi \cdot r \cdot L) \cdot \frac{dT}{dr} \quad (2.3.1-7a)$$

and, in integrated form, Equation (2.3.1-7b):

$$\dot{Q}_{cond., tube} = 2 \cdot \pi \cdot \lambda \cdot \frac{L}{\ln\left[\frac{r_a}{r_i}\right]} \cdot (T_i - T_a). \quad (2.3.1-7b)$$

r_a = outside diameter of the tube
r_i = inside diameter of the tube
L = length of tube.

For the practically important case of conduction through n parallel layers; typical examples:

- Internal fouling layer (e.g., coke deposit), thickness Δx_i
- External steel jacket, thickness Δx_W
- External fouling layer (e.g., algae), thickness Δx_a

Equation (2.3.1-8) results:

$$\dot{Q}_{cond.} = \frac{1}{\sum_{i=1}^{n} \frac{\Delta x_i}{\lambda_i}} \cdot A_W \cdot (T_0 - T_n), \quad (2.3.1-8)$$

and for the special case of Figure 2.3.1-1, Equation (2.3.1-9):

$$\dot{Q}_{cond.} = \frac{1}{\frac{\Delta x_i}{\lambda_i} + \frac{\Delta x_W}{\lambda_{steel}} + \frac{\Delta x_a}{\lambda_a}} \cdot A_W \cdot (T_H - T_K). \quad (2.3.1-9)$$

Convection

In convective heat transport heat transfer takes place via a fluid flowing on a wall. If the fluid flows solely as the result of buoyancy forces, then this is known as free convection, as opposed to forced convection, for example, by pumps or compressors. Since effective heat exchange is only possible in the case of turbulent flow in or around the tube, the heat transfer can be described by a two-film model, in which the wall is regarded as a hydrodynamic boundary layer of thickness δ and a heat-transfer coefficient α is defined (Equation 2.3.1-10):

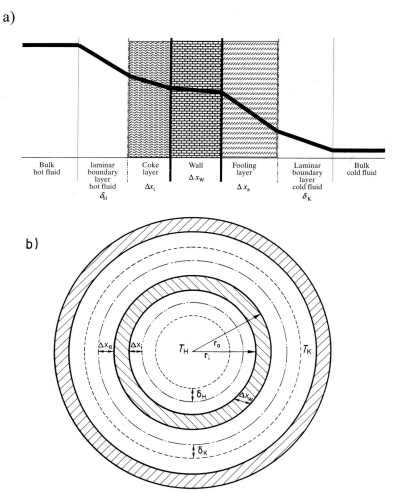

Fig. 2.3.1-1 Stationary heat transfer from a hot liquid (H) to cold liquid (K). a) Through three planar walls (inner coke layer, steel wall, and outer fouling layer). b) Through three cylindrical walls (inner coke layer, steel wall, and outer fouling layer).

$$a = \frac{\lambda_{Fluid}}{\delta(Re)}, \tag{2.3.1-10}$$

where λ_{Fluid} is the thermal conductivity coefficient of the fluid. Thus, Equation (2.3.1-11) results for the description of the heat transfer by convection:

$$\dot{Q}_{Konvektion} = a(Re, \lambda_{Fluid}) \cdot A_W \cdot (T_H - T_K). \tag{2.3.1-11}$$

The practically important case of heat transfer from a hot liquid (index H) through a wall with n layers to a cold fluid (index K) can be treated as a combination of convection (Equation 2.3.1-11) and conduction (Equation 2.3.1-8) according to Equation (2.1.3-12) (Figure 2.3.1-1a):

$$\dot{Q}_W = \frac{1}{\frac{\delta_H}{\lambda_H} + \sum_{i=1}^{n} \frac{\Delta x_i}{\lambda_i} + \frac{\delta_K}{\lambda_K}} \cdot A_W \cdot (T_H - T_K). \qquad (2.3.1-12)$$

By introducing a heat-transfer coefficient k_W (Table 2.3.1-1):

$$\frac{1}{k_W} = \frac{\delta_H}{\lambda_H} + \sum_{i=1}^{n} \frac{\Delta x_i}{\lambda_i} + \frac{\delta_K}{\lambda_K}, \qquad (2.3.1-13)$$

a simplification can be made:

$$\dot{Q}_W = k_W \cdot A_W \cdot (T_H - T_K). \qquad (2.3.1-14)$$

This equation is only precise for the case of a plane wall. For the technically important case of heat transfer through tubes (cylindrical geometry), the different geometry must be taken into account by a mean tube surface area $\overline{A_W}$ for heat transfer (Equations 2.3.1-15 and 2.3.1-16; Figure 2.3.1-1b):

Tab. 2.3.1-1 Approximate k_W values for different types of heat exchangers.

Type of heat exchanger	From	To	Approx. k_W – value $W\ m^{-2}\ K^{-1}$
Doubletube	low-pressure gas	low-pressure gas	10–40
	high-pressure gas	low-pressure gas	25–60
	high-pressure gas	high-pressure gas	200–400
	high-pressure gas	liquid	250–600
	liquid	liquid	400–1800
Tube-bundle	low-pressure gas	low-pressure gas	5–40
	high-pressure gas	high-pressure gas	200–450
	high-pressure gas	liquid	250–700
	Saturated steam	liquid	500–4000
	liquid	liquid	200–1500
Spiral	liquid	liquid	600–2800
Plate	cooling water	gas	25–80
	liquid	liquid	500–4000

$$\overline{A_W} = \frac{A_W^a - A_W^i}{\ln \frac{A_W^a}{A_W^i}} = 2\pi \cdot L \cdot \frac{r_a - r_i}{\ln \frac{r_a}{r_i}} \qquad (2.3.1-15)$$

$$\frac{1}{k_W \cdot \overline{A_W}} = \frac{\delta_H}{\lambda_H \cdot A_{WH}} + \frac{\Delta x_i}{\lambda_i \cdot A_{WH}} + \frac{\Delta x_W}{\lambda_W \cdot \overline{A_W}} + \frac{\Delta x_a}{\lambda_a \cdot A_{WK}} + \frac{\delta_K}{\lambda_K \cdot A_{WK}}. \qquad (2.3.1-16)$$

For the calculation of the k_W value or the α value according to Eqaution (2.3.1-10) the thickness of the hydrodynamic boundary layer must be known. However, this can only be estimated, for example, by Equation (2.3.1-17) (see also Equation 2.4-14 in Section 2.4.2).

$$\delta \approx \frac{d}{\sqrt{Re}}. \qquad (2.3.1-17)$$

A more accurate method is the use of characteristic numbers such as the Nusselt number Nu:

$$(Nu)sselt^4\text{-no.} = \frac{a \cdot d}{\lambda} \qquad (2.3.1-18)$$

which can be expressed as functions of the Reynolds number Re, the Prandtl number Pr, and the geometry number (Equation 2.3.1-21):

$$(Re)ynolds\text{-no.} = \frac{u \cdot d \cdot \rho}{\eta} \qquad (2.3.1-19)$$

$$(Pr)andtl^5\text{-no.} = \frac{c_p \cdot \eta}{\lambda} \qquad (2.3.1-20)$$

$$Geometry\text{-no.} = \frac{d}{L}. \qquad (2.3.1-21)$$

In the literature numerous empirical equations can be found with which the Nusselt number and the corresponding heat-transfer coefficients can be reliably calculated (e.g. Equation 2.3.1-22):

$$Nu = const. \cdot Re^n \cdot Pr^m \cdot \left(\frac{d}{L}\right)^k. \qquad (2.3.1-22)$$

An example is the equation for heat-transfer coefficients for turbulent longitudinal flow in smooth tubes (Equation 2.3.1-23):

$$Nu = 0.02 \cdot Re^{0.80} \cdot Pr^{0.43}. \qquad (2.3.1-23)$$

---
4) Ernst Kraft Wilhelm Nusselt, (1882–1957).
5) Ludwig Prandtl, German physicist (1875–1953).

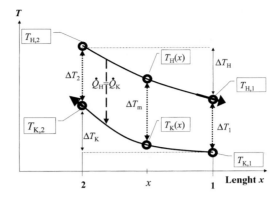

Fig. 2.3.1-2 Heat exchange along a pipe (for discussion, see text).

2.3.1.2
Dimensioning

Up to now heat transfer was treated at a particular point x (Figure 2.3.1-2) of a tube. However, for dimensioning a heat exchanger, the heat transfer along the whole tube has to be considered, since the driving force for heat transfer, that is, the temperature difference, varies along the tube. The mean driving temperature difference is given by Equation (2.3.1-24) (see Figure 2.3.1-2 for nomenclature):

$$\Delta T_m = \frac{\Delta T_2 - \Delta T_1}{\ln \frac{\Delta T_2}{\Delta T_1}} \qquad (2.3.1-24)$$

and hence:

$$\dot{Q} = k_W \cdot A_W \cdot \Delta T_m. \qquad (2.3.1-25)$$

This is the key equation for the design of a heat exchanger. For a given amount of heat to be transferred (Equation 2.3.1-26) and a known value of k_W, the area A_W required for heat exchange can be calculated. This area must be distributed over the tubes according to appropriate design criteria (see Example 2.3.1-1):

$$\dot{Q}_H = \dot{m}_H \cdot cp_H \cdot \Delta T_H = \dot{Q}_K = \dot{m}_K \cdot cp_K \cdot \Delta T_K \qquad (2.3.1-26)$$

$\dot{m}_H$ = mass flow of the hot stream in kg h^{-1}
$\dot{m}_K$ = mass flow of the cold stream in kg h^{-1}.

A heat exchanger can be operated in co-, counter-, or cross-current mode. In practice the countercurrent mode is used almost exclusively since it is the most effective. Cocurrent heat exchangers are used only under special conditions (thermally labile

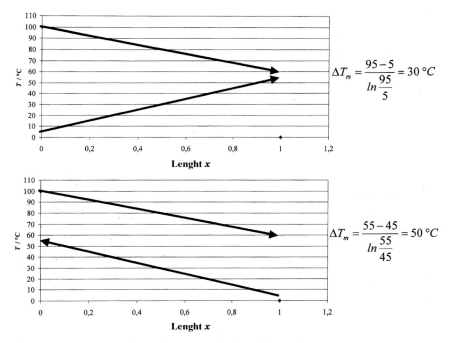

Fig. 2.3.1-3 Comparison of co- and countercurrent flow (schematic).

products, fouling behavior, etc.). The cross-current heat exchanger, a combination of the co- and countercurrent principles, is less efficient than pure countercurrent designs but is often unavoidable for constructional reasons.

The most important designs are the tube-bundle, spiral, and plate heat exchangers (Figure 2.3.1-4).

The *tube-bundle heat exchanger* (Figure 2.3.1-4a), in various construction variations, can be regarded as the standard heat exchanger.

The *spiral heat exchanger* is the ideal countercurrent apparatus. It consists of two or four metal sheets wrapped around a central cylindrical core tube and is divided into two parts by a middle sheet. The two media are introduced into the appropriate spirals through peripheral inlets. Bolts, welded to the spiral sheets, transfer the pressure to the next outermost sheet and maintain a constant distance between the sheets. This heat exchanger is built into a cylindrical jacket equipped with flanges (Figure 2.3.1-4b).

Advantages	Disadvantages
• Ideal countercurrent	• laborious maintenance
• Very compact design (150 m² m⁻¹), with up to 600 m² in one unit	• difficult to clean
• Pronounced self-cleaning effect due to centrifugal force	• unsuitable for high pressure (max. 25 bar)
• High k value due to strong turbulence	
• No risk of mixing	
• Suitable for high temperatures (up to 400 °C)	

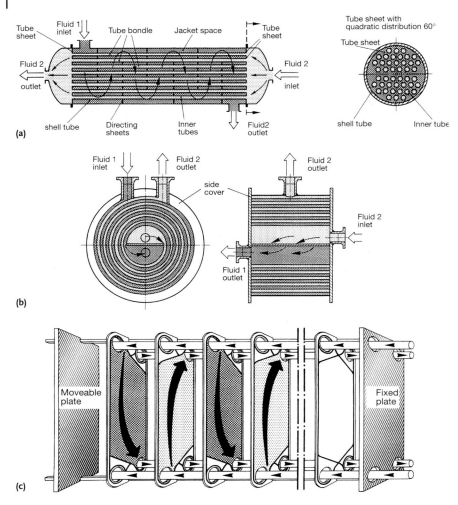

Fig. 2.3.1-4 The most important types of heat exchangers. a) Tube-bundle, b) spiral, and c) plate heat exchanger.

The *plate heat exchanger* consists of a number of profiled cold-pressed metal plates and gaskets which are arranged between a fixed plate and a moveable plate, mounted on guide rails, and held together by tie rods. The hot and cold circuits are separated from one another and the surroundings by elastic gaskets and flow in countercurrent through the spaces formed by two plates (Figure 2.3.1-4c). For high pressures and strict demands on tightness, welded designs are used instead of gasketed ones.

Advantages	Disadvantages
• Compactness: large heat-exchange area in a small space	• unsuitable for high pressure (max. 50 bar)
• Flexibility: number and type of plates are easily changed	• unsuitable for high temperatures (gaskets)
• Simple maintenance (can be dismantled)	
• High k value due to strong turbulence	
• Good antifouling behavior due to strong turbulence	
• Low cost	

Design criteria for tube-bundle heat exchangers

The following design criteria are intended to give a rough guide to assessing the suitability of heat exchangers designed with the aid of computer programs (see below):

Tube length: 1.5–6 m. If a greater length is required, multiple tubes must be used.

Tube diameter: standard tubes have diameters of 16, 20, or 25 mm.

State of flow in the tubes: to achieve intense turbulence and hence good a values, a flow velocity of 1–2.5 m s^{-1} should be aimed for. A design with lower velocity has a lower pressure drop, whereas one with a higher velocity is less prone to fouling and has a higher a value.

Example 2.3.1-1

A volume flow of 100 m^3 h^{-1} in the tubes (d = 20 mm) is to be cooled. At the chosen flow velocity of 2 m s^{-1}, ca. 2.3 m^3 h^{-1} can be passed through each tube (see Equation 2.4-16), and hence 100/2.3 = 44 tubes are required. To ensure the required area the tubes must have a certain length (Equation 2.3.1-27):

$$A = \pi \cdot d \cdot L \cdot N = 100 \ m^2 = \pi \cdot 20 \cdot 10^{-3} \cdot L \cdot 44, \qquad (2.3.1-27)$$

from which a tube length of 36 m results. For a standard length of 6 m a sixfold construction of the heat exchanger is required.

Heat transfer without a change of phase rarely causes problems in process design. Sufficiently accurate methods of calculation are available for dimensioning heat exchangers. The influence of fouling, for example polymer deposits or biological fouling when cooling with river water, is more critical, and each case must be examined individually. The usual remedial measures are smooth surfaces, high flow velocities (1.5–5 m s^{-1}), and switching to noncritical coolants such as recycled cooling water. In critical cases the heat exchangers may be built larger, or multiple heat exchangers may be installed to allow interruption-free cleaning (see Section 6.1.7). Fouling leads to changes in the k_W value with time (Figure 2.3.1-5), which must be taken into account in the design of the heat exchanger. Therefore, it is important to gather as much information as possible on the fouling mechanism between the medium and the wall material during the process-development stage [Scoll 1996].

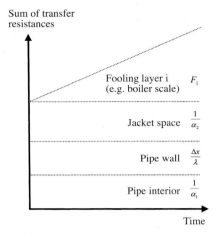

Fig. 2.3.1-5 Influence of fouling on the total heat-transfer resistance.

When thermally sensitive materials are to be handled, special techniques must be used in the experimental inveatigation. In the small test apparatus the ratio of the heat-exchange surface area to the volume is always larger than in a production-scale apparatus; hence, the temperature differences in the test apparatus are smaller than in the subsequent production plant. Therefore, potential product damage due to high wall temperatures in the production plant could remain undetected in the experimental test phase. In such cases the test apparatus must be modified by equipping it with smaller heat-exchange surfaces.

Example 2.3.1-2

In a stirred tank heat is to be introduced through the vessel wall. If the entire vessel jacket is heated, the test apparatus must be operated with a smaller temperature difference. By segmentation of the double jacket a temperature representative for the industrial plant can be attained in the test apparatus.

Evaporators and condensers can also be reliably designed with the available methods of calculation and generally do not require investigations in representative test apparatus.

Evaporators

In terms of process technology the most easily realized evaporators are the natural-convection evaporator with heating elements and the thin-film evaporator with rotating wiper blades. When the exhaust vapors of a wiped thin-film evaporator are withdrawn at the bottom, for small heights, a falling-film evaporator can be simulated (Figure 2.3.1-6). Exceptions are cases, especially with evaporators, in which material-, temperature-, and residence-time-dependent changes in the product are ex-

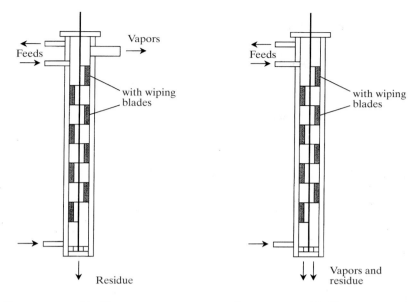

Fig. 2.3.1-6 Co- and countercurrent thin-layer evaporators.

pected. Here an experimental treatment is unavoidable. The appropriate type of apparatus – natural- or forced-convection evaporator, falling-film evaporator (which even in small test apparatus can have a problematic height of more than 3 m), thin-film evaporator, or, in extreme cases, short-path evaporator – must be selected first, with consideration of the costs, and experimentally tested. Even for evaporations in which formation of solids does not occur, an experimental investigation is unavoidable; this is usually performed in a rotary flash evaporator. For viscous media and media that cause extensive fouling, a helical-tube evaporator with cylindrically wound tubes is an appropriate solution (Figure 2.3.1-7). In this design the high vapor velocity of about 100 m s^{-1} and the induced eddies result in effective cleaning of the heating surfaces and high heat transfer [Casper 1986, 1970]. This type of evaporator is readily

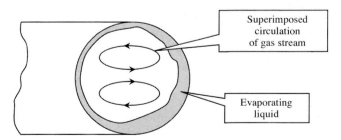

Fig. 2.3.1-7 Helical-tube evaporator with cylindrically wound tubes (cross section).

Tab. 2.3.1-2 Operating properties of some heat-transfer media [Hänßle 1984].

Trade name	Producer	Chemical structure	Operating range
Malotherm S	Hüls	dibenzyltoluene	−14 to 350
Malotherm L	Hüls	benzyltoluene	−55 to 350
Diphyl	Bayer	diphenyl/diphenyl oxide	20 to 400
Syltherm 800	Dow Corning	polydimethylsiloxane	−40 to 400

tested in pilot plants. Short-path evaporators are used under high vacuum (< 0.1 mbar). Small test units are commercially available.

The following mean temperature differences can be regarded as commercially viable:

- Liquid/liquid   10–20 °C
- Gas/liquid   20–30 °C
- Gas/gas   50–80 °C
- Evaporation   10–30 °C
- Thin-film eveporation   30–100 °C
- Low-temperature processes   2–10 °C.

Condensers

For particularly critical products (e.g., monomers) methods with direct cooling (quenching) have proved suitable (Figure 2.3.1-8a). Here condensation takes place in the cold medium itself. The heat is removed at a lower temperure in a liquid/liquid heat exchanger. Combinations of direct and indirect heat removal are also possible in which the condenser is additionally quenched (Figure 2.3.1-8b).

As a heat-transfer medium water has a limited operating range between 0 and about 200 °C. For applications outside this range a wide variety of organic heat-transfer media, covering operating temperatures from –55 to 400 °C (with nitrogen blanketing), is available (Table 2.3.1-2).

A disadavntage of all organic heat-transfer media is their potential flammability. Salt melts, which can be used between 150 and 550 °C, do not have this disadvantage. Typical mixtures [Albrecht 2000] are:

- A ternary eutectic mixture of $NaNO_2$, $NaNO_3$, and KNO_3 (m.p. 142 °C, bulk density 1200 kg m^{-3}, density of the solidified salt 2100 kg m^{-3}, heat capacity 1.56 kJ kg^{-1}K^{-1}, operating range 200–500 °C).
- A binary mixture of 45% $NaNO_2$ and 55% KNO_3 (m.p. 141 °C, bulk density 1200 kg m^{-3}, density of the solidified salt 2050 kg m^{-3}, heat capacity 1.52 kJ kg^{-1}K^{-1}, operating range 200–500 °C).

The main difference between the two mixtures is the thermal conductivity. It is considerably higher for the three-component mixture, which results in better heat transfer. Because of the pressureless operation, conventional materials such as St 35.81 steel can be used in spite of the higher temperatures.

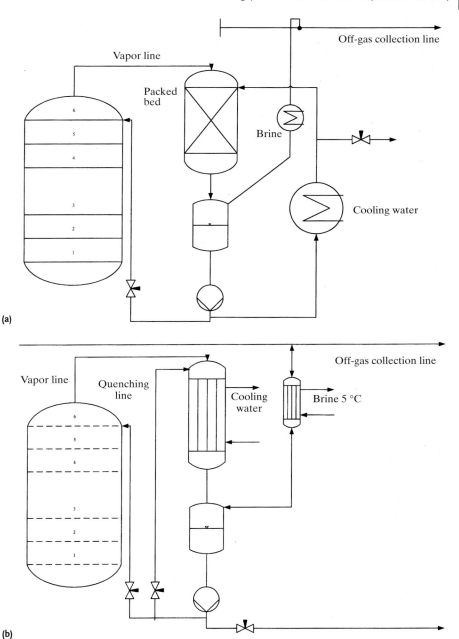

Fig. 2.3.1-8 Examples of indirect heat exchenge. a) Direct cooling with recirculating quench. b) Combination of direct and indirect heat exchange.

2.3.2
Distillation, Rectification

2.3.2.1
Fundamentals of Gas–Liquid Equilibria

Phase equilibria are the fundamental basis of many steps in chemical processes. Quantitative information on these equilibria and data on the substances involved are therefore a prerequisite for process development and design of apparatus.

Today the design of separation processes is performed almost exclusively with the aid of process simulators by solving the balance equations. In the case of separation processes, apart from the data of the pure substances, reliable information on the phase equilibria of the multicomponent system that is to be separated is required.

Whereas about 20 years ago the design of a thermal separation process required numerous time- and cost-intensive pilot plant tests and laborious measurements of phase equilibria, modern thermodynamic models (state equations or G^E models) allow, for the case of nonelectrolytic systems, reliable calculation of the phase-equilibrium behavior of multicomponent systems if the behavior of the two-component systems is known. Therefore, we will briefly summarize the most important relations for describing binary mixtures.

The basis for the separation of a binary mixture A/B by distillation is the difference in concentration (mole fraction) of component A in the vapor (y) and in the liquid phase (x). The relationship is described by the vapor/liquid equilibrium (Raoult's law[6]). In the following, component A is always the low-boiling and component B the high-boiling component. For the sake of simplification, $x_A + x_B = 1$ applies for the liquid phase, and $x_A = x$ or $y_A = y$ for the gas phase. Figure 2.3.2-1 shows schematically the corresponding $P/T/x/y$ phase diagram. This 3D depiction demonstrates that the phase diagram can be highly complex, even for simple mixtures. Therefore, in practice 2D diagrams are widely used for quantitative descriptions, as is discussed briefly below.

2.3.2.1.1 Pure Substances

Gas phase
The $P/V/T$ behavior of gases can be described mathematically by means of thermal state diagrams [Gmehling 1992]. The oldest and simplest relationship is the ideal gas law (see Chapter 3). A series of state equations are available for describing real gases, for example:

---

6) Francois Marie Raoult, French Chemist (1830–1901).

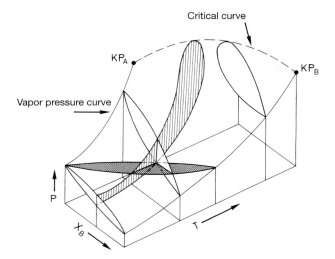

Fig. 2.3.2-1 P–T–x–y phase diagram of a simple binary mixture (schematic) with three sections: isotherms (P–x diagram, light), isoplethes (P–T diagram, hatched), and isobars (T–x diagram, dark).

Virial equation (Equation 2.3.2-1)

$$P = z \cdot \frac{RT}{\bar{V}} \qquad (2.3.2-1)$$

where $z = 1 + B(T) \cdot P + C(T) \cdot P^2 + ...$,

where the virial coefficients $B(T)$, $C(T)$,... can be described by potential functions of the forces of the intermolecular interactions.

Cubic state equations

The semi-empirical relationships are all derived from the van der Waals[7] equation (Equation 2.3.2-2), which dates back more than 100 years:

$$P = \frac{RT}{\bar{V} - b} - \frac{a}{\bar{V}^2} \qquad (2.3.2-2)$$

where $a = 27 \cdot b^2 \cdot P_{krit}$ and $b = \frac{RT_{krit}}{8 \cdot P_{krit}}$.

A widely used modification is the Redlich–Kwong equation (Equation 2.3.2-3) [Gmehling 1992]:

$$P = \frac{RT}{\bar{V} - b} - \frac{a}{\sqrt{T} \cdot \bar{V} \cdot (\bar{V} + b)} \qquad (2.3.2-3)$$

---

7) Johannes D. van Waals, Dutch physicist (1837–1923).

where $a = \dfrac{R^2 \cdot T_{crit}^{5/2}}{9 \cdot (\sqrt[3]{2} - 1) \cdot P_{crit}}$ and $b = \dfrac{1}{3} \cdot (\sqrt[3]{2} - 1) \cdot \dfrac{RT_{crit}}{P_{crit}}$.

Gas/liquid equilibrium of pure substances

Isochores: P/T diagrams (vapor pressure curves)
The vapor pressure curves of the pure components A and B at $x_B = 0$ and 1, respectively, are shown schematically in Figure 2.3.2-1. The vapor pressure curves of the pure components begin at the triple point and end at the critical points CP_A and CP_B. Over a wide range they can be described by the two-parameter Clausius[8]–Clapeyron[9] equation (Equation 2.3.2-4):

$$P = P_0 \cdot exp\left(-\dfrac{\Delta_V H}{RT}\right). \qquad (2.3.2-4)$$

If the assumption of constant enthalpy of evaporation made in the introduction does not hold, then equations with three or more parameters are used, such as the Antoine equation (Equation 2.3.2-5; see also Appendix 8.13 and Section 3.3.2):

$$ln\, P = A + \dfrac{B}{C + T}. \qquad (2.3.2-5)$$

2.3.2.1.2 Binary Mixtures [Ghosh 1999]

For the design of distillation processes the important sections through the $P/T/x/y$ diagram are:
Isotherms: $P/x/y$ diagrams (Raoult[10] diagram, Figure 2.3.2-2; Equations 2.3.2-6 and 2.3.2-7)

$$P(x) = P_A + P_B = x \cdot P_A^0 + (1 - x) \cdot P_B^0 \qquad (2.3.2-6)$$

$$P(y) = \dfrac{P_A^0 \cdot P_B^0}{P_A^0 + (P_B^0 - P_A^0) \cdot y}. \qquad (2.3.2-7)$$

Isobars: $T/x/y$ diagrams (boiling curve, Figure 2.3.2-3; Equation 2.3.2-8).

$$x(T) = \dfrac{P - P_B^0}{P_A^0 - P_B^0} \quad dew-point\ line \qquad (2.3.2-8a)$$

$$y(T) = \dfrac{P_A^0}{P_A^0 - P_B^0} \cdot \left(1 - \dfrac{P_B^0}{P}\right) \quad boiling\ line \qquad (2.3.2-8b)$$

---

8) Rudolf E. Clausius, German physicist (1822–1888).
9) Benoit P. E. Clapeyron, French physicist (1799–1864).
10) Francois M. Raoult, French chemist (1830–1901).

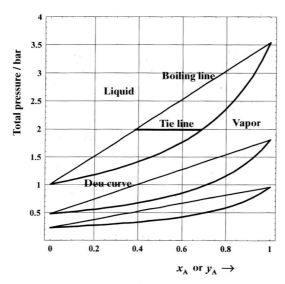

Fig. 2.3.2-2 P–x–y diagram (Equations 2.3.2-6 and 2.3.2-7) of a methanol–water mixture (assumed to be ideal) at three temperatures (from top to bottom: 63, 80, 100°C).
$p^0_{MeOH}/bar = \exp[11.96741 - 3626.55/(238.23 + T/°C)]$
$p^0_{H2O}/bar = \exp[11.78084 - 3887.20/(230.23 + T/°C)]$.

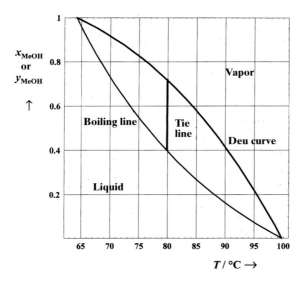

Fig. 2.3.2-3 T–x–y diagram (Equations 2.3.2-8a and 2.3.2-8b) of a methanol–water mixture (assumed to be ideal) at 1 bar total pressure with:
$p^0_{MeOH}/bar = \exp[11.96741 - 3626.55/(238.23 + T/°C)]$
$p^0_{H2O}/bar = \exp[11.78084 - 3887.20/(230.23 + T/°C)]$.

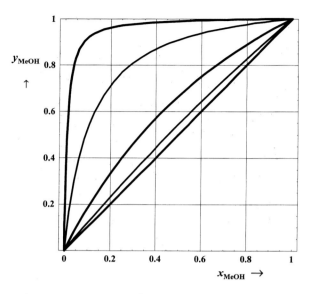

Fig. 2.3.2-4 x–y diagram (Equation 2.3.2-9) of an A–B mixture (assumed to be ideal) at 1 bar total pressure and four relative volatilities (from top to bottom: 1, 1.2, 2, 10, 100).

For the design of rectification columns, the equilibrium concentrations of the gas and liquid phases at constant pressure are of decisive importance. The McCabe–Thiele diagram (Figure 2.3.2-4), based on Raoult's law, describes this relationship (Equation 2.3.2-9):

$$y(T) = \frac{a(T) \cdot x}{1 + (a(T) - 1) \cdot x} \qquad (2.3.2-9)$$

with the relative volatility $a(T) = P_A^0(T)/P_B^0(T)$.

Types of mixtures

Ideal A/B mixtures that can be described by Raoult's law do not exist in practice. Generally, mixtures that have a more or less pronounced real behavior are encountered. The deviations from ideal behavior are taken into account by correcting factors known as activity coefficients γ in the case of the liquid phase or fugacity coefficients ϕ in the case of the gas phase (Equation 2.3.2-10; see also Section 3.3.2):

$$y \cdot \varphi(x, T) \cdot P = x \cdot \gamma(x, T) \cdot P_A^0. \qquad (2.3.2-10)$$

The deviations can also be expressed in terms of excess quantities (with superscript E, Equations 2.3.2-11 to 2.3.2-13):

$$S^E = \Delta_{mix} S - \Delta_{mix} S^{id} = \Delta_{mix} S - R \cdot \sum_{i=1}^{2} x_i \cdot \ln x_i \qquad (2.3.2-11)$$

$$H^E = \Delta_{mix}H - \Delta_{mix}H^{id} = \Delta_{mix}H - 0 \qquad (2.3.2-12)$$

$$G^E = \Delta_{mix}G - \Delta_{mix}G^{id} = \Delta_{mix}G - RT \cdot \sum_{i=1}^{2} x_i \cdot \ln x_i. \qquad (2.3.2-13)$$

The term *athermal mixtures* refers to those in which the heat of mixing $\Delta_{mix}H$ is very small but S^E differs considerably from zero, for example, polymer solutions. In regular mixtures H^E differs considerably from zero but S^E is negligible, for example, mixtures of highly polar low-molecular substances such as nitriles/esters.

To achieve a reliable fitting of the required parameters over the entire range of concentration and temperature, a simultaneous fitting to all reliable thermodynamic data (vapor/liquid equilibrium, azeotropic data, excess enthalpy, actuivity coefficients, liquid/liquid equilibria) should be performed.

If experimental data are not available, then *group-contribution methods* such as UNIFAC and ASOG [Ochi 1982] can be used. These methods were developed in particular for calculation of vapor/liquid equilibria. By modification of the models (mod. UNIFAC), definition of new main groups, introduction of temperature-dependent group-interaction parameters, and the use of a broad database (Dortmund databank) for simultaneous fitting of the required group-interaction parameters, the range of application and reliability of the results of the group-contribution methods could be considerably increased. In particular, the description of the temperature dependence and the behavior at infinite dilution was improved.

Highly promising models were obtained by using so-called G^E models for cubic equations of state. Apart from the description or precalculation of the behavior of highly polar systems, these models allow supercritical components to be taken into account. A further advantage of these models is that, in addition to phase equilibria, other important quantities such as densities and caloric data can be calculated. The required G^E values can be obtained either by fitting the parameters of proven G^E models (e.g., Wilson, NRTL, or UNIQUAC equation) to experimental phase equilibrium data, or with the aid of group-contribution methods such as UNIFAC.

2.3.2.2
One-Stage Evaporation

By continuous, one-stage, closed evaporation of an A/B mixture the lower boiling component A can be enriched in the gas phase in accordance with its gas/liquid equilibrium (Figure 2.3.2-5). When the evaporation takes place under adiabatic conditions, it is known as flash evaporation.

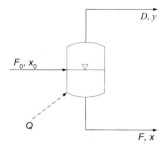

Fig. 2.3.2-5 Continuous one-stage closed evaporation.

The design is performed by equating the mass-balance relationship (lever rule, Equation 2.3.2-14):

$$D \cdot (y - x_0) = F \cdot (x_0 - x) \rightarrow y = -\frac{F}{D} \cdot x + \left(\frac{F}{D} + 1\right) \cdot x_0 \qquad (2.3.2-14)$$

with a given vapor/liquid equilibrium equation (e.g., Equation 2.3.2-9). The nonlinear system of equations can be solved graphically or numerically (Figure 2.3.2-6).

Example 2.3.2-1

For a given liquid/vapor ratio of 1, an inlet concentration of $x_0 = 0.5$, a known value of $\alpha = 5$, and constant pressure, the equation system (Equations 2.3.2-14 and 2.3.2-9) can be set up:

$$y = -1 \cdot x + 1$$

$$y_{gl} = y(T) = \frac{5 \cdot x}{1 + 4 \cdot x}$$

Graphical or numerical solution gives vapor and bottoms concentrations of $x = 0.31$ and $y = 0.69$, that is, an enrichment of A in the overhead product from 0.50 to 0.69 (Figure 2.3.2-7).

2.3 Product Processing (Thermal and Mechanical Separation Processes) | 101

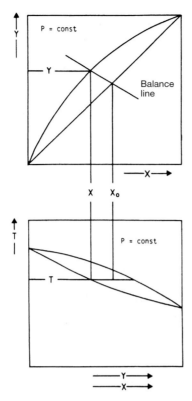

Fig. 2.3.2-6 Determination of the equilibrium composition of vapor and liquid phase and the temperature for a given liquid/vapor ratio and pressure P.

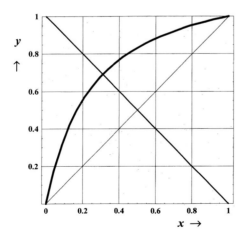

Fig. 2.3.2-7 Graphical solution of the system of equations in Example 2.3.2-1.

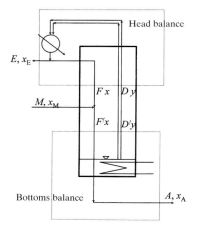

Fig. 2.3.2-8 Schematic of a rectification column with the flows M = feed, E = product, A = outlet, F = liquid, D = vapor.

2.3.2.3
Multistage Evaporation (Rectification)

Only in the case of very large differences in boiling point does simple continuous evaporation lead to effective separation (see Example 2.3.2-1). A marked improvement is obtained by rectification, in which the vapor and the liquid flow in countercurrent (coupled evaporation and condensation, see Figure 2.3.2-8). The liquid flows downwards and transfers low boiler A into the upward-flowing vapor phase, which passes high boiler B to the liquid phase. This countercurrent flow is achieved by generating vapor at the bottom of the column and adding liquid at the top, usually the condensate from the uppermost plate. Intensive mass transfer between the vapor and the liquid takes place in the individual stages (plates) of the column.

For the separation of the liquid feed M (composition x_M) into two streams of composition x_E (product, distillate) and x_A (bottom product), the required number of theoretical plates can be determined graphically with the aid of the McCabe–Thiele[11] method, which was used in the past because no computers were available to solve the extensive systems of equations for mass balances and equilibrium relationships. This method is no longer of practical importance, but it is an excellent didactic aid for understanding the basic principles of rectification.

The McCabe–Thiele method makes the following assumptions:

- The column consists of theoretical plates, that is, equilibrium exists at each plate.
- The vapor (D in moles per unit time) and liquid (F in moles per unit time) streams in the rectification and stripping sections are constant (prerequisite: the molar enthalpies of evaporation of A and B are almost equal and adiabatic conditions are present).

---

11) L. McCabe and E. W. Thiele.

- The vapors are completely condensed at the boiling temperature, and the return stream returns to the top of the column as a boiling liquid.
- The pressure drop in the column is negligible.

Total mass balance around the rectification section:

$$D = E + F \qquad (2.3.2-15)$$

Balance of low boiler A around the rectification section:

$$D \cdot y = E \cdot x_E + F \cdot x \qquad (2.3.2-16)$$

$$y = \frac{F}{D} \cdot x + \frac{E}{D} \cdot x_E. \qquad (2.3.2-17)$$

With the reflux ratio $v = F/E$ this gives the balance line around the rectification section:

$$y = \frac{v}{v+1} \cdot x + \frac{1}{v+1} \cdot x_E. \qquad (2.3.2-18)$$

For $x = x_E$ this equation gives the value $y = x_E$, and for $x = 0$ the expression $y = x_E/(v+1)$ (Figure 2.3.2-9).

An analogous treatment for the stripping section under the assumption that the feed stream is a boiling liquid, that is, $M + F = F^!$ gives:

Total mass balance around the stripping section:

$$F^! = D^! + A \qquad (2.3.2-19)$$

Balance of low boiler A around the stripping section:

$$F^! \cdot x = D^! \cdot y + A \cdot x_A \qquad (2.3.2-20)$$

$$y = \frac{F^!}{D^!} \cdot x - \frac{A}{D^!} \cdot x_A. \qquad (2.3.2-21)$$

With the overall column balance $M = E + A$ and the definition of the feed ratio $u = M/E$, the following applies for the balance line of the stripping section:

$$y = \frac{v+u}{v+1} \cdot x - \frac{u-1}{v+1} \cdot x_A. \qquad (2.3.2-22)$$

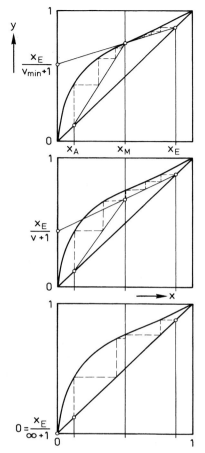

Fig. 2.3.2-9 Balance lines and staircase construction. a) Minimum reflux ratio, infinite plate number. b) Finite reflux ratio, resp. finite plate number. c) Total reflux, minimum plate number.

For $x = x_A$ this equation gives the value $y = x_A$ (Figure 2.3.2-9).
The intersection of the two lines lies at:

$$x_{intersection} = \frac{x_E + (u-1) \cdot x_A}{u}. \qquad (2.3.2-23)$$

Under the above assumptions (boiling liquid feed), this intersection lies at the point $x = x_M$ (Figure 2.3.2-9). For under- or overheated feeds Equation 2.3.2-24 applies:

$$y = \frac{q}{q-q} \cdot x + \frac{1}{q-1} \cdot x_M \qquad (2.3.2-24a)$$

where

$$q = \frac{H_D - H_M}{H_D - h_F} \qquad (2.3.2-24b)$$

H_D = enthalpy of the vapor
f_F = enthalpy of the liquid
H_M = enthalpy of the feed.

Then the balance lines are successively intersected with the equilibrium curves (Figure 2.3.2-9). The number of theoretical plates required for separation results from the number of points lying on the equilibrium line of a staircaselike construction between the rectification line and the equilibrium curve from x_E to x_M and between the stripping line and the equilibrium curve from x_A to x_M. Since the lowest plate is the evaporator plate, the true number of plates of the column is one lower.

The reflux ratio can be chosen in a range from the lower limit v_{min} up to ∞:

- One limit to solving the separation task is to increase the number of theoretical plates N_{theo} to ∞. This is the case when the rectification line intersects the equilibrium line at x_M. Then v_{min} can be calculated from the ordinate intersection of the rectification line $x_E/(v_{min} + 1)$ (Figure 2.3.2-9, top).
- For v going to infinity the rectification line becomes a diagonal with $y = x$. In this case, the smallest number of plates is required, as shown by the staircase construction (Figure 2.3.2-9, bottom).

By determining for various $v > v_{min}$, one obtains for a given separation task a relationship between the number of theoretical plates and the reflux ratio (Figure 2.3.2-9, middle), that is the N/v diagram (Figure 2.3.2-10).

With increasing number of plates the investment costs and hence the depreciation of the column increases; with increasing reflux ratio the operating costs for evaporation and condensation increase, as do the investment costs for the evaporation and condensation equipment. This economic consideration provides the optimum reflux ratio. In many cases the rule of thumb $v_{opt} \approx (1.2-2)v_{min}$ applies.

From the optimal reflux ratio the theoretical number of plates and, with the plate efficiency (Equation 2.3.2-25), the practical plate number and therefrom the column height result.

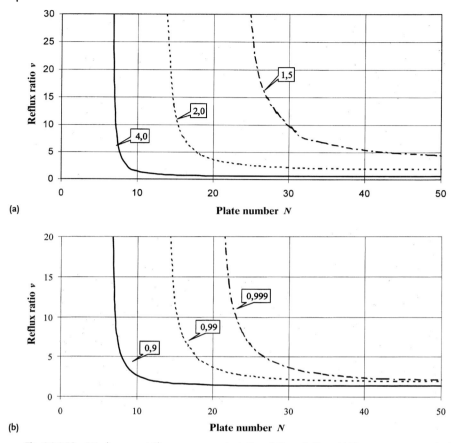

Fig. 2.3.2-10 N/v diagram. a) The curve parameter is the relative volatility α. b) The curve parameter is the separation yield σ.

$$\eta = \frac{N_{theo}}{N_{pract}}, \qquad (2.3.2-25)$$

The column diameter follows from hydrodynamic considerations and is independent of the type of separation internals. The design is largely determined by the vapor loading factor F. At high liquid loadings (pressure columns) the permissible F factor decreases. The required design documents are provided by the suppliers of the column internals (Figure 2.3.2-11). Standard F factors are:

- 2.0–2.5 $Pa^{0.5}$ for ordered packings
- 1.5–2.0 $Pa^{0.5}$ for random packings
- 1.2–1.8 $Pa^{0.5}$ for plate columns.

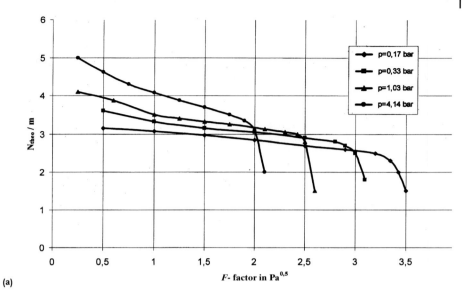

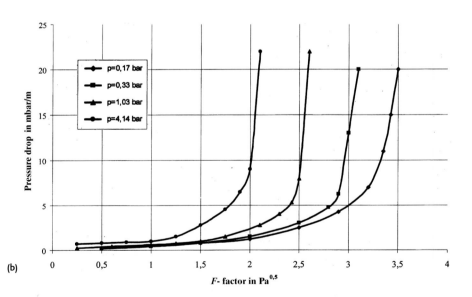

Fig. 2.3.2-11 a) Separation performance in the system cyclohexane/n-heptane at four system pressures in terms of the number of theoretical plates per meter as a function of the F factor of a uniform material-exchange packing (Montz-Pak B1-350M). b) Ditto for the pressure drop.

2.3.2.4
Design of Distillation Plants

Rectification is by far the most widely used separation process in the chemical industry. It is characterized by robust technology that is not prone to disturbances and has comparitively low investment costs. The energy requirement is, however, considerable. About 40% of energy consumption in the chemical industry is used for distillative separation. Therefore, integrated energy measures for multiple use of the thermal energy are of special importance and are widely used (see Section 6.1.6.2) [Kaibel 1990].

The theoretical design of distillation columns is performed wih the aid of computer programs. Commercially available programs include ASPEN and HYSIM, which permit both thermodynamic and fluid-dynamic design (see Section 4.2). Nonideal boiling behavior can be described by various mathematical approaches with sufficient accuracy for practical applications.The Wilson and NRTL models are widely used, and the latter is also suitable for two-component systems with phase separation. The mathematical modeling of distillation processes has reached a high standard. In about 50% of practical applications the design of distillation plants can be performed soley on the basis of calculations.

Short-cut methods and graphical procedures such as the McCabe–Thiele method have declined in importance, since the combination of computer programs with material databanks allows rapid treatment of distillation processes.

For an overview approximation and dimensioning, the relationships of Fenske and Underwood can be used to determine the minimum plate number N_{min} and the minimum reflux ratio v_{min} (Equations 2.3.2-26 and 2.3.2-27):

$$N_{min} = \frac{\ln\left(\frac{y}{1-y} \cdot \frac{1-x}{x}\right)}{\ln a} \qquad (2.3.2-26)$$

$$v_{min} = \frac{1}{a-1} \cdot \left(\frac{y}{x_M} - a \cdot \frac{1-y}{1-x_M}\right) \qquad (2.3.2-27)$$

y = concentration of the low boiler in the distillate
$1-y$ = concentration of the high boiler in the distillate
x = concentration of the low boiler in the bottoms
$1-x$ = concentration of the high boiler in the bottoms
x_M = concentration of the low boiler in the feed
$1-x_M$ = concentration of the high boiler in the feed

By definition the Fenske equation for the minimum plate number only applies for plate columns with a plate efficiency of unity. The use of this equation for packed columns leads to considerable errors, especially at high relative volatilities. The minimum plate number of packed columns shoul be calculated with Equation 2.3.2-28:

$$N_{min} = \ln\left(\frac{x}{1-y}\right) + \frac{1}{a-1} \cdot \ln\frac{y \cdot (1-x)}{x \cdot (1-y)} \qquad (2.3.2-28)$$

The Gilliland diagram depicts the relationship between the plate number and the corresponding reflux ratio (Figure 2.3.2-12).

Economic designs result for plate numbers that are about 1.3 times the minimum plate number or 1.2–1.3 times the minimum thermal power. With regard to controllability and the minimum trickle density, the reflux ratio should not be less than about 0.3–0.5.

Nowadays, distillative separations are readily accessible to mathematical modeling, and therefore in many cases experimental treatment can be greatly limited or even dispensed with.

Experimental treatment of a distillative processing step is avoidable when:

- A production process for the product is already in operation and the changes to the process are restricted to the processing section, for example, switching to a different type of column or introduction of heat-coupling measures.
- Only moderate demands are made with regard to the product specifications.
- No chemical reactions are expected, for example, product damage due to excessive temperatures, discolorations, or polymerizations.

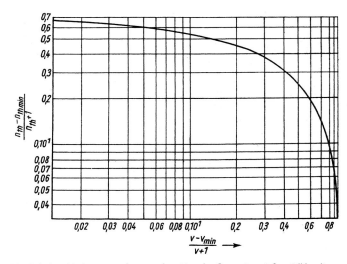

Fig. 2.3.2-12 Relationship between plate number N and reflux ratio v (after Gilliland).

Experimental treatment of a distillative processing step is unavoidable when:

- Certain required product properties can not be mathematically modeled, for example, discolorations due to thermal loading, olfactory properties of fragrances, taste of flavors, complex product properties in subsequent processing steps, for example, polymerization properties and the strength and spinnability of synthetic fibers.
- Recycling of unconverted reactants to the synthesis stage takes place and damage to the catalyst can not be ruled out.
- Temperature- and residence-time-dependent side reactions can take place.
- Reactive distillations, in which the chemical reaction is integrated in the processing step, are to be designed.
- Suitable construction materials for critical mixtures of substances must be selected.
- Extreme substance properties are present, for example, high viscosity of liquids.

The design of industrial columns is performed via the parameters:

- Pressure and temperature in the column
- Permissible vapor velocity
- Attainable number of theoretical plates per meter of column height
- Pressure drop
- Liquid holdup.

Favorable investment cost are generally obtained for a pressure range of 1–4 bar. In the case of heat-sensitive substances the pressure must be sufficiently lowered, with the penalty of higher investment costs, that a tolerable temperature is achieved. Pressures above 4 bar are made neccessary by the costs for cooling agents. For example, the separation of ammonia from aqueous mixtures is generally performed in the pressure range of 17–20 bar so that condensation can be performed with cooling water instead of brine. The minimum temperatures for cooling with river water, recycled cooling water, and air are 30, 40, and 55 °C, respectively.

The permissible vapor velocity is determined via the vapor loading factor (Equation 2.4.15), often known as the F factor. The permissible vapor loading factors for the various column internals are readily available from manufacturers' specifications.

The determination of the attainable theoretical plate number is more difficult. Initially, manufacturers' specifications can be consulted here. The values must be checked with regard to the material properties. For example, high viscosities such as occur in extractive distillations reduce the efficiency of plates to about 40 % as opposed to the values of 60–70 % that usually can be achieved.

2.3.2.4.1 Batch Distillation

For production volumes that lie below about 1000 t a^{-1}, distillative separation is preferably performed batchwise. For small product quantities, batch distillation has the advantage of lower investment costs, since the individual fractions can be separated one by one in the same plant. It is highly flexible, since it can easily be combined with other process steps. If the distillation vessel is designed as a stirred tank, then other reaction steps such as dissolution of solids, chemical reactions, distillative solvent change, liquid–liquid extraction, evaporative crystallization, crystallization with cooling, and precipitation can be carried out in the distillation apparatus.

Batch distillation is generally carried out in the upward operating mode, in which the initial mixture is heated in the distillation vessel and the individual components are collected overhead one after the other in order of their volatilities, starting with the lowest boiling fraction (Figure 2.3.2-13).

The reflux ratio in a batch distillation should not be kept constant during the separation of a fraction, but should gradually be increased to ensure constant purity of the overhead product with time. At a constant reflux ratio the purity of the overhead product gradually decreases with time, since the content of the component being separated in the distillation vessel decreases and the separation thus becomes more difficult (Figure 2.3.2-14). The mixing of overhead products with different compositions in the distillate receptacle results in an entropy of mixing, which can be avoided by having a distillate composition that does not vary with time. The appropriate increase in the reflux ratio with time is generally calculated. It can usually be achieved with sufficient accuracy by a process control system using a ramp function or temperature signals. The resulting savings are considerable. In comparison to operation with constant reflux ratio, a reduction in the distillation time and energy requirement of about 30–60% can be expected.

When high purities of the individual fractions are required, it is necessary to collect intermediate fractions, which are stored separately and added to the next batch. To minimize mixing enthalpies it is advantageous to have a separate vessel for each intermediate fraction and, instead of adding all fractions together at the start of the next batch, to add each fraction at the beginning of the fraction at the end of which it was

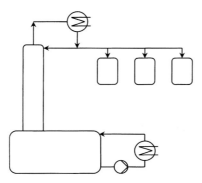

Fig. 2.3.2-13 Batch distillation with upward operating mode.

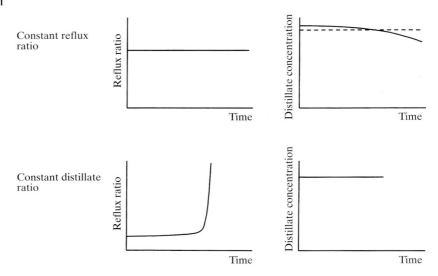

Fig. 2.3.2-14 Temporal course of distillate concentration and reflux ratio for different control of the reflux ratio.

collected. The individual intermediate fractions are also best collected with increasing reflux ratio. To reduce the number of intermediate fractions, the separating internals should have as low a liquid holdup as possible. Bubble-cap and valve plates are unfavorable. The most favorable values are attained with ordered sheet packings.

To limit the temperature, batch distillations are often operated with decreasing pressure. The decrease in pressure can be carried out automatically at an arbitrary point in time. Its rate must be chosen such that, if the energy supply is interrupted, hydraulic damage to the column or the condenser can not occur.

In comparison to continuous distillation batch distillative separations have the disadvantage of higher thermal loading of the product due to the longer residence time, and the energy requirement is also higher.

This disadvantage can partially be compensated by special batch operating modes, of which the upwards mode is already used in some industrial processes. Here the distillation vessel containing the mixture to be separated is located at the top of the column. The individual components are collected one after the other in order of their volatilities, starting with the lowest boiling fraction, at the bottom of the column (Figure 2.3.2-15).

A combination of upwards and downwards operation is especially advantageous. Usually it begins in the downward mode for the first fraction, which has the advantage of saving the heating time and the energy required for heating the initial mixture. At the bottom of the column only the minimum amount required by the evaporator is introduced. Depending on the required purity, the appropriate upwards or dowwards operating mode for the following fractions can be selected.

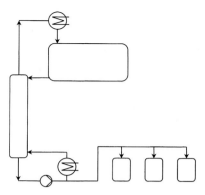

Fig. 2.3.2-15 Batch distillation with downward operating mode.

The suitability of the different operating modes can be determined with the aid of programs for modeling batch distillations. The treatment of batch distillations is much more laborious than that for continuous distillations. However, the decision can be simplified by using Equations (2.3.2-29) and (2.3.2-30), which were determined for the different operating modes with the aid of differential equations. The minimum vapor quantity G is used for comparison:

Batch distillation, upward operating mode:

$$G = n_A \cdot \frac{a}{a-1} \cdot \frac{\sigma_A + \sigma_B - 1}{1 - \sigma_B} \cdot \ln \frac{1}{\sigma_B} + n_B \cdot \frac{1}{a-1} \cdot \frac{\sigma_A + \sigma_B - 1}{\sigma_A} \cdot \ln \frac{1}{1 - \sigma_A} \quad (2.3.2-29)$$

Batch distillation, downward operating mode:

$$G = n_A \cdot \frac{a}{a-1} \cdot \frac{\sigma_A + \sigma_B - 1}{\sigma_B} \cdot \ln \frac{1}{1 - \sigma_B} + n_B \cdot \frac{1}{a-1} \cdot \frac{\sigma_A + \sigma_B - 1}{1 - \sigma_A} \cdot \ln \frac{1}{\sigma_A} \quad (2.3.2-30)$$

Here it is assumed that the control of the reflux ratio is optimal and the removed fractions have constant concentrations with time. For comparison, Equation (2.3.2-31) applies for continuous operation:

$$G = n_A \cdot \frac{a}{a-1} \cdot (\sigma_A + \sigma_B - 1) + n_B \cdot \frac{1}{a-1} \cdot (\sigma_A + \sigma_B) \quad (2.3.2-31)$$

Simpler reationships apply for pure overhead product with upward operating mode (Equation 2.3.2-32) and pure bottom product with downward operating mode (Eq. 2.3.2-33):

$$G = n_A \cdot \frac{a}{a-1} \cdot \sigma_A + n_B \cdot \frac{1}{a-1} \cdot \ln\frac{1}{1-\sigma_A} \qquad (2.3.2-32)$$

$$G = n_A \cdot \frac{a}{a-1} \cdot \ln\frac{1}{1-\sigma_B} + n_B \cdot \frac{1}{a-1} \cdot \sigma_B \qquad (2.3.2-33)$$

G The minimum amount of vapor required for separation in a column with infinite plate number (mol)
n_A Initial amount of low boiler (mol)
n_B Initial amount of high boiler (mol)
a Relative volatility
σ_A Separation yield of the low boiler (ratio of the fraction of low boiler in the distillate to the fraction of low boiler in the bottoms)
σ_B Separation yield of the high boiler (ratio of the fraction of high boiler in the bottoms to the fraction of high boiler in the distillate).

The continuous operating mode always has the lowest energy requirement. The following rules can be given for the suitable operating ranges of the upwards and downwards modes: The downwards mode is preferred when Equation (2.3.2-34) applies:

$$a \cdot \frac{n_A}{n_B} < \frac{\frac{1}{\sigma_A} \cdot \ln\left(\frac{1}{1-\sigma_A}\right) - \frac{1}{1-\sigma_A} \cdot \ln\frac{1}{\sigma_A}}{\frac{1}{\sigma_B} \cdot \ln\left(\frac{1}{1-\sigma_B}\right) - \frac{1}{1-\sigma_B} \cdot \ln\frac{1}{\sigma_B}} \qquad (2.3.2-34)$$

or, when $\sigma_A = \sigma_B$, Equation (2.3.2-35):

$$a \cdot \frac{n_A}{n_B} < 1 \qquad (2.3.2-35)$$

This applies for:

- Difficult separations (low relative volatility)
- Small amounts of low boiler
- Higher purity requirements for the high boiler than for the low boiler.

Figure 2.3.2-16 gives an overview of suitable ranges of application of the upwards and downwards operating modes.

A further possibility for reducing the distillation time is to operate with intermediate storage [Kaibel 1998]. In many cases this achieves a heating power within a few per cent of that of a continuous distillation. In this operating mode the separation of a fraction is divided into two partial steps. In the first partial step, in addition to the distillate the liquid running from the column is removed and stored in an additional vessel (Figure 2.3.2-17). In the second partial step, this stored liquid is fed to the middle of the column to compensate for the deficiency in low boiler that occurs towards the end of the fraction and thus prevent the pronounced increase in reflux ratio that would otherwise occur. The potential savings are highest for difficult separations and high purity requirements.

2.3.2.4.2 Continuous Distillation

For production capacities exceeding about 2000–5000 t/a distillative separations are prefereably performed continuously. Generally, the workup of a reaction product requires a combination of several rectification steps. Particularly extensive separation processes are required in steam cracking, for which about ten distillation columns are required for separating the products.

Because of the numerous possibilities, optimization of the separation sequence with regard to investment costs and energy requirement is a difficult task that has still not been satisfactorily solved. For the simplest arrangement of columns, in which only a overhead and a bottom fraction is obtained from each column, for separaration into an increasing number of fractions $n_{\text{fractions}}$, a strongly increasing number of column circuits n_{circuits} results (Equation 2.3.2-36):

$$n_{\text{circuits}} = \frac{(2 \cdot \{n_{\text{Fractions}} - 1\})!}{n_{\text{Fractions}}! \cdot (n_{\text{Fractions}} - 1)!}. \qquad (2.3.2-36)$$

If no side fractions are removed, the required number of columns is given by Equation (2.3.2-37).

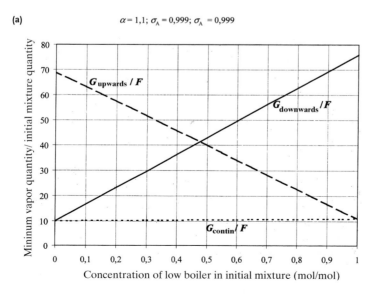

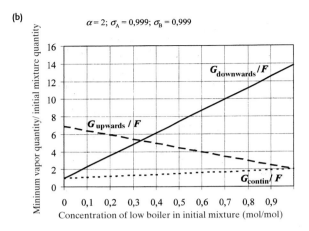

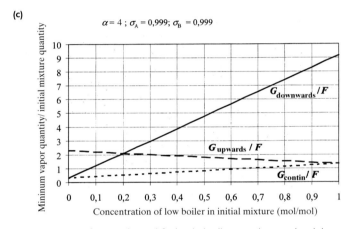

Fig. 2.3.2-16 Comparison of energy demand for batch distillation with upward and downward operating modes as a function of the required purity (the parameter is the relative volatility α, and the constant is the separation yield: $\sigma_A = 0.999/\sigma_B = 0.999$, minimum vapor quantity G, amount of starting mixture F): a) $\alpha = 1.1$, b) $\alpha = 2$, c) $\alpha = 4$.

$$n_{columns} = n_{Fractions} - 1. \qquad (2.3.2-37)$$

Thus, for the separation of a feed mixture into three fractions there are two column circuits with two columns. Figure 2.3.2-18 shows the two possibilities. In the direct separation sequence the low-boiling components are separated one after the other overhead, and in the inverse sequence the high-boiling components are removed at the bottoms of the columns.

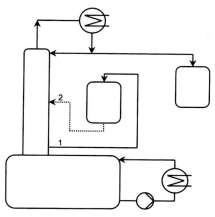

Fig. 2.3.2-17 Batch distillation with upward operating mode and intermediate storage.

For separations into more than three fractions, an arbitrary combination of direct and inverse sequences can be used. Furthermore, it is not only possible to separate the highest or lowest boiling fraction in each case; instead, the separation steps can be arranged at will. Figure 2.3.2-18 shows the different sequences for separation into four fractions.

For the sake of simplicity, column circuits can be used in which a main column is combined with a side column that operates as a rectification or stripping column. This arrangement saves an evaporator or a condenser (Figure 2.3.2-20).

A more widely used and simpler design are columns with side outlets (Figure 2.3.2-21). In this arrangement one distillation column can be saved for each side outlet. To

Tab. 2.3.2-1 Number of possible column circuits as a function of the number of fractions.

$n_{fractions}$	$n_{circuits}$
2	1
3	2
4	5
5	14
6	42
7	132
8	429
9	1430
10	4862

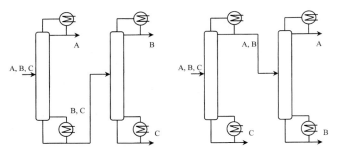

Fig. 2.3.2-18 Direct and inverse separation sequences in the separation of a feed mixture into three fractions.

improve the purity, the side products are removed from the rectification section in liquid form, and from the stripping section in gaseous form. Side columns have the disadvantage that the product removed at the side outlet is always contaminated, either with the lower boiling components when the outlet is in the rectification section, or with the higher boiling components when the side stream is removed from the stripping section. Its use is therefore restricted to cases with moderate requirements for the purity of the middle-boiling fraction. This is a major disadvantage, since in practice the middle fraction is often the valuable product. Therefore, when high purity of the middle fraction is required, it is necessary to use several columns and to dispense with side streams.

The disadvantage of contaminated side fractions can be avoided by using columns with a dividing wall. In this type of column, which has only been in industrial use for a few years, a vertical dividing wall in the region above and below the feed and take-off

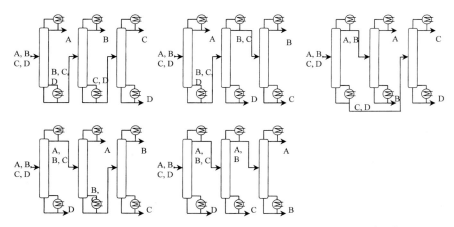

Fig. 2.3.2-19 Possible separation sequences for the separation of a feed mixture into four fractions.

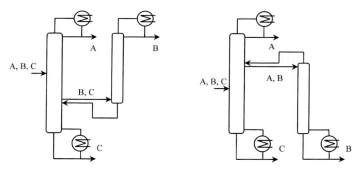

Fig. 2.3.2-20 Distillation columns with side columns.

points divides the column into feed and take-off sections and thus prevents cross-mixing of liquid and vapor streams [Kaibel 1987]. Three or four pure fractions can be obtained from a divided column (Figure 2.3.2-22). In comparison to conventional columns for separation into pure fractions, such columns have 20–40% lower energy requirements and about 30% lower investment costs.

An alternative to columns with dividing walls are the thermally coupled distillation columns (Figure 2.3.2-23). With regard to energy requirements they are equally good, and they are especially suited for retrofitting, because the evaporators and condensers already present can be retained. In contrast to columns with dividing walls, the individual columns can be operated at different pressures.

The various separation sequences can exhibit considerable differences in investment costs and energy consumption. Expert systems have been developed to simplify the time-consuming design of the individual variants and of the most favorable separation sequence [Erdmann 1996a, Trum 1986]. They consider criteria such as feed composition, relative volatilities, the absolute positions of the boiling points, thermal stability of the individual components, purity rquirements, and corrosive properties.

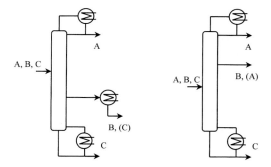

Fig. 2.3.2-21 Distillation columns with side outlets.

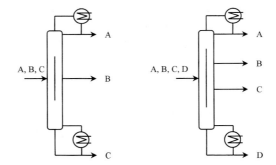

Fig. 2.3.2-22 Distillation columns with a dividing wall.

A guideline for the development of suitable separation sequences can be obtained by using heuristic rules, the best known of which are:

- Always carry out the simplest separation step.
- Use the direct separation sequence.
- Position the separation step such that approximately equimolar quantities of overhead and bottom products are obtained.
- Separate the products that are present in the largest amounts first.
- Choose the separation sequence so that the streams of the nonkey components are minimized.

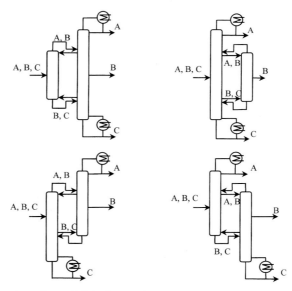

Fig. 2.3.2-23 Thermally coupled distillation columns.

Since the individual heuristic rules are related to various parameters (material properties and quantity ratios) they are ambiguous and can also contradict one another.

As an alternative or supplement to the heuristic rules the reversible separation sequence given by physics can be used as a guideline for developing favorable separation sequences. Figure 2.3.2-24 shows this general separation scheme for the example of the separation of a four-component mixture [Kaibel 1989a]. In each separation step the simplest separation is performed, and components that have very similar boiling points are not separated. It can be shown analytically that the separation of such components, as is always the case in direct and inverse sequences, unavoidably leads to mixing entropies on the feed plate, which result in a higher energy requirement. Columns with separating walls and thermally coupled columns are in accordance with this thermally optimal separation scheme. The general separation scheme can be used for developing favorable conventional separation schemes by making certain simplifications. Connections between the individual partial columns are removed in such a manner that as many of the advantageous properties of the general separation scheme as possible are retained.

The individual distillation columns can be further optimized by means of an energetic analysis [Kaibel 1990a]. Here the extent too which mixing entropies or energy losses occur at individual locations in the distillation plant is analyzed. Mixing entropies at the feed points of the columns are an important source of heat losses. In the separation of three- and multicomponent mixtures these mixing entropies can only be avoided by taking the thermodynamic separation into account. If a distillation column has several feeds with different compositions, then these should be introduced at different points in the column. Mixing entropies in the column also occur during the separation of easily separable mixtures when the concentrations and temperatures differ widely from plate to plate. A remedial measure here is the use of intermediate evaporations and condensations.

Example 2.3.2-2

Figure 2.3.2-25 *shows the separation of two feed mixtures of methanol and water with different compositions. The two streams are fed separately into the column to avoid mixing entropies. In the column without intermediate evaporation* (Figure 2.3.2-25a), *pronounced mixing entropies are found in the stripping section, where the vapor quantity is too high to meet the requirements of the rectification section. By means of an intermediate evaporation* (Figure 2.3.2-25b) *about 60% of the total energy can be introduced at a lower temperature of about 75°C, and the stripping section can be constructed with a smaller diameter. The lower temperature facilitates the exploitation of waste heat from neighboring plants. When a vapor compressor is used, the stripping performance is lower, since the temperature difference between head and intermediate evaporator (65 vs 75°C) is considerably smaller than that between head and bottom (65 vs 100°C).*

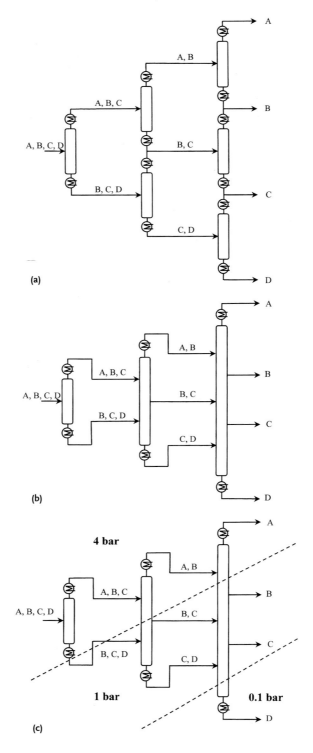

2.3 Product Processing (Thermal and Mechanical Separation Processes)

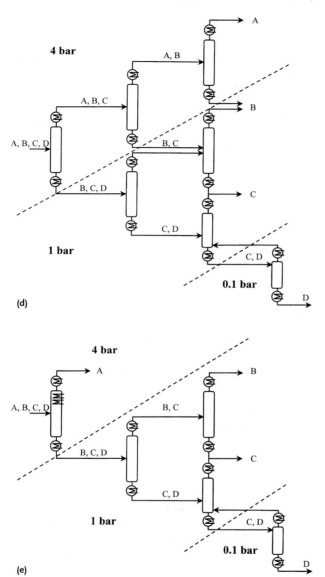

(d)

(e)

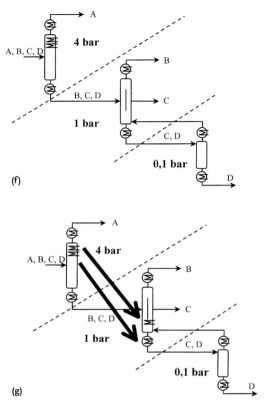

Fig. 2.3.2-24 General separation scheme for splitting a four-component mixture (a) and development of a thermodynamically optimal separation scheme (a–g) [Kaibel 1989a].

The actual amount of vapor required at individual points in the column can easily be calculated with balance equations by assuming an infinitely high number of plates [Kaibel 1989a]. For a two-component mixture the minimum vapor quyntity G at any position in a distillation column with a contration x of the light boiler in the liquid is given by Equation (2.3.2-38), where the quantities are in moles:

$$G = \frac{D_A \cdot (1-x) - D_B \cdot x}{x} \cdot \frac{1 + (a-1) \cdot x}{(a-1) \cdot (1-x)} \qquad (2.3.2-38)$$

G Minimum vapor quantity
D_A Quantity of low boiler moving upwards in the column
D_B Quantity of high boiler moving upwards in the column
a Relative volatility
x Concentration of low boiler in the liquid.

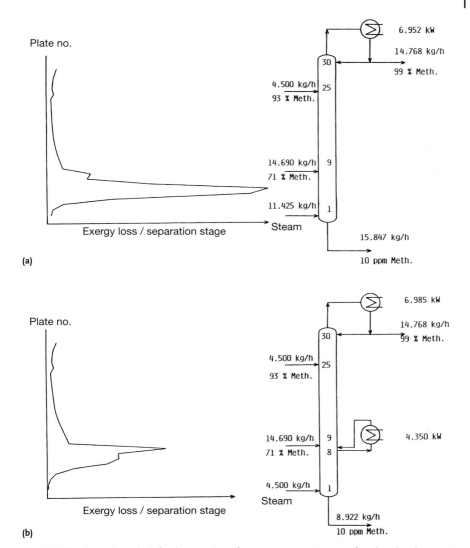

Fig. 2.3.2-25 Exergetic analysis for the separtion of two-component mixtures of methanol and water. a) Without intermediate evaporator. b) With intermediate evaporator.

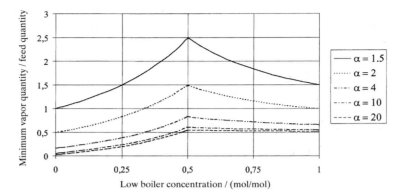

Fig. 2.3.2-26 Minimum vapor quantity for the separation of equimolar two-component mixtures with different relative volatilities α. The high boiler in each case is n-octane.

Figure 2.3.2-26 shows the minimum vapor quantity relative to the feed quantity M as a function of the low-boiler concentration for the separation of two-component mixtures with different relative volatilities. It can be seen that the stripping section in particular offers good possibilities for the application of intermediate evaporators. Deviations from ideal boiling behavior strongly influence the course of the minimum vapor quantity curve. Figure 2.3.2-27 shows as an example the separation of an equimolar mixture of acetone and water [Kaibel 1989b]. Because of the almost azeotropic boiling behavior, the majority of the heat is required in the rectification section of the column, near the

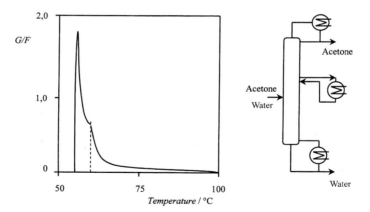

Fig. 2.3.2-27 Separation of an equimolar acetone/water mixture of. Possibility of using an intermediate evaporator in the rectification section of the column due to the nonideal boiling behavior.

head of the column. This suggests using an intermediate evaporator in the rectification section of the column.

2.3.2.4.3 Internals

Internals can be divided into plates and packings. With plates the liquid largely forms the continuous phase, while the gas phase rises through the liquid in the form of bubbles. This leads to intensive mixing of the two phases and effective mass transfer, albeit at the expense of a relatively high pressure drop. In columns with random or ordered packings, the gas phase is the continuous phase, and the liquid phase trickles over the surface of the packing elements, generally in the form of a laminar layer. In this case a high mass transfer performance requires a large contact area between the gas phase and the liquid. An advantage is the low pressure drop, which makes distillation pressures as low as 2 mbar possible.

Plates are available in the following designs: bubble-cap plates, plates with fixed or moving valves, sieve plates, and dual-flow plates.

Bubble-cap plates (Figure 2.3.2-28) have a high liquid capacity and can be used for reactive distillations which require long residence times. The height of the liquid on the plate can be up to 0.5 m. In some applications it is advantageous that the liquid remains on the plates when the plant is off-line. Bubble-cap plates have a particularly wide operating range.

Valve plates are a cheaper design of plate, which are also suitable for high loadings and have small pressure drops.

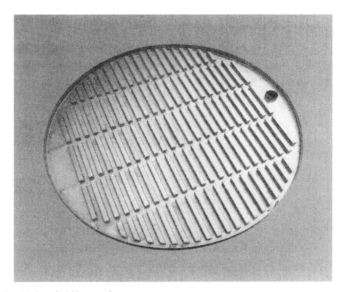

Fig. 2.3.2-28 Modern bubble-cap plate.

The cheapest plates are the simple *sieve plates*, but their loadability is limited.

The disadvantage of limited loadability especially applies for *dual-flow plates*. These have no dowcomers for the liquid, and gas and liquid pass together through the holes in the plate. The main application is separations that are prone to soiling, for example, substances that tend to polymerize.

At low operating pressures, due to the large gas volumes, plate columns exhibit pronounced droplet entrainment, which lowers the separation performance due to backmixing of the liquid. The throughput can be increased by introducing into the spray layer a 5–10 high packed layer that acts as a mist collector.

High-performance plates, which are available in various designs, must be caraefully adapted to a particular application and have only a limited range of loadability.

Distillations at a pressure of more than about 4 bar are preferably carried out in plate columns. Because of the high liquid trickle densities in pressure distillations, the liquid film in packed columns can tend to exhibit instabilities.

The major advantage of plates are the low costs when used in columns with a diameter of more than about 1.5 m. Plates can not easily be mounted in small columns, and hence packings are preferred here. In the case of low pressures, as are needed for the separation of highly heat sensitive substances, packed columns are the better alternative.

In production plants the plate columns with diameters up to about 1.5 m that were formerly widely used have in many cases been replaced by columns with random or ordered packings. Exceptions are separations involving two liquid phases, separation of low-boiling hydrocarbons, columns operated at a pressure above about 4 bar, and the separation of substances that tend to polmerize.

In the region of very low pressures, that is, head pressures that can be as low as 1–2 mbar, ordered packings of wire mesh are mainly used. Typical industrial designs are the types Montz A3 and Sulzer BX with specific surface areas of $500\,m^2\,m^{-3}$. For pressures between about 10 mbar and 4 bar sheet packings (e.g., Montz B1, Sulzer Mellapak) or packings made of expanded metal with specific surface areas of 100–$750\,m^2\,m^{-3}$ can be used. The packings usually have a cross-channel structure that effects mixing of the gas and liquid streams over the column cross section (Figures 2.3.2-29 and 2.3.2-30). The angle of inclination of the bends to the horizontal is usually 45°, but for higher throughput at the expense of separation performance it can be increased to about 60°. Modern packings have a steep angle of inclination, at least at the lower end of the bends. This facilitates runoff of the liquid onto the underlying packing layer and increases the loading limit by about 30% in comparison to packings with straight course (Figures 2.3.2-29 and 2.3.2-30).

Random packings in the form of Raschig rings, Pall rings, saddles, and other geometries are also still in widespread use but are increasingly being replaced by ordered packings. They can be used in applications with high liquid loads (e.g., absorptions) and low demands on the pressure drop.

The attainable number of plates depends on the surface area per unit volume of the packing. It lies in the range of $3\,m^{-1}$ for $250\,m^2\,m^{-3}$, $5\,m^{-1}$ for $500\,m^2\,m^{-3}$, and up to $10\,m^{-1}$ for $750\,m^2\,m^{-3}$ packings. In order to compare random packings with structured packings, the specific surface area of a packing element can be estimated as 5.7 divided

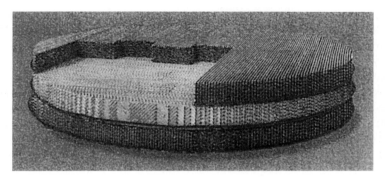

Fig. 2.3.2-29 Cross-channel packing with straight course of the bends.

by the size of the element. The specific surface area of structured packings is usually specified by the manufacturer. Otherwise it can be estimated by 2.8 divided by the breadth of a single layer.

When high product purities are to be maintained, then in random-packed columns the liquid must be collected and redistributed after about ten plates. Inhomogeneities in the liquid distribution, for example, due to peripheral flow of the liquid in the stripping section of the column or too low a trickle density in the rectification section, can make it impossible to maintain high product purity without intermediate redistributors, even when highly overdimensioned packing heights are used. For this reason plate columns are still used today when high product purities are required.

Fig. 2.3.2-30 Packing with inclined bends.

Liquid distributors should preferably use the cumulative operating principle, in which the liquid accumulates to a height of about 15 cm in a distributor box and flows out through precisely dimensioned bores (Figure 2.3.2-31). This design is insensitive to faulty mounting. For typical separation requirements, about 200 drainage points per square meter of cross-sectional area are required. Very small amounts of liquid, such as those that occur in high-vacuum distillations, result in bores with very small diameters that are prone to blockage. A possible remedial measure is the use of designs in which the required number of drainage points is achieved by means of distribution tongues operating according to the capillary principle (Figure 2.3.2-32). Distributors that operate according to the overflow principle, for example, in which the liquid flows over a toothed weir, are highly sensitive to faulty horizontal mounting.

In the case of columns with ordered packings, the manufacturers specified separation performance usually refers to the chlorobenzene/ethylbenzene system, for which particularly high plate numbers can be achieved because of the high liquid density and low viscosity. Simply changing to alkane mixtures lowers the attainable plate number by about 30%. For high viscosities, a further decrease of about 50% can be expected. Similar drops in performance can occur with systems that exhibit phase decomposition [Siegert 1999]. In the case of poor wetting, for example, as occurs with highly polar liquids and plastic packings, formation of streams can lead to drastic declines in performance. The vapor loading factor, the liquid trickle density, and the operating pressure also have an influence. The separation performance of columns with ordered packings strongly depends on their loading (Figure 2.3.2-11). Approximate values for the performance of column packings follow:

Fig. 2.3.2-31 Liquid distributor with cumulative operating principle.

Fig. 2.3.2-32 Liquid distributor with cumulative operating principle and additional distribution tongues operating according to the capillary principle.

- Bubble-cap plates, valve plates, sieve plates: 2–3 plates per meter, plate efficiency 60–70%.
- Dual-flow plates: 2–3 plates per meter, plate efficiency 40–50%.
- Fabric packings with $750/500/250 \, m^2 \, m^{-3}$: 10/5/3 plates per meter.
- Sheet packings with $750/500/250 \, m^2 \, m^{-3}$: 6/4,5/3 plates per meter.
- Packing elements with dimensions of 50/25/15 mm: 1.5/2/4 plates per meter.

The pressure drop can be taken from the manufacturer's specifications. For a given vapor loading factor, it increase with increasing liquid trickle density. Values of about 5 and about 0.1–0.5 mbar for plate and packed columns, respectively, can be assumed.

The liquid content in plate columns is about 5–10% of the column volume. For packed columns, values of 2.5 (high vacuum or coarse sheet packing) to 8% can be assumed. If, in the case of heat-sensitive substances, small liquid contents are aimed for, ordered packings can be used in which the inclinations to the verticalare is only small, so that the liquid can run off more quickly. Acccordingly, high liquid holdups can be achieved in homogeneously catalyzed reactive distillations by using greater inclinations. If these measures are not sufficient, then plate columns with high liquid heights on the downcomers or external residence-time vessels are used.

2.3.2.4.4 Experimental Design

In the experimental design of distillative separations both bubble-cap columns with a minimum diameter of 30–40 mm and columns with random or ordered packings and a minimum diameter of 30 mm can be used. With regard to scale-up of the plate number, columns with bubble-cap plates are most favorable, since their performance depends least on loading, system pressure, and the system to be separated. Columns made of glass are preferred.

The separation performance of columns with random packings of, for example wire-mesh rings with a diameter of 3–5 mm, is strongly dependent on loading (10–40 plates per meter), and therefore it is very difficult to transfer results directly to large-scale plants. With decreasing liquid trickle density, smaller film thicknesses become established on the packing elements, which facilitates mass transfer and thus leads to higher plate numbers. This behavior is also exhibited by ordered packings of wire mesh which have a specific surface area of more than $1000-1900\,m^2\,m^{-3}$ and thus allow small heights of test apparatus to achieved, such as the types Sulzer DX (15–30 plates per meter) and EX (15–40 plates per meter). Small dependences of the plate number on the throughput are exhibited by ordered sheet packings such Kühni Rombopak 9M and graphite packings such as Sulzer Mellacarbon. The use of the same type of packing as in the large-scale plant – provided this makes sense in the small diameter of the test apparatus – is only of limited help here, becase the shorter channels in the cross-flow structures lead to more frequent changes in flow direction and hence to higher pressure drop and about 20% higher plate number. Only above a column height of about 0.8 m does this dependence of the separation performance no longer exist.

Because of these fundamental deficiencies of currently available experimental methods, experimental design of distillative separations is generally combined with a mathematical model to decrease the uncertainty in determining the plate number, which is decisive for the design of the large-scale plant.

2.3.4.5
Special Distillation Processes

For the separation of azeotropes or mixtures with relative volatilities that lie below about 1.4, which are difficult to separate, special distillation processes are available, such as two-pressure distillation and extractive and azeotropic rectification.

Two-pressure distillation is especially promising when the azeotrope-forming components have very different heats of evaporation. The thus resulting differing slopes of the vapor-pressure curves leads to a pressure dependence of different azeotropic compositions.

Example 2.3.2-3

At normal pressure tetrahydrofuran forms an azeotrope with water at a water content of 5.3%. At a pressure of 7 bar the azeotrope has a water content of 12%. These different azeotropic compositions can be used to separate the mixture into its pure components (Figure 2.3.2-33).

Extractive distillation exploits the differing solubilities of the azeotrope-forming components in a higher boiling extractant. Aids to the selection of suitable extractants (e.g., expert systems) are available [Erdmann 1986a, Trum 1986]. Recovery of the extractant requires an additional downstream column.

Example 2.3.2-4

A azeotropic mixture of isopropyl alcohol and water is separated by addition of ethylene glycol. Extractive distillation leads to pure isopropyl alcohol as overhead product, and water and ethylene glycol as bottom product. This mixture is split into water and ethylene glycol under reduced pressure in a downstream distillation column. The ethylene glycol is then cooled and recycled to the extractive distillation (Figure 2.3.2-34).

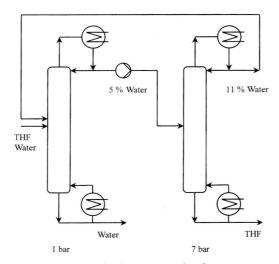

Fig. 2.3.2-33 Two-pressure distillation for dewatering tetrahyrofuran.

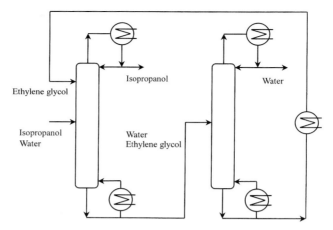

Fig. 2.3.2-34 Extractive distillation for dewatering isopropyl alcohol.

In azeotropic distillation, mitures that are difficult to separate are separated by addition of a substance that forms an azeotrope with only one of the components of the mixture. This process is rarely used. A widely used special case is heteroazeotropic distillation, in which water is present as azeotrope component and forms a miscibility gap.

Example 2.3.2-5

Phthalic acid is discontinuously esterified with n-butanol. The released water of reaction is distilled off together with n-butanol, separated in a phase separator, and eliminated via a stripping column (Figure 2.3.2-35).

In reactive rectification a rectification process is coupled with a chemical reaction. A simple example is the combination of a stirred tank in which an esterification takes place with a column for separation of the water of reaction [Stichlmair 1998, Frey 1998a].

Reactive rectifications without a separate reactor, in which the reaction takes place in the distillation column, are increasingly being used in industry. Typical classes of reactions are esterification, transesterification, acetal formation and cleavage, etherification, oxidation, and hydrogenation. With regard to the design of such processes, three cases can be distinguished:

1. The reaction is autocatalytic
2. The reaction is homogeneously catalyzed
3. The reaction is heterogeneously catalyzed.

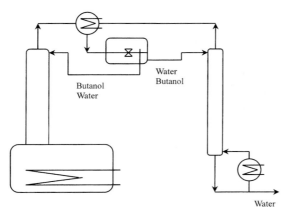

Fig. 2.3.2-35 Hetero-azeotropic distillation.

For autocatalytic reactions with moderate reaction rates, technical measures must be taken to ensure a sufficiently long residence time. A typical example is catalysis of the hydrolysis of an ester by the released acid.

Homogeneously catalyzed reactions also often require prolonged residence times. The catalysts are often high-boiling mineral acids or alkalis. In some cases, volatile catalysts that are held in the reaction zone by the distillative action can be successively used, for example, nitric or hydrochloric acid.

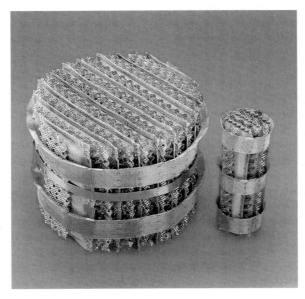

Fig. 2.3.2-36 Internals for holding catalysts.

Often the required residence time of the liquid in the column can not be achieved with conventional distillation internals. Here special residence-time plates can be used, generally bubble-cap plates with liquid heights of up to 0.5 m (Figure 2.3.2-28).

For heterogeneously catalyzed reactions numerous catalytically active internals have been developed in which the catalyst lies on the packing or, for example, is present in wire-mesh pockets through which liquid can flow (Figure 2.3.2-36). In plate columns the catalyst can be incorporated in the downcomers for the liquid.

2.3.3
Absorption and Desorption, Stripping, Vapor-Entrainment Distillation

2.3.3.1
Fundamentals

Absorption resembles extractive distillation. However, the component to be absorbed is removed from a gas stream that essentially consists of a gas that is not condensable under the operating conditions:

A (substance to be absorbed) in noncondensable gas (G, Y_0) + absorption liquid (F, X_0) → loaded solvent (F, X_e) + noncondensable gas with residual A (G, Y_e)

The reverse process, the removal of an absorbed component, is known as desorption. The transition from chemisorption, for example, the absorption of SO_3 in sulfuric acid

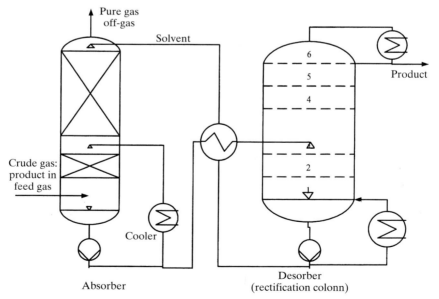

Fig. 2.3.3-1 Schematic of an absorber/desorber separation unit.

production, to physisorption, for example, the removal of residual solvent from air with a high-boiling solvent, is fluid. If the absorbed component is subsequently to be recovered by desorption, the absorbent must be carefully selected with regard to selectivity and capacity in order to achieve a favorable overall process (Figure 2.3.3-1).

The choice of extractant can be made with aid of expert systems, in similar manner as for extractive distillations or extractions [Erdmann 1986a, Trum 1986].

Important selection criteria for absorption liquid follow:

- High capacity for A
- High selectivity for A
- Easy regeneration (A/F separation)
- Suitable boiling point (not too high, so that it can be regenerated by distillation, but not too low, so that the loading in the waste gas stream is not too high
- Noncorrosive
- Chemically and thermally stable
- Nontoxic and biodegradable
- Low viscosity
- Ready available and cheap.

2.3.3.2
Dimensioning

Formally, the design of an absorber with respect to plate number and cross-sectional area can be performed analogously to a rectification by means of the staircase method. Henry's[12] law is often used as equilibrium relationship (Equation 2.3.3-1):

$$P_A = H_A \cdot x \text{ resp. } y = \frac{H_A}{P} \cdot x \qquad (2.3.3-1a)$$

$x/$ mol mol^{-1} = mole fraction of A in the liquid
$y/$ mol mol^{-1} = mole fraction of A in the gas
P_A = partial pressure of A
P = total pressure

or in the form

$$A/(\text{mol L}^{-1}) = K_H/(\text{mol bar}^{-1} \text{ L}^{-1}) \cdot P_A/bar$$

where

$$K_H \approx \frac{P}{RT} \cdot \frac{1}{H_A}, \text{ if } x_A \ll x_B \qquad (2.3.3-1b)$$

(R = 0.082052 L atm K^{-1}).

---

12) William Henry, British physicist and chemist (1774–1836).

A rough guideline is that the Henry coefficient should be less than about 10 bar mol mol^{-1} for an economically viable absorption process. Using the loadings X and Y instead of the mole fractions y and x gives the equilibrium curve (Equation 2.3.3-2):

$$Y = [H_A/P] \cdot \frac{X}{1 + (1 - [H_A/P]) \cdot X} \qquad (2.3.3-2)$$

X = loading of liquid with A
Y = loading of gas with A
P = total pressure.

The *balance equation* can be most easily formulated when the loadings are introduced as the measure of concentration. Initially, the carrier gas is regarded as insoluble in the solvent, and it is assumed that the vapor pressure of the solvent is negligible. The loading of the gas stream (G in moles per unit time) or of the absorption liquid (F in moles per unit time) with A is given by Equations (2.3.3-3) and (2.3.3-4), respectively:

$$Y = \frac{n_{AG}}{n_G} = \frac{n_{AG}}{G} = \frac{y}{1-y} \text{ and } y = \frac{n_{AG}}{n_G + n_{AG}} = \frac{Y}{1+Y} \qquad (2.3.3-3)$$

or

$$X = \frac{n_{AF}}{n_F} = \frac{n_{AF}}{F} = \frac{x}{1-x} \text{ and } x = \frac{n_{AF}}{n_F + n_{AF}} = \frac{X}{1+X} \cdot \qquad (2.3.3-4)$$

With the above-mentioned assumptions, G and F are constants, and for the balance space shown in Figure 2.3.2-2, Equation (2.3.3-5) applies:

Tab. 2.3.3-1 Henry-constants K_H of some gases at 20 °C [Bliefert 2002].

Gas	K_H / 10^{-3} mol bar^{-1} L^{-1}
N_2	0.64
H_2	0.74
CO	0.95
O_2	1.27
C_2H_6	1.94
C_2H_4	5
CO_2	36
C_2H_2	42
Cl_2	93
SO_2	1620

$$G \cdot Y_0 + F \cdot X = G \cdot Y + F \cdot X_e. \qquad (2.3.3-5)$$

From which Equation (2.3.3-6) follows:

$$Y = \frac{F}{G} \cdot X + \left(Y_0 - \frac{F}{G} \cdot X_e\right). \qquad (2.3.3-6)$$

The same applies for the overall balance (Equation 2.3.3-7):

$$G \cdot Y_0 + F \cdot X_0 = G \cdot Y_e + F \cdot X_e. \qquad (2.3.3-7)$$

Inserting Equation (2.3.3-7) into Equation (2.3.3-6) for $X = X_0$ gives $Y = Y_e$, that is, all balance lines must pass through the point (X_0, Y_e). The slope of the balance lines corresponds to F/G and is a minimum when the straight line passes through the intersection of the Y_0 parallel with the equilibrium curve (Figure 2.3.3-3a). This slope gives the minimum quantity of absorption liquid $(F/G)_{min}$, that is, the mass flux of F which, at infinite plate number, is just sufficient to ensure the performance Y_e. By plotting different balance lines with different F/G ratios in a staircase process, one obtains a N/F diagram that gives the economically optimal plate number N and amount of absorption liquid F (Figure 2.3.3-3b).

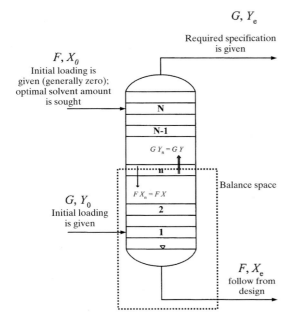

Fig. 2.3.3-2 Schematic of an absorber.

By plotting the balance equation (2.3.3-6) and the equilibrium realtionship (Equation 2.3.3-2) in a staircase process, one obtains the following quantities:

- The minimum amount of absorption liquid (amount of solvent that is needed to just achieve the separation at infinite plate number) from the intersection of the balance line with the point (X_0, Y_e) (Figure 2.3.3-3a).
- The economically optimal combination of plate number N and amount of absorption liquid F by plotting an N/F diagram (Figure 2.3.3-3b).

A prerequisite for construction of this staircase is that the column operates isothermally. However, since almost the entire heat of absorption ($\approx$ heat of condensation) is liberated on the first plate and thus increases the temperature of the lower plates, it is favorable to remove the energy from the first plate by a recycle quench via an external heat exchanger (Figure 2.3.3-1).

For absorption the same plate designs as for rectification are used. However, the efficiency of a practical stage is lower in absorption than in rectification. Therefore, besides the conventional columns, equipment with highly intensive contact of the phases, like in a jet washer or spray tower, is used. Hence, in contrast to rectification, the mass-transfer concept (Figure 2.3.3-4), in which gas and liquid transfer units

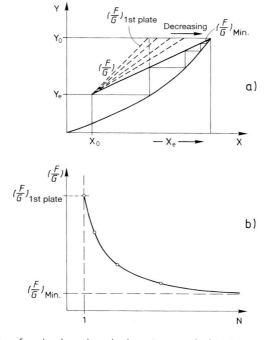

Fig. 2.3.3-3 Design of an absorber column by the staircase method. a) Determination of the minimum amount of absorption liquid. b) Determination of the optimum amount of absorption liquid.

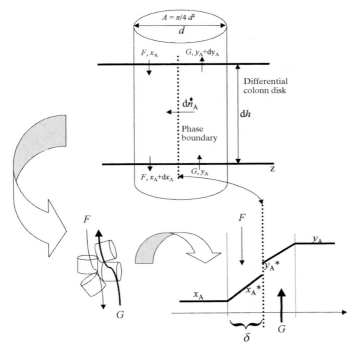

Fig. 2.3.3-4 Mass-transfer model (HTU–NTU method):
- A = Column cross section = $\pi/4\, d^2$
- dh = height of differential column disk
- $A_{Ges} = a\, A\, dh$ = total inner exchange area of differential column disk
- Mass transfer of A from gas phase to boundary phase:

$$\dot{n}_A = D_G \cdot A_{tot} \cdot \frac{dc_A}{dz} = \frac{D_G}{\delta} \cdot A_{Ges} \cdot dc_A$$

with diffusions coefficients D_G. Convesion of molarity c_A to mole fraction y_A:

$$c_A \approx \frac{\rho_{Mixture}}{M_{inert\,gas}} \cdot y_A \quad \text{(s. Appendix 8.6)}$$ and introduction of the mass
transfer coefficient: $\beta_G^* = \frac{D_G}{\delta}$

in [m s^{-1}] respectively $\beta_G = \frac{\rho_{Mixture}}{M_{inert\,gas}} \cdot \frac{D_G}{\delta}$ in [m s^{-1} mol m^{-3}] give:

$$\dot{n}_A = \beta_G \cdot A_{tot} \cdot dy_A = \beta_G \cdot a \cdot A \cdot dh \cdot (y_A - y_A^*)$$

- Mass transfer of A to liquid phase analogous to above:
$$\dot{n}_A = \beta_F \cdot A_{tot} \cdot dx = \beta_F \cdot a \cdot A \cdot dh \cdot (x_A^* - x_A)$$
- Mass transfer balance:
$$\dot{n}_A = G \cdot dy_A = F \cdot dx_A$$

$$G \cdot dy_A = \beta_G \cdot a \cdot A \cdot dh \cdot (y_A - y_A^*)$$
$$F \cdot dx_A = \beta_F \cdot a \cdot A \cdot dh \cdot (x_A^* - x_A)$$

$$H = \int_0^H dh = \frac{G}{\beta_G \cdot a \cdot A} \cdot \int_{Y_{top}}^{Y_{bottom}} \frac{dy_A}{(y_A - y_A^*)} = \frac{F}{\beta_F \cdot a \cdot A} \cdot \int_{X_{top}}^{X_{bottom}} \frac{dx_A}{(x_A^* - x_A)}$$

H = HTU $\cdot$ NTU

HTU = height of a theoretical unit
NTU = number of theoretical units.

and the height H thereof are determined (Equation 2.3.3-8), is often used in preference to the concept of theoretical plates [Sattler 2001]:

$$H = \frac{F}{\beta_F \cdot a \cdot A} \cdot \int_{x_{over}}^{x_{under}} \frac{dx}{(x^* - x)} = \frac{G}{\beta_G \cdot a \cdot A} \cdot \int_{y_{over}}^{y_{under}} \frac{dy}{(y - y^*)} \qquad (2.3.3-8)$$

One reason for this is that mass transfer in absorption often depends much more strongly on the system, and hence on the dimensions of the apparatus, than in rectification. Corresponding calculation programs are commercially available. Since it only known in advance for a few systems to what extent mass-transfer resistance lies on the gas or the liquid side, one is highly dependent on experiments, usually in columns. Packed columns are generally used for large amounts of liquid, and also plate columns for small amounts. The strong tendency to foaming that is often encountered may force the use of large packed columns. If a single theoretical plate is sufficient for separation, Venturi washers and bubble columns can also be used.

From Equation (2.3.3-8) it results that for dimensioning the effective exchange area a_{eff} and the mass-transfer coefficients in the two phases must be known. In general, only the product of the exchange area a_{eff} and the overall transfer coefficient β can be determined (see also Section 2.1.3.1, Equation 2.1-17) [Last 2000].

2.3.3.3
Desorption

Desorption (Figure 2.3.3-1) is generally carried out by heating in a column. Depending on the details of the overall process, stripping with another gas stream or desorption by pressure reduction can be used. Stripping can also be carried out at lower temperatures (e.g., room temperature). A typical application is the removal of malodorous components from heat sensitive substances.

2.3.3.4
Vapor-Entrainment Distillation

A special case of desorption is vapor-entrainment distillation, which is used for the thermally mild separation of high-boiling substances that are immiscible with water. Steam is generally used as entraining vapor (steam distillation) since it can subsequently simply be condensed, but nitrogen is also used. The maximum temperature is limited to the boiling point of water at the operating pressure (Equation 2.3.3-9):

$$P = P^0_{H2O}(T) + P^0_A(T) \text{ (1 bar, STP)} \qquad (2.3.3-9)$$

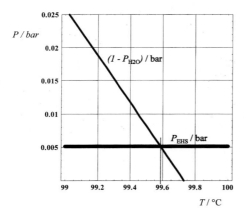

Fig. 2.3.3-5 Example for the determination of the azeotropic temperature T_{AZ} for the system 2-ethylhexanoic acid/water:

$$\ln P^0_{EHS}/bar = 11.3053 - \frac{4533.73}{173.87 + T/°C}$$

$$\ln P^0_{H2O}/bar = 11.78084 - \frac{3887.20}{230.23 + T/°C}$$

For the chosen example, Equation (2.3.3-10) shows that 25 t of water is needed to strip 1 t of ethylhexanoic acid.

The design involves determining the azeotropic temperature at which Equation (2.3.3-9) is fulfilled. In the simplest case this can be performed graphically (Figure 2.3.3-5) or by solving nonlinear eqauations. The amount of entraining vapor required to strip 1 mol of A is then given by Equation (2.3.3-10):

$$n_{H2O}/n_A = \left(\frac{P - P_A}{P_A}\right) \approx \frac{P}{P_A}. \qquad (2.3.3-10)$$

2.3.4
Extraction

In liquid/liquid extraction, a dissolved component (E, extracted component, solute) is transferred from a loaded liquid (R, raffinate phase) to another liquid (L, extract phase, solvent). A prerequisite for this is immiscibility of raffinate and extract phases. If the driving force for mass transfer is not only attainment of the physical liquid/liquid equilibrium, but there is also a chemical driving force such as solvation, chelation, and anion or cation exchange, then the term reactive extraction is used:

(E in R) + $L \rightarrow$

(E in L, saturated with R) + (residual E in R, saturated with L)

Thus, direct separation of E and R is not achieved; the separation problem is simply displaced. In addition, due to the finite solubility, R becomes saturated with L, and vice versa. Hence extraction must be followed by at least two more separation steps, generally rectifications (Figure 2.3.4-1) [Schierbaum 1997].

Therefore, extraction is preferred to rectification only under certain boundary conditions:

- Extraction of heat-sensitive, high-boiling, or nonvolatile substances
- Separation of azeotropic mixtures
- Separation of substances with similar boiling points (low separation factors, i.e., relative volatilities of the components to be separated of less than 1.1)
- Presence of inorganic salts (incrustation of heating surfaces in rectification).

In comparison with distillation, the most widely used process, knowledge of the fundamentals of liquid/liquid extraction is limited. A sufficiently accurate description of the hydrodynamics and mass-transfer rates of liquid systems for the design of apparatus is currently not possible for many practical applications. The development of an extraction apparatus generally requires cost-intensive and time-consuming tests, tailored to the system to be processed, in laboratory and pilot plants. Tests with original solutions, for example, from an integrated miniplant, are especially important here (see Section 4.5).

2.3.4.1
Fundamentals

The fundamentals of liquid/liquid extraction are provided by the thermodynamic theory of equilibrium. Two immiscible liquid partial systems 1 and 2 are in equilibrium when all mass-, energy-, and impulse-transfer processes have come to a stop, that is, when the chemical potential, temperature, and pressure are the same in both phases. If Equation (2.3.4-1) is set up for a component E in phase 1 or 2, then the chemical potential describes the state of the pure component E with the properties of the ideal dilute solution:

$$\mu_E^{(1)} = \mu_{E,0}^{(1)}(T, P) + RT \cdot \ln\left(\gamma_E^{(1)} \cdot x_E^{(1)}\right) \tag{2.3.4-1}$$

In the partial systems 1 and 2 the chemical potentials of component E can each be described by Equation (2.3.4-1). When equilibrium is being established, Equation (2.3.4-2) applies for the distribution of the component E between the two phases 1 and 2:

$$\exp\left[\frac{\mu_{E,0}^{(1)} - \mu_{E,0}^{(2)}}{RT}\right] = \frac{\gamma_E^{(2)} \cdot x_E^{(2)}}{\gamma_E^{(1)} \cdot x_E^{(1)}} = K_E(T). \tag{2.3.4-2}$$

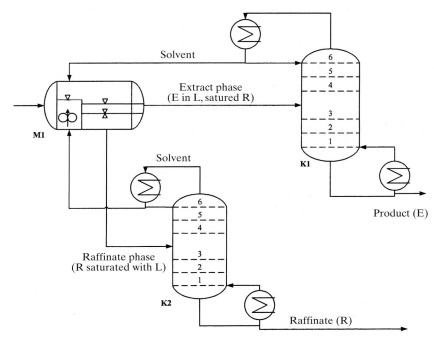

Fig. 2.3.4-1 Extraction with downstream separation units (schematic) [Schierbaum 1997]. M1: mixer – settler stage; K1: rectification column, separation of solvent from product; K2: stripper, separation of solvent from raffinate.

At sufficient dilution, $\gamma_E \to 1$ and Equation (2.3.4-2) simplifies to the Nernst distribution law (Equation 2.3.4-3), which has the same significance for extraction as Raoult's law for rectification:

$$K_E(T) = \frac{x_E^{(2)}}{x_E^{(1)}}, \qquad (2.3.4-3)$$

However, for real systems the distribution coefficient K_E is strongly concentration dependent. Thus, one important requirement for the extract phase is established: it should have as high a K_E value as possible. Further important boundary conditions for the extract phase are [Schierbaum 1997]:

- High selectivity $S = \dfrac{[E]/[R]|_{extract\ phase}}{[E]/[R]|_{raffinate\ phase}}$;

- Low solubility in the raffinate
- No azeotrope formation with the solute

- High chemical and thermal stability
- Low viscosity
- Suitable values of the boundary surface tension (risk of emulsion formation)
- Noncorrosive
- Low toxicity and good biodegradability
- Cheap and readily avaialable.

For the choice of suitable solvents, guidelines and expert systems are available. The determination of the extraction temperature is an optimization task with respect to:

- Choosing the temperature of the feed so that no additional heat exchange is necessary
- Choosing a high temperature so that the separation behavior is improved
- Choosing a low temperature so that the miscibility gap is as large as possible.

2.3.4.2
Dimensioning

A number of methods are available for dimensioning extraction apparatus with regard to theoretical plate number:

McCabe–Thiele Method

If the mutual solubility of the two phases is negligible, the theoretical number of stages can be determined by means of a staircase diagram (steps between balance lines and equilibrium curves), in a similar manner to rectification or absorption. In this case, too, it is more convenient to work with loadings X, Y than with molar fractions x, y (Equations 2.3.4-4, 2.3.4-5):

$$Y = \frac{E \text{ in (mol/h or kg/h)}}{L \text{ (mol/h or kg/h) extract}} \qquad (2.3.4-4)$$

$$X = \frac{E \text{ in (mol/h or kg/h)}}{R \text{ (mol/h or kg/h) raffinate}}. \qquad (2.3.4-5)$$

From Figure 2.3.4-2a the balance line is obtained as Equation (2.3.4-6):

$$R \cdot (X_0 - X) = L \cdot (Y_e - Y) \qquad (2.3.4-6)$$

or, after rearrangement, Equation (2.3.4-7):

$$Y = Y_e + \frac{R}{L} \cdot (X - X_0). \qquad (2.3.4-7)$$

The equilibrium curve can often be described by the Nernst distribution law (Equation 2.3.4-3), but because of the concentration dependence it must often be determined experimentally:

$$y = K_E \cdot x \text{ (Mole fractions)} \qquad (2.3.4-8a)$$

$$Y = \frac{\dfrac{K_E \cdot X}{1+X}}{1 - \dfrac{K_E \cdot X}{1+X}} \text{ (loadings)}. \qquad (2.3.4-8b)$$

The measurement of the phase equilibria can readily be carried out in thermostatically controlled stirred vessels. These preliminary investigations already provide valuable indications of the dispersibility and the time required for phase separation. The required number of stages is calculated on the basis of the phase equilibria.

Example 2.3.4-1

given: X_0: initial loading of E in R (0.14)
X_e: required final loading (0.04)
Y_0: purity of the solvent (E in R is zero)
$K = 5$.
sought: number of stages N for a given ratio R/L.
Figure 2.3.4-2b gives a theoretical number of stages of 7.5.

Pole point construction

If mutual immiscibility is no longer given, then design with the McCabe–Thiele method is no longer possible. In practice, systems with partial miscibility are more common. In this case the pole point construction is used, in which the mutual solubility of the three components solvent L (extract phase), raffinate R, and solute E is given by binodal curves in a triangular diagram (Figure 2.3.4-3). The binodal curve can be determined experimentally relatively easily by means of turbidity titration. The equilibrium relationships are given by the tie lines, which become shorter on going upwards towards the critical point (Plait point PP), which must not necessarily lie at the maximum. The experimental determination of the tie lines, for which many points are required, is laborious and time-consuming. Therefore, only a limited number of tie lines are determined, and the intermediate lines are extrapolated graphically. The lever rule for the phase quantities also applies in the triangular diagram. In Figure 2.3.4-3 mixtures P_1 and P_3 give mixture P_2, or P_2 splits into P_1 and P_3. The mass and component balance for each component gives Equation (2.3.4-9):

$$\begin{aligned} P_1 \cdot (x_{R1} - x_{R2}) &= P_3 \cdot (x_{R2} - x_{R3}) \\ P_1 \cdot (x_{L1} - x_{L2}) &= P_3 \cdot (x_{L2} - x_{L3}) \\ P_1 \cdot (x_{E1} - x_{E2}) &= P_3 \cdot (x_{E2} - x_{E3}). \end{aligned} \qquad (2.3.4-9)$$

According to the law of similar triangles P_1, P_2, and P_3 are joined by a straight line $[P_i\text{–}P_j]$ and Equation (2.3.4-10) applies:

$$P_1 \cdot [P_1 - P_2] = P_3 \cdot [P_2 - P_3]. \qquad (2.3.4-10)$$

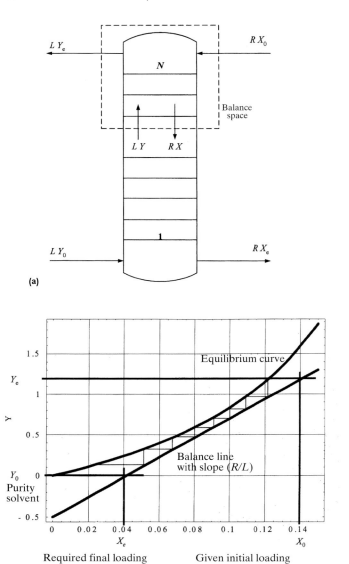

Fig. 2.3.4-2 Mass balance of an extraction (a) and application of the McCabe–Thiele method (b) for determining the theoretical number of stages N and the minimum amount of solvent L_{min}.

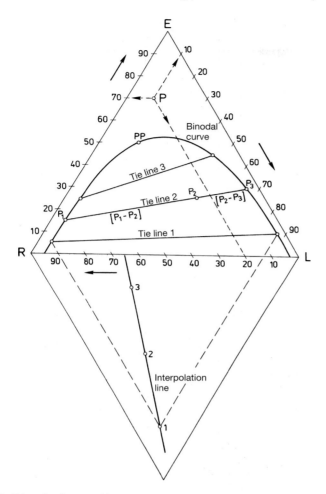

Fig. 2.3.4-3 Triangular diagram with:
- Point P (70% E, 10% L, 20% R)
- Binodal curve
- Three experimentally determined tie lines
- Interpolation line
- Lever rule: mixtures P_1 and P_3 give mixture P_2, or P_2 splits into P_1 and P_3 in accordance with Equation (2.3.4-10).

Single-stage extraction

The raffinate is set in equilibrium with the solvent L. The given quantity ratio of R_0 and L is gives the mixing point M in accordance with the lever rule. This mixture splits in accordance with the tie line into homogeneous mixtures R_1 and L_1 lying on the binodal curve.

Countercurrent extraction according to the pole point construction

Apart from the initial streams R_0 and L_0, all streams result from the establishment of equilibrium. Therefore, all mixture compositions apart from R_0 and L_0 and lie on the binodal curve.

For example, if the target composition of the raffinate phase relative to the binary system R/E is given, then R_E is obtained by connecting R_{soll} and L_0. Furthermore, if the ratio of R_0 (initial raffinate phase) and solvent L is given, Then the mixing point M is given. Then L_E is obtained from the total balance ($R_0 + L_0 = R_E + L_E$). With R_0, R_E, and L_0, the pole P of the balance line is given (Figure 2.3.4-5).

For stage 1 the mixtures L_E and R_1 are connected by the tie line. The tie line passing through L_E gives R_1 on the raffinate branch of the binodal curve. L_2 is obtained in accordance with the balance ($R_1 x_1 + P x_P = L_2 y_2$, straight line through R_1 and P).

The procedure for determining the plate number N for a given quantity of solvent can be summarized as follows:

- Given enrichment from R_0 to R_{soll}.
- Given L_E, which results from the boundary cases of minimum solvent quantity $L_{\min}$ and minimum stage number $N_{\min}$ from economic considerations analogous to rectification (Section 2.3.2.3) or absorption (Section 2.3.3) by plotting an L/N diagram.

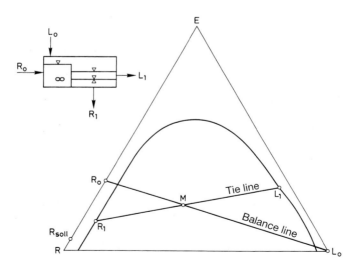

Fig. 2.3.4-4 Simple one-stage extraction.

- R_E via R_{soll}.
- R_1 via tie line.
- L_2 via balance stage 1 (straight line through P).
- R_2 via tie line.
- L_3 via balance stage 2 (straight line through P), etc.

The required number of stages is equal to the number of tie lines required to reach R_E. By plotting an L/N diagram analogous to Figure 2.3.3-3b, the optimal solvent quantity L_{opt} and the corresponding plate number N_{opt} are obtained.

Boundary cases
Minimum amount of solvent
The minimum amount of solvent L_{min} is the quantity that just fulfills the given task (enrichment from R_0 to R_{soll}) at infinite plate number (Figure 2.3.4-6). Infinite theoretical plate number occurs when the pole P coincides with the innermost intersection of a tie line. This pole P_{min} is determined by extending several tie lines. The straight line R_0–P_{min} intersects the binodal line at $L_{E,min}$. The two straight lines R_0–L_0 and R_E–$L_{E,min}$ then intersect at the point M_{min}. The minumum feed ratio results from the lever rule as $R_0/L_0 = (R_0-M_{min})/(M_{min}-L_0)$.

Amount of solvent for one theoretical stage
The given task of enriching the raffinate phase from R_0 to R_E can often be achieved in a single stage, albeit with a disproportionately large amount of solvent. The straight line R_E–L_E then coincides with a tie line. If the point M then lies on the binodal curve or outside of the miscibility gap, phase separation no longer takes place, and the solvent dissolves the entire mixture.

Up to now only the number of theoretical stages was determined for dimensioning an extraction. The connection between the theoretical stage model and real separation column was established through the plate efficiency for a plate column or the height of a theoretical plate HETP for a packed column. Hereby, the actual mass-transfer pro-

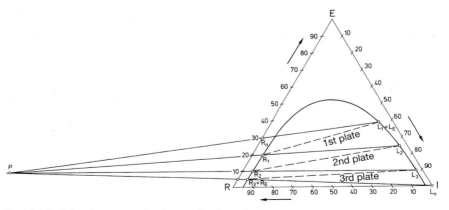

Fig. 2.3.4-5 Pole point construction (see text for discussion).

cess between the two phases was not considered, that is, the rates of exchange were expressed indirectly in term of the empirically determined efficiency or the HETP. By using mass-transfer models (HTU–NTU method, Figure 2.3.3-4), the kinetics of the mass-transfer process are taken into account. Mass transfer between two phases is the result of three steps: transport from the raffinate to the phase boundary, passage through the phase boundary, and transport into the extract. A description of mass transfer in liquid/liquid extraction therefore requires knowledge of phase equilibria, extraction kinetics, and the apparatus-specific hydrodynamics. Knowledge of material-specific properties such as diffusion coefficients is essential for model calculations. This method provides a deeper understanding of the true processes but is often not applicable in practice due to the lack of reliable data.

The dimensioning (diameter) of extraction columns is performed as follows: After determining the optimum plate number and the corresponding optimal solvent quantity for a given process, the ratio of the volumetric flow rates of the continuous and disperse phases $\dot{V}_{cont}/\dot{V}_{disp}$ is fixed, and one defines an area-specific volumetric throughput (Equation 2.3.4-11) and, in a pilot column, determines the flood point as a function of the pulsation $a \cdot f$, where a is the amplitude (typically $\pm\,8\,\mathrm{mm}$) and f is the frequency (typically ca. 60 Hz):

$$\dot{v}/(\mathrm{m}^3\ \mathrm{h}^{-1}\ \mathrm{m}^{-2}) = \frac{(\dot{V}_{cont} + \dot{V}_{disp})}{column\ cross\text{-}section\ area}. \qquad (2.3.4-11)$$

The flooding of an extraction column can be recognized by a phase inversion or by the appearance or disappearance of several phase boundaries in the column. The pressure difference across the column as a funtion of the specific throughput can be used as a quantitative measure. From the obtained optimum $\dot{v}_{opt}$ (Figure 2.3.4-8) and the

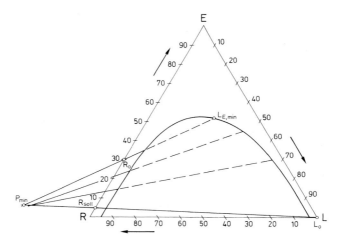

Fig. 2.3.4-6 Determination of the minimum amount of solvent L_{min}.

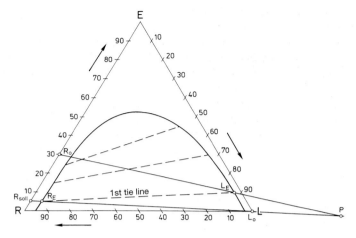

Fig. 2.3.4-7 Determination of the amount of solvent for a theoretical stage.

scale-up law (Equation 2.3.4-12) the diameter of the production column can be calculated (Equation 2.3.4-13).

$$\dot{v}_{lab} = \dot{v}_{prod} \text{ or } u_{lab} = u_{prod} \text{ in m s}^{-1} \tag{2.3.4-12}$$

$$d_{prod} = \sqrt{\frac{4 \cdot \dot{V}_{prod}}{\pi \cdot (\dot{v}_{opt}/_{lab})}}. \tag{2.3.4-13}$$

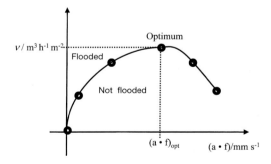

Fig. 2.3.4-8 Column loading as a function of pulsation (schematic).

2.3.4.3
Apparatus

Batch extractions are often performed in stirred vessels following synthesis steps. They can readily be investigated in test apparatus. In particular, the following are to be tested:

- The dispersion performance in the extraction.
  In industrial plants large stirrer diameters are favorable to avoid the peaks in shear force that occur at small stirrer diameters. For large plants drives with speed control are recommended.
- The settling velocity.
  Demixing is favored by approximately equal quantities of the two liquid phases. It can be favorable to run the stirrer at a lower speed at the beginning of the settling phase. Introduction of large amounts of energy in the mixing stage prolongs the phase-separation time.
- Possible formation of mud can be caused by small amounts of impurities, which sometimes only become apparent after closure of the recycle streams. If necessary, the mud layer must be removed separately and separated.

Continuous separations are performed in mixer–settlers (most common apparatus), in extraction columns with or without pulsation, in special constructions such as rotating-disk separators.

In the case of mixer–settlers (Figure 2.3.4-9) test apparatus is commercially available, and the results can readily be transferred to industrial plant. For very different phase fractions the rate of separation can be increased by recycling of the phase present in excess to the mixing stage. Mixers can be static (e.g., a bed of spheres), small mixing pumps, or stirred vessels. Stirred tanks are equipped with flow breakers. In both test apparatus and large-scale plant, blade and impeller stirrers are favorable due to the uniform distribution of the shear forces. The stage efficiency is high (0.8-0.9).

Example 2.3.4-2

Typical static mixer in a test apparatus:
Pipe diameter 10 mm, Filling of glass beads of 1–2 mm diameter, length of packing 100 mm, liquid throughput 0.5-3 L h^{-1}, max. diameter of the feed tube for the two liquids 2 mm.

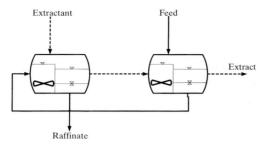

Fig. 2.3.4-9 Two-stage mixer–settler (schematic).

In phase separators the level of the separation layer can best be achieved with internal or external syphons with free overflow. Coalescence aids can be incorpoarted in the feed zone. The material (metal, glass, PTFE) should be chose such that it is preferably wetted by the disperse phase.

Investment costs for industrial mixer–settlers are high. Extraction columns (Figure 2.3.4-10), and in some cases also special constructions, are much cheaper. However, in this case scale-up is problematic. For example, in large equipment widespread back-mixing or poor initial distribution of the disperse phase can occur, although they are not observed in the small dimensions of the test apparatus. Such effects can lead to dramatic drops in separation performance.

2.3.5
Crystallization

The separation of an ordered solid phase from a solution or supercooled melt is known as crystallization. It is one of the most effective separation processes in the chemical industry. For the initiation of crystallization a finite supersaturation or supercooling is always necessary, which is largely determined by the activation energy of crystal seed formation [Mersmann 2000a]. Industrial crystallizations should yield a crystals with certain product properties (e.g., grain size distribution, crystal morphology). These depend on numerous influencing factors such as temperature, pressure, wall material, pH value, concentration of impurities, additives, and so on. The degree of local supersaturation or supercooling, which also influences these properties, must be precisely controlled in crystallizers. Supersaturation is generally achieved by cooling (Figure 2.3.5-1a) or evaporation (Figure 2.3.5-1b) of the solvent (cooling and evaporative crystallization).

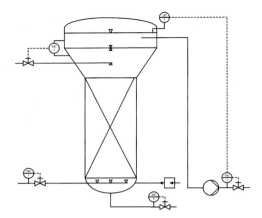

Fig. 2.3.4-10 Extraction column (schematic) with a pulsation device (amplitude ca. ± 8 mm, frequency ca. 60 Hz) for improving the HTU.

2.3.5.1
Fundamentals

Thermodynamics
Solid/liquid equilibria are calsssified as two types: mixed crystals (Figure 2.3.5-2a) and eutectic mixtures (Figure 2.3.5-2b). In multicomponent mixtures, mixed-crystal and eutectic systems can co-exist. From eutectic mixtures in the equilibrium state, one component crystallizes in pure form. In systems that form mixed crystals, the product crystals consist of a mixture of sustances.

Kinetics
Crystals can only form from a supersaturated solution or a supercooled melt [Mersmann 2000]. The initial formation of a solid phase requires a sufficiently large degree of supersaturation or supercooling because of the increased solubility of very small crystals. When small and large crystals are introduced into a saturated solution, the large crystals grow while the small crystal dissolve. This phenomenon is described by the Gibbs–Thomson[13] equation, which describes how the solubility of a crystal depends on its thickness l (Equation 2.3.5-1):

---

13) William Thomson (later Lord Kelvin), British physicist (1824–1907).

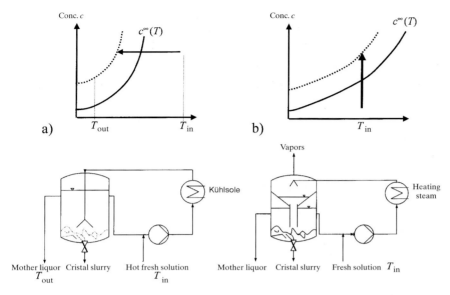

Fig. 2.3.5-1 Schematics of cooling (a) and evaporative crystallization (b). a) Solubility curve $c^\infty(T)$ with "steep" course and the corresponding schematic apparatus for cooling crystallization. b) Solubility curve $c^\infty(T)$ with "flat" course and the corresponding schematic apparatus for evaporative crystallization.

$$\ln \frac{c_S(l)}{c^\infty} = \frac{2 \cdot \sigma \cdot M}{\rho_K \cdot l \cdot f_A \cdot RT} \qquad (2.3.5-1)$$

$c_S(l)$ saturation concentration of a crystal with thickness l
c^∞ saturation concentration of a very large crystal
l mean crystal thickness
σ interfacial energy of the crystal in the saturated solution
ρ_K density of the crystal
M molar mass of the crystallizing substance
f_A form factor correlating the size l^2 and the crystal surface area $A_K = f_A l^2$

A given supersaturation thus corresponds to a crystal seed of length l in labile equilibrium. Aggregation of molecules to form clusters (ca. 20–100 molecules) with the critical crystallite size is only possible in the case of a local deviation in the supersaturation of the solution. Figure 2.3.5-3 shows a schematic plot of the supersolubility curves, which lie above the saturation curve and run approximately parallel to it. They indicate to which degree of supersaturation spontaneous crystallization is not observed in a technically acceptable time (20–30 min). This experimentally determined time is an important quantity in the design of industrial crystallizers.

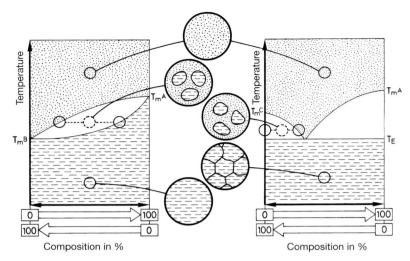

Fig. 2.3.5-2 Solid–liquid phase diagrams.
a) Mixed-crystal formation.
b) Eutectic system.

In general, three types of crystal seed formation are distinguished:

- Homogeneous primary seed formation in clear supersaturated solutions.
- Heterogeneous seed formation on the walls of apparatus and so on.
- Secondary seed formation in the presence of seed crystals.

In industrial crystallizers, spontaneous crystallization, which is difficult to control, must be avoided [Herden 2001]. Therefore, one attempts to establish supersaturation concentrations that still lie below the supersolubilty curve. Only the third type of seed formation can guarantee a reproducible particle size distribution under technical conditions. It also requires the smallest degree of supersaturation of the three mechanisms of seed formation. Thus, one operates in the region between saturation and

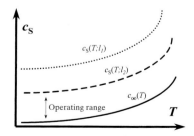

Fig. 2.3.5-3 Schematic depiction of a solubility curve $c^\infty(T)$ and two supersolubility curves $c_S(T; l_i)$.

supersolubilty and, instead of spontaneous crystallization, exploits secondary seed formation, which is caused by abrasion and breakage of crystals that are already present. In this way, part of the already formed crystalline product is recycled and used to achived well-defined frequency of secondary seed formation and growth rate.

2.3.5.2
Solution Crystallization

In solution crystallization a pure substance R is crystallized from a solution and thus separated from a foreign substance F, which remains in solution [Birmingham 2000]. The solubilty realtionships of this three-component system can be depicted in a triangular diagram (Figure 2.3.5-4).

Three types of process can be distinguished:

- Simple evaporation and subsequent crystallization, either continuous (Figure 2.3.5-4) or batch.
- Batch fractional crystallization.
- Fractional crystallization with continuous reycling of the mother liquor. In the case of simple evaporation and subsequent crystallization, the following mass balances apply:

Total mass balance:

$$L_0 = L_E + R + D. \tag{2.3.5 – 2}$$

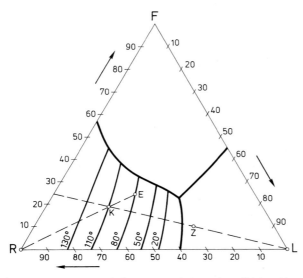

Fig. 2.3.5-4 Three-component system (solvent, pure substance, impurity) in a triangular diagram (Z = feed, E = mother liquor).

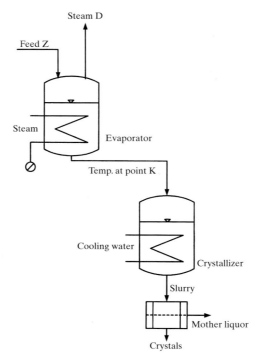

Fig. 2.3.5-5 Continuous evaporative crystallization.

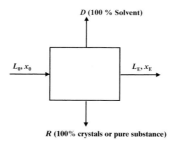

Fig. 2.3.5-6 Mass balance of a one-stage crystallization. The vapor D and the pure substance R are assumed to be 100% pure.

Solids balance:

$$L_0 \cdot X_0 = L_E \cdot X_E + R \cdot 1 + D \cdot 0. \qquad (2.3.5-3)$$

From which Equation (2.3.5-4) follows for the mass of isolate pure substance:

$$R = \frac{L_0 \cdot (X_0 - X_E) + D \cdot X_E}{1 - X_E}. \qquad (2.3.5-4)$$

A simple one-stage crystallization is designed with aid of the triangular diagram (Figure 2.3.5-4) as follows:

- The feed to the evapoartor (point Z) is usually given.
- Determine the end point of the evaporation (point K = slurry, i.e., hot solution from the evaporation stage). This is determined by:
  a) The evaporation temperature, which is limited by, for example, the thermal stability of the pure material.
  b) The temperature of the heating steam, which has an upper limit.
  c) The yield.
- Determine the composition of the mother liquor (point E). This is determined by:
  a) The yield. The closer the point E to the eutectic line or the sides (R, F), the higher the yield.
  b) The maximum amount of deposited crystals for which the slurry (crystalline product plus mother liquor) is still transportable or pumpable (ca. 0.2 t m^{-3}).
  c) The distance from the eutectic line (should be ca. 5 °C).

The amount of R or E follows from the lever rule.

2.3.5.3
Melt Crystallization

In any crystallization process seed formation is an indispensible, individually adjustable step that decisively influences process control and the properties of the crystalline product (e.g., purity, crystal size, crystal habit). It is known from industrial practice that seed formation is a problem for certain systems, especially in film crystallization processes. Here seed formation requires large temperature differences between cooling surface and melt. These then initially grow rapidly due to the high degree of supercooling. The very high growth rates can result in the incorporation of large amounts of impurities into the forming layer and thus lower product purity. Only after seed formation has ceased can layer growth proceed in a controlled fashion.

For optimum process control and good economics, seed growth must be initiated without a large degree of supercooling. Theories describe the relationship between all the quantities and the work of seed formation [Volmer 1931]. One of these quantities is the interfacial tension between suface and melt. Thus seed formation in melt crystallization can be improved by optimizing the material combination cooling surface/

melt. Then a combination of a small degree of supercooling and good wettability will lead to high crystal purity at lower energy consumption.

Suspension crystallization
In suspension crystallization the melt is cooled to below the stauration temperature, and crystals grow under adiabatic conditions. The degree of supersaturation is the driving force. Specialist knowledge is required to obtain crystals with a certain purity, structure, and particle size distribution. The residual melt, which contains the impurities, must be mechanically separated from the crystals.

Film crystallization
In film crystallization the crystals grow on a cooled wall. Therefore, the crystals are colder than the melt (nonadiabatic process), and the driving force is the temperature gradient. Thus, rates of crystal growth that are 10–100 times higher than in suspension crystallization are attainable.

The two most widely used systems are:

- Static crystallization, in which the crystals grow on a cooled wall in a nonmobile liquid.
- Falling film crystallization, in which the liquid is circulated as a falling film over the crystal layer growing on the cooled wall.

Falling film crystallization was developed by Sulzer Chemtech. Figure 2.3.5-7 shows a schematic of the process, which is characterized by three operating phases:

- Phase 1: crystallization (Figures 2.3.5-7 and 2.3.5-8a). At the beginning of the first phase the collecting tank is filled with molten starting material, and the pumps for product and heat-transfer medium are switched on. A crystal layer forms on the

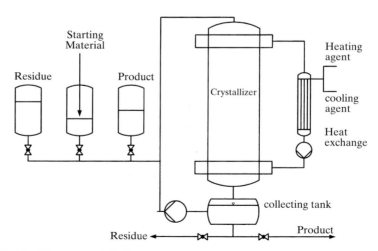

Fig. 2.3.5-7 Falling film crystallization.

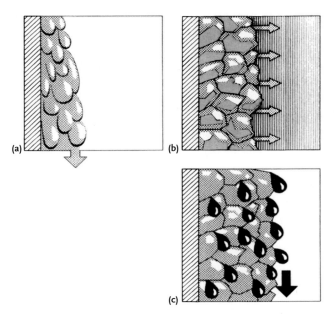

Fig. 2.3.5-8 The three stages of falling-film melt crystallization.
a) Crystallization.
b) Partial melting.
c) Melting.

inner wall of the falling tower, and the level in the collecting tank decreases with increasing thickness of the crystal layer. When a certain level is reached, the crystallization process is interrupted, and the residual melt in the collecting tank is removed via the residue pipe or pumped to a storage tank.
- Phase 2: partial melting (sweating; Figures 2.3.5-7 and 2.3.5-8b). In the second phase, the crystal layer is tempered, so that impurities that adhere to it or are incorporated therein can drip off. The dripping liquid, also known as partial melt, is collected in the collecting tank, and when a certain quantity is reached is removed via the residue pipe or pumped to a stage tank.
- Phase 3: Melting (Figures 2.3.5-7 and 2.3.5-8c). In the third phase, the remaining crystal layer is melted. The resulting melt is either the desired pure product or an intermediate stream that is recrystallized in the following stage.

The product purity depends on the amount of product remaining on or in the crystal layer in the second phase. However, any desired purity can be achieved by repeated crystallization, sweating, and melting. In the same way, a higher yield is achieved by repeated crystallization of the residue from the first phase. Figure 2.3.5-9 shows the mass-flow scheme of a seven-stage process, depending on the nature of the product and the required purity and yield, up to nine stages are possible.

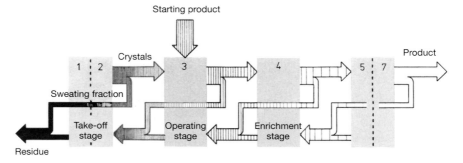

Fig. 2.3.5-9 Example of a seven-stage melt crystallization.

2.3.5.4
Dimensioning

Crystallizations and precipitations are processes of material separation and shaping that are not readily accessible to mathematical modeling [Franke 1999]. Apart from problems in describing the thermodynamic relationships, hydrodynamic effects, and changes in solids, another reason for this is that traces of impurities or additives influence the crystallization process in an unpredictable way. These are preferentially adsorbed on particular crystal faces and thus strongly influence crystal growth. For example, crystals that formerly formed cubes can become needles and vice versa. An experimental treatment is indispensable. Some guidelines are:

- Improvement of mass transfer by relative motion between crystals and mother liquor.
- As high a supersaturation as possible, but only so far that crystals that are too small are not formed.
- Favorable introduction or recycling of seed crystals.
- Sufficiently dimensioned heating and cooling surfaces.

Scale-up is also difficult [Bermingham 2000]. Because solids are handled, the minimum size of test apparatus is more strongly limited than in the case of, for example, distillation. The minimum size for evaporative and cooling crystallizers is about 5 L. Process steps in which solids are produced must be very carefully designed as they directly influence downstream steps such as filtration, drying, storage of solids, and compaction. If possible, tests in a miniplant in which the total process is mirrored should be supplemented by individual tests in a larger semitechnical plant.

Investigations of crystallization and precipitation are usually performed in stirred vessels. This type of apparatus has the fundamental problem that the individual functions exhibit different dependences on scale-up.

The mixing intensity, which is decisive for the supersaturation, is directly proportionla to the stirring speed, which must be reduced on scale-up. In the case of precipitations this leads to different supersturations, rates of seed formation, and particle sizes.

Heat transfer is proportional to stirring speed to the power 5/3 and also changes on scale-up. For example, for a tenfold increase in size of the crystallizer, the introduced stirring power must be increased by a factor of $10^{5/3}$ (risk of crystal abrasion) to remove the heat of crystallization at the same driving force ΔT for heat exchange.

Furthermore, on scale-up the ratio of surface area to volume also changes and thus influences the temperature differences in heat transfer and seed formation, as well as the vapor velocity at the liquid surface in evaporative crystallization. In scale-up the energy introduced per unit volume is usually kept constant, and hence this specifies the stirring speed. This results in almost comparable conditions for the suspension of the solids and the abrasion behavior.

Preliminary tests must be carried out to determine which of the process quantities – mixing intensity, temperature differences, supersaturation (metastable region with weak seed formation and labile, highly supersaturated region with spontaneous seed formation), seeding, mechanical loading – are of decisive influence for a particular application. If necessary, modified test apparatus must be used in individual cases.

Example 2.3.5-1

If the mixing intensity proves to be the main influencing factor, then a stirred tank with premixing is used for a targeted investigation of this quantity.

For high-capacity continuous crystallizations, various types of crystallizer are used. Here additional tests on a semitechnical scale are recommended (Table 2.3.5-1).

Crystallization is a highly cost intensive separation process since it involves numerous mechanical separation units such as filters, centrifuges, dryers, and storage facilities (Figure 2.3.5-10).

2.3.6
Adsorption, Chemisorption

Adsorption is used for the fine purification of liquids and gases when the loading with the components to be removed is low. The adsorbent, for example, activated carcon, molecular sieves, zeolite, clay, or silica gel, is generally used in granular form in a fixed bed and less often as a suspension in a liquid. Desorption is usually carried out by physical methods such as temperature increase, pressure drop, or exchange against another medium. In the case of high binding energies (> 50 kJ mol^{-1}, chemisorption), physical desorption is no longer possible.

For suspended adsorbents, stirred vessels are generally used, and the process is easily designed. For scale-up the erosion behavior of the adsorbent and its separability from the liquid are important. The amount of energy to be introduced is determined experimentally and should be high enough that suspension just takes place. Slow-running stirrers with propeller or inclined blades of large diameter are favorable.

Fig. 2.3.5-10 Crystallization and downstream units.

Tab. 2.3.5-1 Process engineering parameters that can be used as guidelines for the design of crystallizers.

Mechanical energy introduced Energy for suspension is proportional to the content of crystallizate	0.2–1.0 kW m^{-3}
Final solids concentration	0.2–0.5 kg L^{-1}
Residence time	1–10 h
Cooling rate in batch crystallization	2–20 K h^{-1}
Vapor velocity at the surface in evaporative crystallization	2–5 m s^{-1}

In most cases adsorbents are used in the form of fixed beds. A number of theoretical treatments are available for the adsorption equilibrium and the adsorption kinetics. However, it is recommended to determine these parameters, the possible loading, and the breakthrough behavior experimentally. This also applies for the desorption process.

In continuous processes at least two adsorbers are used in parallel (adsorption, desorption). Because of the complicated operation, fully automatic apparatus is preferred (Figure 2.3.6-1).

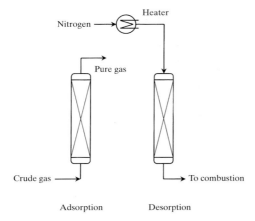

Fig. 2.3.6-1 Adsorber setup.

2.3.7
Ion Exchange

The design of ion exchange plants is similar to that of adsorption plants. Because of the sometimes pronounced swelling behavior of the ion exchanger (up to 50%), fixed beds are often operated in a sump mode so that the ion exchange bed can expand vertically. The limited mechanical stability of the ion exchange resin often does not permit a more highly fluidized operating mode. Often the liquid velocity must be limited.

2.3.8
Drying

The design of drying processes is strongly dependent on experiments. The fundamental investigation is the study of the sorption isotherm in a sorption balance. The result allows an initial evaluation of the extent of the three drying steps:

1. Surface liquid. Here the vapor pressure of the liquid corresponds to the saturation vapor pressure.
2. Capillary-bound liquid. Here a lowering of the vapor pressure is expected for capillary diameters smaller than 0.1 mm.
3. Dissolved or chemically bound liquid whose vapor pressure is established via the osmotic pressure or the chemical bonding.

In the choice of a suitable dryer type, a series of aspects must be considered:

- Type of sorption isotherm, course of drying, final moisture content.
- Nature of the moisture (water, flammable sovent).

- Consistency of damp material (solvent, paste, crystal slurry, filter cake).
- Demands on the dry product (grain size and grain size distribution, freedom from dust, pourability, tendency to cake, abrasion resistance, bulk density, rate of dissolution).
- Safety requirements (dust explosions, ignition energy).
- Permissible product temperature (formation of side products, melting, sublimation).
- Tendency to form encrustations or highly viscous phases in the course of drying.

On the basis of the results of the sorption investigation and with consideration of the demands on the properties of the end product, a preliminary selection of dryer types can be made, the suitability of which is then checked experimentally.

If the product is already in a given grain form, for example, resulting from a crystallization or precipitation, then various dryers of the contact or convective types can be chosen. For large capacities and exclusively surface moisture the pneumatic dryer is particularly favorable with regard to investment cost. Fluidized-bed and flowing-bed dryers also allow longer drying times to be achieved. Blade dryers are suitable for a wide range of product properties.

The choice of dryer is more difficult when the drying step is also used to shape the product. Typical designs here are the spray dryer and the sprayed fluidized bed. If necessary, drying can be followed by a classification step in which undesired particles, for example, fines, are recyled to the feed stream for the dryer. When the desired product properties, for example, freedom from dust, pourability, rate of dissolution, and bulk density, are not attainable in the drying step, additional steps such as compaction and granulation must be used.

Convective drying is carried out where possible with air. In the case of an explosion hazard due to a flammable solvent or an explosible dust, cost-intensive inertization is necessary.

The choice of co- or countercurrent drying is determined by the product properties. Cocurrent drying leads to high drying rates and limits the maximum product temperatures, while countercurrent drying allows lower final moisture contents to be achieved.

For the experimental design of dryers, much information can be obtained from the sorption isotherm and simple tests in a drying oven. Ultimately, however, tests must be performed in a test apparatus of the same design, because only here can certain properties such as caking, damage to the product, and attainable particle size distribution be determined.

Tab. 2.3.8-1 Checklist for dryer selection.

Method of heating	Dryed material	Type of dryer	Residence time
Convection dryer	mobile hoap	fluidized bed	long
		sprayed fluidized bed	long
		drum dryer	long
		convection dryer	very short
		omization dryer	short
	immobile heap	tunnel dryer	long
		circulating drying chamber	long
		belt dryer	long
Contact dryer	mobile product	paddle dryer	long
		tumble dryer	long
		plate dryer	long
		disk dryer	long
		screw dryer	long
		thin-film dryer	short
	immobile material	drying chamber	long
		belt dryer	long
		cylinder dryer	short

2.3.9
Special Processes for Fluid Phases

Membrane separation [Knauf 1998].

Pervaporation is occasionally used for azeotrope separation but has not become well established. It can be readily scaled up from a test plant to production scale.

Gas permeation is finding its first industrial applications. Scale-up is straightforward. The test apparatus is adapted to the available membrane geometries.

Extraction with supercritcal media is used in food technology, but in the chemical industry it is only used for valuable products because of the high investment costs.

Reverse osmosis is hardly used and only in special cases is it to be considered in process development.

Dialysis is not used in the chemical industry. Applications are limited to the food industry.

Electrodialysis has industrial applications where salts must be separated from high-boiling substances. Reliable scale-up is possible.

Chromatographic separation [Strube 1998].

2.3.10
Mechanical Processes

In contrast to fluid process technologies such as distillation, absorption, and extraction, mechanical processes are only accessible to a limited extent to mathematical modeling, and process design is dependent on experiment. However, only a few experimental techniques that allow reliable scale-up to be performed are available.

A further difficulty is that the different process steps strongly influence one another. The grain size and shape achieved by crystallization or precipitation determines the ease of separating the solids, the attainable residual moisture content and purity in the washing step, the drying behavior, and the properties of the end product, which may have to be improved by compaction or sieving.

The result of this problematic is that in process development a choice must be made between process design and design of apparatus. It is possible to design the individual solids-handling steps in a small test plant such that the recycle streams that occur in the process are closed, for example, recycling of mother liquor and washings in filtration or of undesired fines in sieving. Thus, the important question of the effect of recycle streams can be answered, but these tests are not sufficient for carrying out design of apparatus. For this, the tests in an integrated test plant must be supplemented by targeted experiments on a larger scale, for example, by collecting corresponding quantities of product. It is advantageous to coordinate these tests with the apparatus manufacturer and to borrow suitable test apparatus from this company or to carry out the tests in the company itself. Because of the influence of aging processes, such as Ostwald ripening of crystals with a decrease in the fraction of fine material, this approach is problematic in many cases. In critical cases, for example, when the attainable product quality strongly depends on the separation performance of a decanter, then apparatus design must be carried out on a 1:1 scale, which means that sufficiently large amounts of product must be available for the design tests.

For the separation of solids from liquids, filtration and sedimention can be considered. The energy required for separation can be introduced by a centrifugal field, gravity, and, in the case of filtration, by application of under- or overpressure.

Filtration

The suitability of a filtration process can be assessed in a small laboratory pressure filter with a filter area of about 20–50 cm^2 and a capacity of 0.3–1 L, in which various metal and plastic filter materials can be tested (Figure 2.3.10-1). The pore width is about 10–100 μm. The filter should be heatable and coolable. The experiment is evaluated in terms of the applied gas differential pressure ΔP, the filter area A, the filtrate volume V, the final filter cake height H, and the filtration time t until gas breakthrough according to Equation (2.3.10-1).

$$a \cdot \eta = \frac{2 \cdot t \cdot A \cdot \Delta P}{V \cdot H}. \qquad (2.3.10-1)$$

The product of the specific filter resistance a and the liquid viscosity η lies in the ranges between $a\eta = 10^{11}$ mPa s m^{-1} for filterable suspensions and $a\eta = 10^{16}$ mPa s m^{-1} for insufficiently filterable suspensions.

In the evaluation, besides the cake thickness, the attainable residual moisture and the consistency of the cake are determined. Furthermore, possible demixing of the solid due to different sedimentation velocities of the particles during the filtration process must also be taken into account. The washing properties of the filter cake and its tendency to cracking can also be estimated in this apparatus in simple preliminary tests.

Vacuum belt filters for installation in integrated test plants have been described [Maier 1990]. These compact filters allow the functions of filtration of the suspension with removal of the mother filtrate, one or more washing zones with recovery of the wash filtrates, drying of the cake, and washing of the recycled belt to be incorporated. The adjustable working length of the belt is about 1 m for a belt width of about 5–10 cm. The solids throughput is about 0.5–10 kg h^{-1}.

Small bowl centrifuges suitable for incorporation in integrated test plants are also available. Evaluation of the individual steps (fillling with suspension, intermediate acceleration if required, cake washing, spinning dry, and scraping out the solid) is possible. Futhermore information is obtained on potential complete coverage with a bottom layer and the required cleaning time.

For suspensions with unfavorable filtration behavior, cross-stream filtration can be used, in which high flow velocities of the suspension (2–10 m s^{-1}) largely prevent formation of a filter layer (Figure 2.3.10-3). Tubular modules with an internal diameter of about 5 mm, which can readily be integrated into test plants, are particularly suitable for experimental investigations. Particular care must be taken in the choice of the membrane, since the suitability of the process is strongly dependent on the pore width and the membrane material. It is recommended to use a pore width that is ten times smaller than the smallest grain size occurring in the process. For removal of the

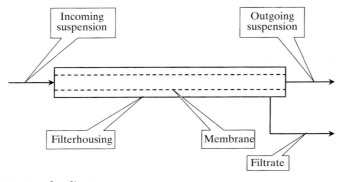

Fig. 2.3.10-1 Cross-flow filtration.

bottom layer that forms in spite of the high suspension speed, it should be possible to initiate backwashing by means of a brief pressure increase on the filtrate side. This is best achieved with an automatic system, which in the case of difficult media, for example, performs backwashing at 1 min intervals for about 2 s at a differential pressure of 3 bar.

Sedimentation
The suitability of sedimentation processes in comparison to filtration processes can be evaluated in a standing cylinder or a laboratory sedimenting centrifuges. For the design of apparatus large-scale investigations are indispensible for both separators and decanters.

2.4
Pipelines, Pumps, and Compressors

2.4.1
Fundamentals of Hydrodynamics

Fluid dynamics, especially hydrodynamics (the study of moving liquids), is of major importance in the design of chemical plant. The hydrodynamics largely determine the energy consumption of the total plant, the separation performance of columns, and the residence-time behavior of reactors; turbulent flow can even destroy pipes. The classical equations for describing the state of flow (= velocity field in space and time) of a fluid are due to Navier[1] and Stokes[2] and date back to the 1800s (Equation 2.4-1):

$$\frac{\partial \vec{u}}{\partial t} = -\vec{u} \cdot div(\vec{u}) + \vec{a} - \frac{1}{\rho} \cdot grad(P) + v \cdot grad[div(\vec{u})] \qquad (2.4-1)$$

$\vec{u}$ = velocity
$\vec{a}$ = acceleration (e.g., due to gravity)
ρ = density of fluid
P = pressure
v = kinematic viscosity,

or, simplified for a single direction in space x (Equation 2.4-2):

$$\frac{\partial u_x}{\partial t} = -u_x \cdot \frac{\partial u_x}{\partial x} + a_x - \frac{1}{\rho} \cdot \frac{\partial P}{\partial x} + v \cdot \frac{\partial^2 u_x}{\partial x^2}. \qquad (2.4-2)$$

---

1) Claude L. M. H. Navier, French physicist (1785–1836).
2) George G. Stokes, mathematician and physicist (1819–1903).

The individual terms on the right-hand side of Equation 2.4-2 have the following meaning:

1st term: local change in the velocity field
2nd term: "mass force" (force = mass × acceleration)
3rd term: "surface force"
4th term: "shear force".

This partial differential equation is deterministic by nature. In practice, however, many hydrodynamic phenomena (e.g., transition from laminar to turbulent flow) have chaotic features (deterministic chaos [Stewart 1993]). The reason for this is that the Navier–Stokes equation assumes a homogeneous ideal fluid, whereas a real fluid consists of atoms and molecules. Today highly developed numerical flow simulators (computational fluid dynamics, CFD) are available for solving the Navier–Stokes equation under certain boundary conditions (e.g., Fluent Deutschland GmbH). These even allow complex flow conditions, including particle, droplet, bubble, plug, and free surface flow, as well as multiphase flow such as that foundin fluidized-bed reactors and bubble columns, to be treated numerically [Fluent 1998].

By assuming a frictionless liquid, the friction term can be omitted to give the so-called Euler[3] equation (Equation 2.4-3):

$$\frac{\partial u_x}{\partial t} = -u_x \cdot \frac{\partial u_x}{\partial x} + a_x - \frac{1}{\rho} \cdot \frac{\partial P}{\partial x}, \qquad (2.4-3)$$

which, for a stationary state can be further simplified to Equation (2.4-4):

$$u_x \cdot \frac{\partial u_x}{\partial x} = a_x - \frac{1}{\rho} \cdot \frac{\partial P}{\partial x}. \qquad (2.4-4)$$

For upward flow ($a_x = -g$, where g is the acceleration due to gravity and the height $h = x$), after separation of the variables and integration, this equation leads to the well-known Bernoulli[4] equation (Equations 2.4-5 and 2.4-6; Figure 2.4-1):

$$\int_{u_1}^{u_2} u_x \cdot du_x = -g \cdot \int_{h_1}^{h_2} dx - \frac{1}{\rho} \cdot \int_{P_1}^{P_2} dP, \qquad (2.4-5)$$

$$\frac{u_2^2 - u_1^2}{2} = -g \cdot (h_2 - h_1) - \frac{1}{\rho} \cdot (P_2 - P_1) \qquad (2.4-6)$$

---

3) Leonhard Euler, mathematician and physicist (1707–1783).
4) Daniel Bernoulli, Renaissance man (1700–1782).

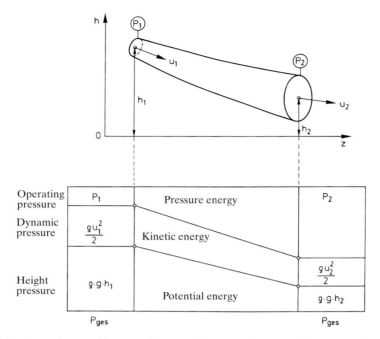

Fig. 2.4-1 Pictorialization of the Bernoulli equation for a pipe with courses of pressure and energy.

or, in general, Equation (2.4-7):

$$\frac{\rho}{2} \cdot u^2 + \rho \cdot g \cdot h + P = P_{ges} = \text{konst.} \qquad (2.4-7)$$

where the terms have the following meanings:

1st term: dynamic pressure (kinetic energy $mu^2/2$).
2nd term: height pressure (potential energy mgh).
3rd term: operating pressure (pressure energy PV).

The aspects of fluid dynamics that are of importance for process development are discussed in the following.

2.4.2
One-phase Flow in Pipelines

The state of flow of a fluid in a pipeline is described by the Reynolds[5] number Re (Equation 2.4-8):

---

5) Osborne Reynolds (1842–1912).

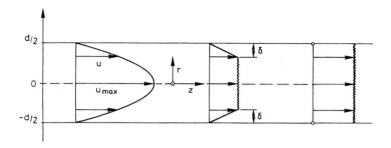

Fig. 2.4-2 Flow profiles.

$$Re = \frac{u \cdot d}{v} \qquad (2.4-8)$$

u = flow velocity
d = pipe diameter
v = kinematic viscosity.

Flow is laminar for $Re < 2300$, and in the ideal case (no disturbances due to pipe roughness, internals, etc.) can be described by the Hagen–Poiseuille law. In the derivation of this law, equating the surface force with the shear force gives a parabolic velocity profile $u(r)$ (Equation 2.4-9), as shown in Figure 2.5-2:

$$u_{laminar}(r) = \frac{\Delta P \cdot d^2}{16 \cdot \eta \cdot L} \cdot \left(1 - 4 \cdot \frac{r^2}{d^2}\right) = u_{max} \cdot \left(1 - 4 \cdot \frac{r^2}{d^2}\right) \qquad (2.4-9)$$

ΔP = pressure difference across the pipe length
d = pipe diameter
η = dynamic viscosity
u_{max} = $2\bar{u}$, where $\bar{u}$ is the mean flow velocity.

The amount of fluid exiting a pipe with cross-sectional area $A = \pi/4 \cdot d^2$ is given by Equation (2.4-10), which is used for the determination of dynamic viscosity with a Ubbelohde viscosimeter:

$$\int_0^{\dot{V}} d\dot{V}' = \int_0^A u(r) \cdot dA' = \int_0^{d/2} u(r) \cdot 2\pi \cdot r \cdot dr \qquad (2.4-10)$$

$$\dot{V} = \frac{\Delta P \cdot \pi}{8 \cdot \eta \cdot L} \cdot r^4 = \frac{\Delta P \cdot \pi}{128 \cdot \eta \cdot L} \cdot d^4.$$

The other limiting case, so-called plug flow, occurs above about $Re = 10\,000$ and is characterized by:

$$u_{plug}(r) = \text{const.} \qquad (2.4-11)$$

In the transition region ($2300 < Re < 10\,000$) a labile flow profile is found that, depending on the roughness of the pipe, can oscillate between these boundary values. For the formation of a constant flow profile, a certain distance l is required, which can be estimated with Equation (2.4-12):

$$\begin{aligned} l_{laminar} &= 0.058 \cdot Re \cdot d \\ l_{turbulent} &= 50 \cdot d. \end{aligned} \qquad (2.4-12)$$

Only after this distance may, for example, flow meters (Section 2.8) be incorporated.

The true state of flow in industrial pipelines lies in the turbulent regime, which can be approximated by the flow profile in Equation (2.4-13):

$$u_{turbulent}(r) = u_{max} \cdot \left(\frac{d - 2 \cdot r}{d}\right)^{1/7}. \qquad (2.4-13)$$

Here a strongly flattened flow profile (Figure 2.4-2) is evident, which drops rapidly to zero only near the wall of the pipe. The thickness δ of the laminar boundary layer can be estimated by using Equation (2.4-14):

$$\delta \approx \frac{d}{\sqrt{Re}}. \qquad (2.4-14)$$

The design of pipelines is an economic optimization between:

- As large a diameter as possible to minimize the pressure drop and thus the energy costs for drive units, especially for pipelines with continuous operation.
- As small a diameter as possible to save material and construction costs, especially for pipelines with short, batchwise operation and pipelines constructed of expensive materials.

As a first estimate the optimum flow velocity can be calculated by means of the F factor (= loading factor = measure of the kinetic energy; Equation 2.4-15):

2.4 Pipelines, Pumps, and Compressors

Tab. 2.4-1 Guideline values of flow velocities of various media for the design of pipelines.

Medium	Flow velocity u/m s^{-1}
Liquids	1–3
Gases (normal pressure)	40
Gases (coarse vacuum, 400 mbar)	70
Gases (fine vacuum, 2 mbar)	80
Low-pressure steam	25
High-pressure steam	40

$$F/(\sqrt{\text{Pa}}) = u_{opt}/(\text{m s}^{-1}) \cdot \sqrt{\rho/(\text{kg m}^{-3})} \approx 50 \qquad (2.4-15)$$

u_{opt} optimal flow velocity in m s^{-1}
ρ density of the flowing medium in kg m^{-3}

In practice the flow velocities listed in Table 2.4-1 have proved suitable.

The pipe diameter is then calculated from the flow velocity according to Equation (2.4-16):

$$d/\text{m} = \sqrt{\frac{\dot{V}/(\text{m}^3 \text{ h}^{-1})}{u_{opt}/(\text{m s}^{-1}) \cdot 2827}}. \qquad (2.4-16)$$

Since standardized pipelines are generally used in plants, the calculated values are rounded up to the next available standardized nominal diameter. Standardized nominal diameters are 15, 25, 40, 50, 80, 100, 150, 250, 300 mm, etc. In miniplants nominal diameters of 2, 4, 6, 10 mm are typical. The corresponding pressure drop is calculated with Equation (2.4-17):

$$\Delta P = \lambda(Re) \cdot \frac{\rho \cdot \bar{u}^2}{2} \cdot \frac{L}{d} = \lambda(Re) \cdot \frac{F^2}{2} \cdot \frac{L}{d} \qquad (2.4-17)$$

$\bar{u}$ = mean flow velocity (= volume flow per unit area)
L = pipe length
d = pipe diameter
F = F factor.

For laminar flow the drag coefficient λ can be equated with:

$$\lambda(Re) = \frac{64}{Re} \qquad (2.4-18)$$

and thus Equation (2.4-18) is transformed into the Hagen[6]–Poiseuille[7] law (Equation 2.4-10):

---

6) Gotthilf H. Hagen, hydrolic engineer (1793–1884).
7) Jean-Louis Poiseuille, French physician (1797–1869).

$$\Delta P = 32 \cdot \eta \cdot \bar{u} \cdot \frac{L}{d^2}. \tag{2.4-19}$$

Formulas are available in the literature for estimating the drag coefficient under turbulent conditions [VDI-Wärmeatlas, Lb1], for example, Equation (2.4-20):

$$\begin{aligned}\lambda &= 0.3165 \cdot Re^{-0.25} &&\text{transition rage (Re from } 3 \times 10^3 \text{ to } 10^5) \\ \lambda &= 0.0054 + 0.3964 \cdot Re^{-0.3} &&\text{turbulent range} \\ &&&(Re \text{ from } 2 \times 10^4 \text{ to } 2 \times 10^6).\end{aligned} \tag{2.4-20}$$

When internals or packings are present in a pipeline, the pressure drop can be estimated with a modified pressure-drop equation (Equation 2.4-21):

$$\Delta P = \frac{1}{\Psi^2} \cdot (\mu \cdot \lambda) \cdot \frac{h}{d'} \cdot \frac{\rho}{2} \cdot u_0^2 \tag{2.4-21}$$

u_0 = mean velocity
ρ = density
h = bed height
$d' = \frac{2}{3} \cdot \frac{\Psi}{1-\Psi} \cdot d_K'$, hydraulic channel diameter
$d_K' = 6\, V_K/A_K$, equivalent packing body diameter
V_K = packing body volume
A_K = packing body surface area
$\Psi = 1 - \frac{\rho_S}{\rho_K}$
ρ_S = packing density
ρ_K = packing body density
$(\mu \cdot \lambda) = f(Re) = \frac{u_0 \cdot d'}{\Psi \cdot \nu}$ effective drag cafficient (= product of path factor and drag coefficient)
ν = kinematic viscosity.

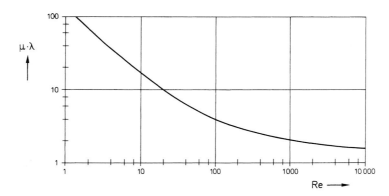

Fig. 2.4-3 Effective drag coefficient as a function of the Reynolds number Re [VDI-Wärmeatlas, Le2].

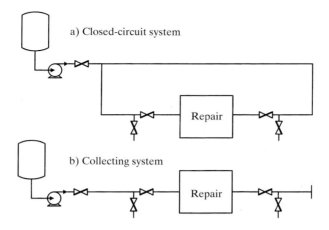

Fig. 2.4-4 Pipeline systems. a) Closed-circuit system. b) Collecting system. When repair work is carried out in the area "Repair", in case a) the plant can still be operated provided the repair site can be closed off. In case b) this is not possible.

The effective drag coefficient is a function of the Reynolds number (Figure 2.4-3).

In planning pipelines for chemical plant, for reasons of safety and ease of maintainence, closed-circuit systems are preferred to collecting systems (Figure 2.4-4).

2.4.3
Liquid Pumps [Sulzer 1987]

The choice of a pump depends on the required pumping performance (volume flow, pressure) and the pumping conditions (temperature, system pressure, properties of the pumped material). Important types of pump are listed in Table 2.4-2.

Tab. 2.4-2 Important types of pump [Poe 1999]. While displacement pumps generate pressure hydrostatically, energy transfer in rotary pumps is based on hydrodynamic processes.

Pumptype	Advantages	Disadvantages
Centrifugal pumps (e.g., rotary pumps)	• large volume therms	• only for low pressures
	• no pulsation	
	• pumps solid particles	
Displacement pumps (e.g., piston pumps)	• high pressure [Maier 1986]	• pulsation problems
		• unsuitable for solid particles
Jet pumps	• no moving parts (low-maintainance)	• low efficiency (ca. 0.1 to 0.2)

The power requirement N of a pump is calculated with Equation (2.4-22):

$$N/\text{kW} = \frac{\dot{m} \cdot g \cdot h}{10^3 \cdot \eta_{mech}} = \frac{\dot{V} \cdot \rho \cdot g \cdot h}{3600 \cdot 10^3 \cdot \eta_{mech}} = \frac{\dot{V} \cdot \Delta P}{36 \cdot \eta_{mech}} \qquad (2.4-22)$$

$\dot{m}$ = mass flow in kg s^{-1}
ρ = density in kg m^{-3}
g = 9.81 m s^{-1}
h = total pump head in m
$\dot{V}$ = usable feed stream in m^3 h^{-1}
η_{mech} = pumping pressure in bar
ΔP = feed pressure in bar.

Of course, the power of the corresponding driving machine must be higher than the power requirement calculated from Equation (2.4-22). As a rule of thumb, an excess power of about 10% is required for very large pumps (> 50 kW), while small pumps require 10–40%.

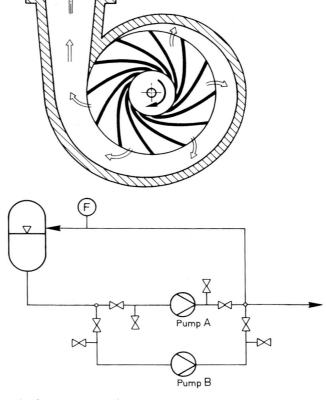

Fig. 2.4-5 Principle of a rotary pump and its incorporation in a pumping station.

The most widely used pump in chemical plant is the rotary pump (Figure 2.4-5). A bladed wheel rotates with constant velocity in a fixed housing, and the liquid is flung from the center to the wall via the channels between the blades. The housing has the shape of a circle whose radius increases until it enters the pressure pipe. The suction pipe enters the center of the bladed pipe axially. Sealing between the pump housing and the drive is achieved with a stuffing box or a floating ring seal.

When started, the rotary pump cannot generate an underpressure to suck in the liquid to be pumped, and the suction pipe and the pump must therefore be filled with liquid. Hence, startup and shutdown are carried out in situ against the closed valve on the pressure side. For reasons of availability, generally two pumps are combined to give a pumping station (Figure 2.4-5).

The pumping height and volume flow are related by the characteristic curve of the pump (Figure 2.4-6), the course of which depends on the shape and size of the bladed wheel and the housing.

Disturbance-free operation is only possible when the pump runs without cavitation, which requires that the pressure at center of the bladed wheel is higher than the vapor pressure of the pumped medium. In particular, this must be taken into account in the case of sump-draining pumps for rectification columns, which operate close to the boiling point of the medium. The sensitivity of a rotary pump to cavitation is quantified by the net positive suction head (NPSH; Figure 2.4-7), which is defined as the total pressure head of the flow at the center of the bladed wheel minus the vapor pressure head of the liquid (Equation 2.4-23).

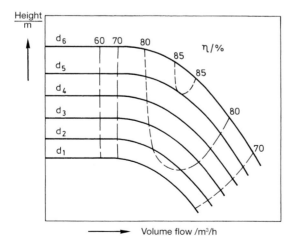

Fig. 2.4-6 Characteristic curve of a rotary pump (pumping height as a function of volume flow; curve parameter d = diameter of the bladed wheel, η = efficiency).

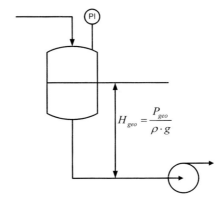

Fig. 2.4-7 Determination of the NPSH in feed operation. Suction operation should be avoided, that is, if possible the pump should be at the lowest point.

$$NPSH/m = \frac{P_{tot}}{\rho \cdot g} - \frac{P^\circ(T)}{\rho \cdot g} - \frac{\Delta P}{\rho \cdot g} \qquad (2.4-23)$$

P_{tot} = total pressure in the suction pipe = $P_{System} + P_{geo}$
$P^\circ(T)$ = vapor pressure of the liquid at the feed temperature T
g = 9.81 m s^{-2}
$\rho(T)$ = density of the liquid
ΔP = pressure drop in the feed line to the pump.

Continuous operation of the pump without dangerous cavitation is only possible if the NPSH at the operating point is at least 0.5 m higher than the required value [Branan 1994].

Vacuum pumps
The choice of a vacuum pump is determined by the required vacuum and pumping performance [Jorisch 1998]. Table 2.4-3 lists a classification of vacuum pumps according to pressure regimes.

Another criterion for the selection of a vacuum pump is the suction speed. In addition to the gas stream form the process, which is generally known, in technical plants the leakage stream, which depends on the quality of the sealing materials and the length of the seals, must also be taken into account. In the case of normal seals a leakage rate of 0.2 kg of air per hour and meter of seal ($P < 500$ mbar) can be expected. In the planning phase, empirical values must be used. The leakage rate Q_L can be determined in a similar or existing plant by means of pressure-rise measurements, in which the plant with volume V is evacuated, the pump is shut off, and the pressure rise ΔP per unit time Δt is measured (Equation 2.4-24).

Tab. 2.4-3 Vacuum ranges and suitable pumps.

Coarse vacuum	• 1000 to 50 mbar	• water ring pumps
	• 1000 to 20 mbar	• water ring pumps with injector
	• 1000 to 1 mbar	• steam jet with water ring pump
Fine vacuum	• down to 0.01 mbar	• steam jet with several prestages
	• down to 0.001 mbar	• sliding vane rotary pump
	• down to 10^{-7} mbar	• diffusion pumps
		• turbo molecular pumps

$$Q_L = V \cdot \frac{\Delta P}{\Delta t}. \qquad (2.4-24)$$

From this the minimum suction speed $S_{\min}$ of the vacuum pump can be obtained (Equation 2.4-25):

$$S_{\min}/(\text{kg air h}^{-1}) = \frac{M/(\text{g mol}^{-1})}{23.1 \cdot T/K} \cdot Q_L/(\text{mbar L s}^{-1}). \qquad (2.4-25)$$

The liquid ring pump [Scholl 1997] is one of the most widely used vacuum pumps because it is suitable for large gas streams (Figure 2.4-8). A rotor equipped with buckets is mounted eccentrically within a stationary housing filled with a working fluid (generally water), which is continually added. On the wall, a ring of liquid is formed that lifts off from the hub of the rotor. The gas to be pumped enters the thus-generated vacuum via the suction opening. After almost a complete rotation, the liquid ring approaches the hub again and drives the compressed gas through the discharge opening together with part of the liquid from the liquid ring, which is reintroduced through an additional opening. The liquid is recycled by separating it from the gas in a separator and, after cooling (the liquid must take up the heat of compression and the heat of condensation of the vapor), returning it to the compressor. Its vapor pressure at the operating temperature determines the resulting underpressure.

This type of pump is highly robust, and maintenance consists simply of lubricating the bearings. Even when the pumped gas contains particles or the operating liquid is not completely clean, prolonged operating times can be achieved. Since no moving metal components are in contact with one another, oil lubrication is unnecessary. The flame front of an ignited gas is stopped by a liquid ring pump, since the water extinguishes the flames (flame barrier). However, the piping and the required auxiliary equipment (Figure 2.4-8) make the use of these pumps highly complicated.

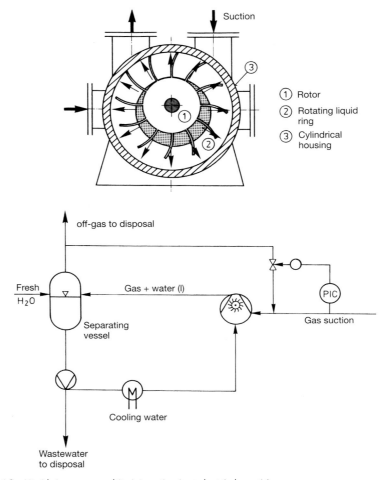

Fig. 2.4-8 Liquid ring pump and its integration in a plant (schematic).

2.4.4
Compressors

Gases are transported with aid of pressure differences. Thus, a characteristic quantity is the compression ratio P_V, that is, the ratio of the pressure P_2 after the compressor to the pressure P_1 before the compressor, which can be used to classify compressors (Table 2.4-4).

Ventilators transport gases in vessels of equal pressure and are used, for example, to circulate the air in drying plants. Ventilators can operate in axial or radial mode.

Turbo compressors for the transport of large volume flows are the most widely used blowers. They transport gases at relatively low pressures. Several radial impellers,

Tab. 2.4-4 Classification and characteristics of compressors.

	Ventilator	Blower	Compressor
Compression ratio	1 to 1.1	1.1 to 3	> 3
Max. pressure in bar (for air)	1.1	4	up to 1000
Specific work in kJ/kg (for air)	< 10	10 to 160	100 to 1300

mounted on a drive shaft, are located in interconnected chambers. When the drive shaft turns the gas is drawn in axially and accelerated by the centrifugal force. Often, up to ten stages are arranged in series, so that for a compression ratio of up to 1.5 per stage, a total compression ratio of 10–15 can be achieved.

Compressors such as reciprocating piston and rotary piston compressors are used to generate the highest pressures.

For the design of compressors the following equations are used:

Case 1: Power for isothermal compression (only possible in very small machines):

$$N/\text{kW} = \frac{P_1 \cdot \dot{V}_1}{\eta} \cdot \ln\left(\frac{P_2}{P_1}\right) \qquad (2.4-26)$$

$\dot{V}_1$ = volume flow on the suction side in m³ s⁻¹
P_1 = absolute pressure on the suction side in bar
η = efficiency (values between 0.6 and 0.85) are possible

Case 2: power for adiabatic or polytropic compression:

$$N/\text{kW} = \frac{P_1 \cdot \dot{V}_1}{\eta} \cdot \frac{\kappa}{\kappa - 1} \cdot \left[\left(\frac{P_2}{P_1}\right)^{\frac{\kappa-1}{\kappa}} - 1\right] \qquad (2.4-27)$$

where κ is the adiabatic or polytropic exponent [VDI-Wärmeatlas], which can be estimated with Equation (2.4-28):

$$\kappa = \frac{c_p}{c_v} = \frac{1}{1 - \left(\frac{8.313}{c_p/(\text{kJ kg}^{-1}\text{ K}^{-1}) \cdot M/(\text{g mol}^{-1})}\right)} \qquad (2.4-28)$$

The compression end temperature T_2 is calculated from Equation (2.4-29):

$$T_2 = T_1 \cdot \left(\frac{P_2}{P_1}\right)^{\frac{\kappa-1}{\kappa}} \qquad (2.4-29)$$

For real gases, especially at high pressure, the design equations must be corrected with the real gas factor (Figure 3.1.2-1, Chapter 3).

2.5
Energy Supply

2.5.1
Steam and Condensate System

Steam is the most widely used energy carrier for heating columns. It has the following advantages:

- Cheap energy source, since the steam can often be generated directly in the synthesis section of the plant.
- No problems with explosion protection (provided the hot surfaces of steam pipes are adequately insulated).
- Coverage of a large temperature range (Table 2.5-1).

The planning and design of the steam and condensate system is based on the mass and energy balance. Startup, shutdown, and operation under partial load must also be taken into account (Table 2.5-2).

The total steam network (Figure 2.5-1) consists of individual steam lines (corresponding to steam pressures). Large chemical installations usually have three steam pressures (e.g., 40, 25, and 4 bar) and hence three temperatures. Each line can take up

Tab. 2.5-1 Temperature and enthalpy of steam pressures widely used in chemical plants.

Steam pressure P_e/bar	Temperature ϑ/°C	Enthalpie of evaporation $\Delta_V H$/(kWh t^{-1})
0	100	627
1.5	127	605
4.0	153	585
16.0	204	534
25.0	226	508
40.0	252	477
100.0	312	361

Tab. 2.5-2 Mass balance (schematic) for a planned steam system consisting of two steam consumers and a steam generator (see example in Fig. 2.3-1).

	Start-up	Full load	Shut down	Partial load 1
Medium steam pressure (e.g., 4 bar)	• consumer W2	• consumer W2	• consumer W2	• consumer W2
	• generator	• generator	• generator	• generator
Sum in t h^{-1}				
High steam pressure (e.g., 25 bar)	• consumer W1	• consumer W1	• consumer W1	• consumer W1
Sum in t h^{-1}				

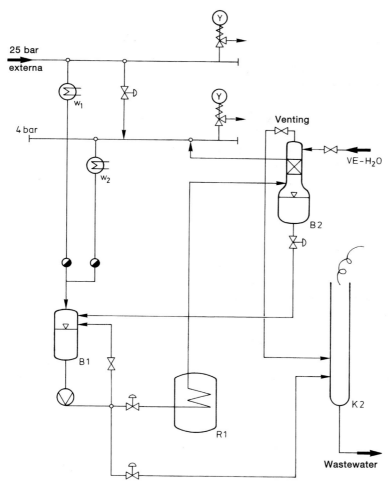

Fig. 2.5-1 Steam network with different steam lines (see also Table 2.5-2).

or release steam and is protected against overpressure by a safety valve. The steam used in a heat exchanger becomes hot condensate, which is collected in condensate collection vessels B1. From here it is pumped to the appropriate steam generators (reactor cooler, product condenser, etc.), so that a closed condensate circuit is formed. Excess condensate is cooled (cooling tower K2) and introduced into the wastewater for treatment.

2.5.2
Electrical Energy

Electrical energy is generally unsuitable as a heat source for columns due to the higher price (more than twice as expensive as steam) and in cases of explosion hazard. It is most important for driving applications (e.g., pumps, blowers, compressors) and for performing mechanical operations such as size reduction and mixing. Another important application is electrical tracing in chemical plants (protection against frost).

The electricity supply of a plant must be designed with sufficient redundancy, so that safety-relevant systems (e.g., pumps for cooling water) are in different independent electrical circuits.

2.5.3
Cooling Water

Surface water or ground water can be used for cooling [Auelmann I 1991, Auelmann II 1991, Auelmann III 1991, Hellmund 1991]. Compared to surface water brackish water and sea water have the advantage of more consistent temperature and less contamination with algae, but they require corrosion-protection measures. A cooling-water circuit (Figure 2.5-2) is preferable to simply passing cooling water through the plant. Sufficient cold river water is introduced to maintain the desired temperature (adavntages: saves river water in cold seasons, a minimum temperature can be maintained to prevent freezing of liquids that crystallize below 20 °C, and high flow rates in the circuit prevent corrosion and fouling).

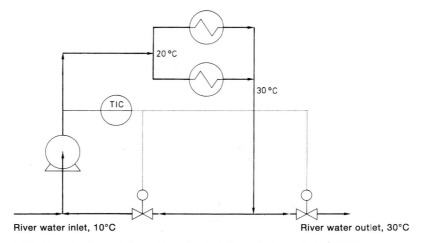

Fig. 2.5-2 Example of a circuit for cooling a chemical plant with river water (schematic).

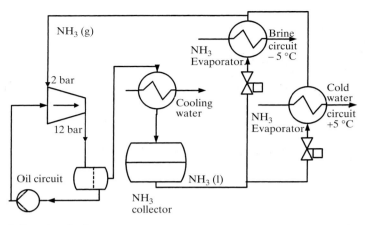

Fig. 2.5-3 Refrigeration circuit.

2.5.4
Refrigeration

If temperatures below 25–30 °C are to be attained in a process, then energy must be consumed to generate cold. Generally this is achieved by compression, cooling, and adiabatic expansion of a suitable refrigerant in the plant itself (Figure 2.5-3), that is, the requirement for cold corresponds to a requirement for heat and cooling water. To remove a certain amount of heat, about 20–50 % of the energy quantity is required in the form of additional electrical energy, depending on the required temperature level. Thus, the amount of heat to be removed increases to about 120–150 %.

2.5.5
Compressed Air

When small amounts of compressed air are required in a plant, it is supplied by pipe. An example is the air required for process control devices (rule of thumb: 1 m³ of control air per hour per control device).

Larger quantities are generated in the plant by air compressors, and hence the requirement can be expressed in terms of electrical energy [Coulson 1990]. The power N_{theo} required to compress a gas is obtained from Equation (2.4-27).

2.6
Product Supply and Storage

From an economic viewpoint, storage facilities have the following roles:
- Lowering costs (e.g., discount can be obtained by making larger orders).
- Price buffering (e.g., by filling the store prior to price rises and vice versa).
- Increasing production stability (minimum reserves can make up for shortages in supply, e.g., when delivery by ship is restricted by high or low water levels).
- Overbridging bottlenecks in turnover (school and public holidays, etc.).

For the operator of a chemical plant, storage is mainly used to achieve independence in the operation of the plant:
- Storage of starting materials to buffer the raw materials supply (independence from suppliers).
- Storage of intermediate products to increase the independence between different plant sections (e.g., synthesis and workup, balancing the capacities of individual plant sections).
- Storage of products to buffer sales (overbridging plant shutdowns and inventories, increased security of supply).

Today, in the case of starting materials that are not produced by the companies thenselves, more and more companies demand tank farms that are built and maintained by the supplier, who ensures that sufficient material is always available. The advantage for the supplier is a long-term contract.

The size of storage facilities is a matter of economic optimization (optimum stock) between availability (as large as possible) and tied-up capital (as small as possible). Depending on the state of aggregation, different storage facilities are required for:

- Solids such as bulk materials (open storage, covered storage, silos, bunkers) and piece goods (sack and drum storage, etc.).
- Liquids (drums, fixed-roof tanks, floating-roof tanks, etc.).
- Gases (spherical pressure gas tanks, low-pressure gas containers such as disk-type gas tanks and floating-bell gas tanks).

In planning tank farms, legally required measures must be taken against the propagation of fires and explosions. Tank farms are built in a closed concrete basin that is sufficiently large to accommodate the contents of the largest tank. A minimum distance must be maintained between the individual tanks. Tanks for liquids are often mounted on double-T beams so that they can be inspected from below (Figure 2.6-2).

The tanks are equipped with fittings for aeration and ventilation, inertization, pressure equalization, and flame barriers (Figure 2.6-2).

With regard to the tank farm, in the course of process development the construction materials and the thermal stability of the stored materials must be considered. Thermally sensitive materials (e.g., monomers such as styrene and acrylates) can be circulated through a cooler.

2.6 Product Supply and Storage

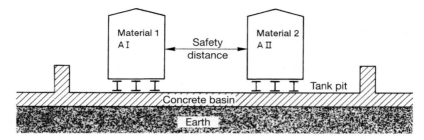

Fig. 2.6-1 Tank farm for liquids (schematic).

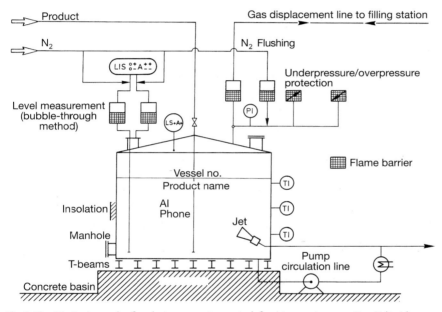

Fig. 2.6-2 Monitoring and safety devices on a storage tank for a temperature-sensitive AI liquid.

2.7
Waste Disposal [Rothert 1992]

2.7.1
Off-Gas Collection System and Flares

To prevent emission of vapors all vessels, columns and so on must be connected to an off-gas collection system. The substances in the collected off-gases are disposed of in a targeted fashion (thermal or catalytic combustion [Hünig 1986, Hüls 1989], absorber, adsorber, biowasher, etc.). The off-gas collection system is flushed from one end with sufficient inert gas (mostly nitrogen or oxygen-depleted air) that a vessel that sucks in gas (e.g., on being emptied) receives pure inert gas and no vapors from neighboring vessels. To avoid contaminations several independent off-gas collection systems may need to be installed. These are combined before the disposal system, generally a flare, in a sufficiently large collecting vessel, the level of which must be monitored by several means for reasons of safety. The off-gas collection system (Figure 2.7-1) must therefore be carefully designed on the basis of chemical knowledge.

In off-gas purification via a flare, the organic constituents of the off-gas are burnt at temperatures higher than 800 °C without recovery of the heat (Figure 2.7-2). Combustion can occur in the open air or a shielded flare. Since the off-gases from the off-gas collection system have only a low calorific value, the flare is operated with a supporting gas (e.g., natural gas). Depending on the type of flare, the combustion air is drawn in from the surroundings or force fed through a blower.

The flare must be equipped with diverse safety systems to prevent flashback of the flame (e.g., water-immersion bath, flame traps, acceleration path as dynamic flashback protector).

In the chemical industry, flares are being used less and less for off-gas purification and more as safety measures, for example, for fast pressure relief or for the disposal of harmful gases during process disturbances.

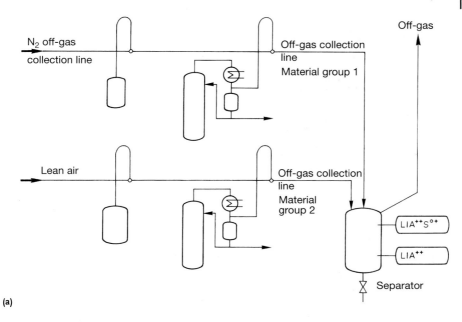

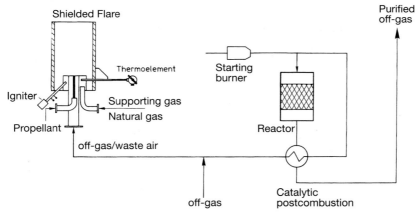

Fig. 2.7-1 Off-gas collection system (a) with two alternative methods of disposal (b): flare or catalytic combustion.

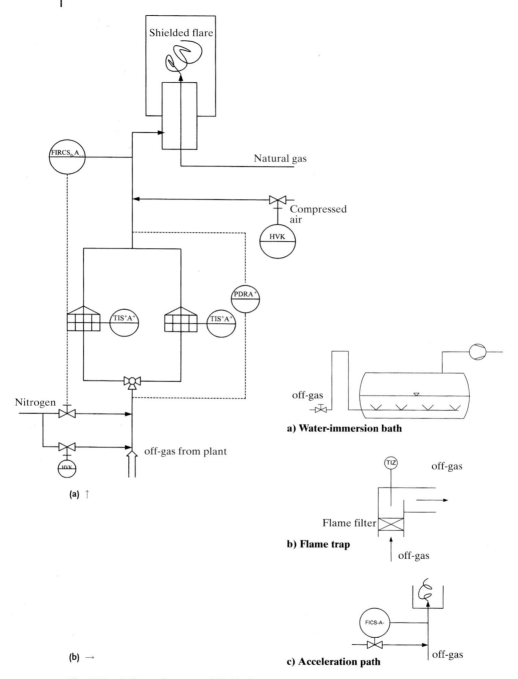

Fig. 2.7-2 a) Flare with integrated flashback protection (acceleration path and switchable flashback protector. b) Three types of flashback protection [Mendoza 1998].

2.7.2
Combustion Plants for Gaseous and Liquid Residues

The thermal combustion of off-gases and liquid residues is carried out at 750–1200 °C [Loo 2000]. The heat of the purified off-gas can be used to preheat the off-gas or the combustion air or for steam generation (Figure 2.7-3). These measures greatly increase the efficiency of a combustion plant and lower the operating costs. Depending on the location, synergies with other combustion plants may be possible (e.g., nearby power stations).

2.7.3
Special Processes for Off-Gas Purification

In the last few decades numerous developments have taken place in the field of off-gas and waste-air purification. They are treated briefly in the following:

Absorption
Waste air that contains only one certain pollutant can be purified by physical or chemical absorption of this contaminant in a suitable washing agent. This method is often economical in those cases in which the loaded washing liquid can be used in a process without workup.

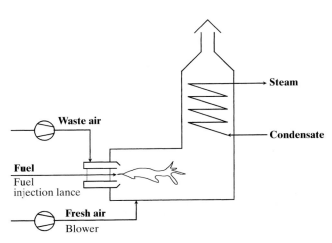

Fig. 2.7-3 Combustion plant (schematic).

Adsorption

Adsorption exploits the capability of, for example, activated carbon to take up large amounts of organic vapors. Very low residual concentrations can be achieved. Since this is a regenerative process, at least two adsorption toweres are required.

Biofilters

Microorganisms that can oxidize organic contaminants to carbon dioxide and water are fixed on a continually moistened carrier material through which the loaded waste air is passed. Prerequisites here are low concentrations and water-soluble pollutants.

Biowashers

An aqueous slurry of activated sludge is passed through a column in countercurrent to the waste air.

Membrane separation (steam permeation)

The prerequisites are the availability of a suitable membrane and an off-gas free of particulate contamination (Figure 2.7-4).

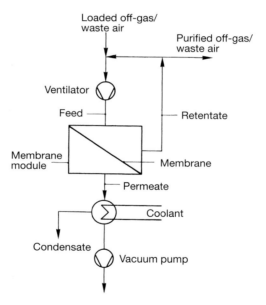

Fig. 2.7-4 Air-purification plant using vapor permeation (schematic).

2.7.4
Wastewater Purification and Disposal

In chemical production processes wastewaters loaded with side products are often formed, and these must be cleaned in a wastewater-treatment plant before they are discharged into a body of water. The amount of wastewater can be reduced, for example, by recycling, recovery of ingredients, and by using low-water processes (integrated environmental protection) [Christ 1999]. In spite of these measures, more than $4 \times 10^9 \, m^3 \, a^{-1}$ of wastewater is produced in Germany.

The German *Wasserhaushaltsgesetz* (WHG: Water Resources Management Act) of 1957 and its revision of 1960 regulates the exploitation of water bodies and the withdrawal and discharge of water, and up to the present day it has been repeatedly been revised to meet changing requirements. Furthermore, in 1978 the *Abwasserabgabengesetz* (law governing wastewater discharge) was introduced with the aim of extending the principle that the polluter pays to the field of water protection. The local water authority, which is responsible for granting permission to discharge wastewater and specifying and monitoring limiting values and conditions, must be involved in the project as early as possible.

Depending on the nature of the wastewater, the following processes are available:

- Biological treatment (favored for dilute wastewaters ($< 5\%$) and readily biodegradable substances).
- Chemical treatment (favored for substances that are not readily biodegraded), for example wet catalytic oxidation, catalytic UV oxidation, oxidation with H_2O_2, supercritical water oxidation (SCWO) [Baur 2001].
- Physical treatment (extraction [Schierbaum 1997], membrane processes, adsorption on resins, activated carbon, etc.).
- Combined processes (e.g., biological/chemical/physical treatment processes).

2.7.4.1
Clarification Plant

Large industrial sites generally have a mechanochemical biological wastewater-treatment plant (clarification plant). The wastewaters of the individual production plants are collected and delivered to the clarification plant (additive environmental protection). It essentially consists of the following steps:

- *Removal of sand and coarse material*
  Mechanical devices such as screening units, sand traps, and sieving plants remove sand and coarse material to ensure smooth operation of the downstream plant.
- *Neutralization stage*
  Since biological treatment of the wastewater is most effective under neutral conditions (pH 6–8), depending on its pH, the wastewater must first be neutralized with slaked lime or sulfuric or hydrochloric acid. Heavy metal ions (Pb, Hg, Cd, Cr, etc.) and colloidal materials are removed prior to neutralization by precipitaion as hydroxides or flocculated by addition of iron(II) sulfate.

- *Preclarification in a settling basin*
  Here the precipitates are allowed to settle by lowering the flow velocity.
- *Biological purification in a bioreactor*
  Adapted microorganisms (bacteria, fungi, etc.) aerobically degrade the contaminants in the water to carbon dioxide and activated sludge (humic acids, bacterial mass). The concentration of activated sludge is 1–5 g of dry matter per liter. The performance of this biological stage depends on the supply of degradable substances, the oxygen supply, the temperature, and the residence time. Bacterial toxins must be excluded, whereby the toxicity depends on the concentration and the degree of adaption of the microorganisms to the substance in question. Therefore, large fluctuations in the quality and quantity of wastewater coming from the production plants should be avoided (buffer tank for wastewater from washing). This can be achieved by buffering with a constant load of communal wastewater. Modern bioreactors differ from one another in the manner in which the bacteria are held (supension, fixed bed, etc.), the way in which oxygen is introduced, and the hydrodynamics.
- *Postclarification in a settling basin*
  Here the activated sludge is separated from the purified wastewater by precipitation with flocculants and partially recycled to the bioreactor.
- *Sludge workup*
  The excess activated sludge is denatured, dewatered, and disposed of, for example, by mixing it with coal dust and burning the resulting mixture in a fluidized-bed or rotary furnace. The laborious treatment and disposal of the sludge are responsible for a large part of the operating costs.

Wastewater analysis and characterization

During process development in the miniplant, representative wastewater samples must be analyzed. The wastewater quantity of the planned industrial plant must be determined, together with the analytical data of the wastewater, so that the loading of the clarification plant, either present at the intended site or to be built, can be determined. Since quantitative analysis of all components is often too laborious (AAS, GC-MS, HPLC, etc.), the following sum parameters are generally used to characterize wastewaters:

- Sedimentable substances.
- Salt content.
- Chemical oxygen demand (COD) according to DIN 38409. The wastewater sample is acidified with sulfuric acid, a known excess of potassium dichromate (oxidizing agent) and silver ions as catalyst are added, and the mixture heated to reflux for two hours. After cooling, the unconsumed oxidizing agent is backtitrated with iron(II) sulfate to give the COD in milligrams of O_2 per liter.
- Biological oxygen demand (BOD_5) according to DIN 38409. The wastewater sample is mixed with oxygen-saturated water containing bacteria, and the oxygen content is determined with an O_2 glass electrode. The sample is sealed and kept at 20 °C (thermostat). After five days the oxygen content is measured, and the BOD_5 is determined from the difference in milligrams of consumed oxygen per liter.

- Total organic carbon (TOC) according to DIN 38409-H3.
- Dissolved organic carbon (DOC) according to DIN 38409 Part 3 (1983).
- Adsorbable organic halogen compounds (AOX) according to DIN 38409-H14.
- Volatile organic halogens (VOX), purgable organic halogens (POX), and extractable organic halogens (EOX) according to DIN 38409-H8.

To avoid a threat to the clarification plant from toxic substances, where necessary wastewaters are subjected to biological test methods to determine their toxic effects (toxicity to bacteria and fish). Each individual component must be sufficiently biodegradable, so that no nondegradable component is masked by the total amount of degradable components. A substance is regarded as sufficiently degradable when the ratio BOD_5/COD is larger than 0.5; readily degradable wastewaters have values greater than 2/3. If necessary, further tests must be performed that better take the adaptability of the microorganisms into account, for example, Zahn–Wellens test. This is a static method for testing ease of elimination or potential biodegradability of of a test substance under aerobic conditions (Figure 2.7-5). In Germany, pure substances are classified according to their biodegradability and their toxicity to aquatic organisms into three *Wassergefährdungsklassen* (WGK; water hazard classes):

- WGK 1: slightly hazardous to water (e.g., acetic acid, maleic anhydride).
- WGK 2: hazardous to water (e.g., aceteonitrile).
- WGK 3: highly hazardous to water (e.g., arsenic(III) oxide).

Discussions concerning wastewater problems should be held with the corresponding waste-disposal department at as early a stage as possible.

2.7.4.2
Special Processes for Wastewater Purification

Laws and regulations concerning wastewater treatment have become increasingly strict, and in many cases today the wastewater must be treated on-site before it is fed to the central clarification plant. Besides destructive processes such as:

- Chemical degradation with oxidizing agents (ozone, nitric acid, or wet oxidation with atmospheric oxygen at high pressure and high temperature with or without a catalyst).
- Pyrolysis in wastewater combustion plants.

today regenerative processes are increasingly being used as a component of integrated environmental protection:

- Adsorption/desorption
- Extraction [Schierbaum 1997]
- Distillation
- Precipitation
- Ion exchange, and so on.

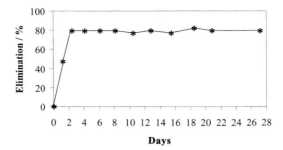

Fig. 2.7-5 Example of a Zahn–Wellens test of wastewater containing acetic acid. The plot shows the relative decrease in DOC.

The test substance is diluted and stirred and aerated with communal activated sludge in a batch vessel for 28 d. From the shape of the elimination curve, degradation, adsorption on the activated sludge, and stripping effects can be distinguished. Purely biological degradation can be recognized by:
- A typical S-shaped elimination curve
- Small contribution of adsorption (elimination after 3 h < 20%)
- No indication of stripping effects
- A wastewater is readily degradable if the elination value exceeds 70% (20–70% moderately degradable, < 20% difficultly degradable).

The type and extent of the chosen method always depends on the task to be solved (wastewater quantity, qualitative and quantitative wastewater loading), the target (reduction, elimination, recycling), and the operating costs of the corresponding wastewater-treatment plant. For example, for the elimination of highly stable organic compounds from small amounts of lightly loaded wastewater streams, an ozone-treatment plant [Ozonia] can be economically viable in comparison with combustion. On the other hand, the higher the COD of the wastewater, the greater the advantages of a thermal wastewater combustion plant.

Once a treatment process has been chosen, tests must be carried out with original wastewater in a miniplant, so that scale-up problems can be solved. Especially in the case of regenerative processes, in which separated substances are recycled as valuable products to the production process, the influence of the new reycle streams must be tested in an integrated test plant.

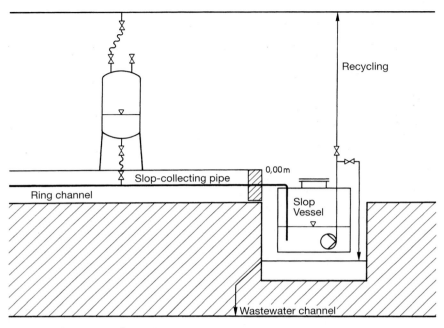

Fig. 2.7-6 Slop system (schematic).

2.7.5
Slop System

The slop system (Figure 2.7-6) is an important auxilliary system for plant startup and shutdown and for controlling emergency situations. The slop-collecting pipe runs in a ring channel around the plant. Containers, columns, and so on, which must be emptied, for example, for repair, are connected to the ring main via outlet fittings and metal hoses, so that their contents can be temporarily stored in the sloptanks, which are the hydrodynamically lowest point of the plant. After completion of the repair work, the contents can be returned to the corresponding plant components by submerged pumps.

2.8
Measurement and Control Technology

For reasons of rationalization and safety modern chemical plants are automated to such an extent that they can be operated with a minimum of personnel. Today an automated plant, regardless of its size, requires a permanent staff consisting of only five shift workers:

- One shift foreman
- Two shift workers in the measuring and control station
- One shift worker on site
- One substitute.

In a properly automated plant, this skeleton crew can concentrate on monitoring the plant and eliminating disturbances. This is made possible by process control engineering (Figure 2.8-1).

2.8.1
Metrology

Sensors are used to measure operating parameters, the most important of which are:

- Temperature T
- Pressure P
- Flow F
- Level L
- Quality Q.

2.8.1.1
Temperature Measurement

The most widely used temperature scale in the German-speaking countries is the Celcius scale, while the Fahrenheit scale dominates in Anglo-Saxon countries. The absolut Kelvin scale is rarely used in the chemical industry. The scales are interconverted as follows:

$$°C = 10/18 \cdot (°F - 32) \quad (0\,°C = 32\,°F)$$
$$°F = 0.18 \cdot °C + 32 \quad (100\,°C = 212\,°F)$$
$$K = °C + 273.2.$$

The by far most widely used temperature-measuring devices are the termoelement and the resistance thermometer, since these give a temperature-dependent electrical signal that can readily be processed.

2.6 Measurement and Control Technology

Level	Location	Activity	Hardware
Process control level	Control room	• Display • Operation • Documentation • Alarm • Report • Depict curves	• Screen • Keyboard • Light pen/Alarm • Printer • Writer
Near-process components	Switching room	• Measure • Monitor • Control • Calculate	• Process stations with power supply • CPU • Bus coupling components • Signal (analog, digital, impuls input/output • Direct digital information from digital field devices
Field level	Plant	• Measure • Adjust	• Sensors • Actuators • Contacts, etc.

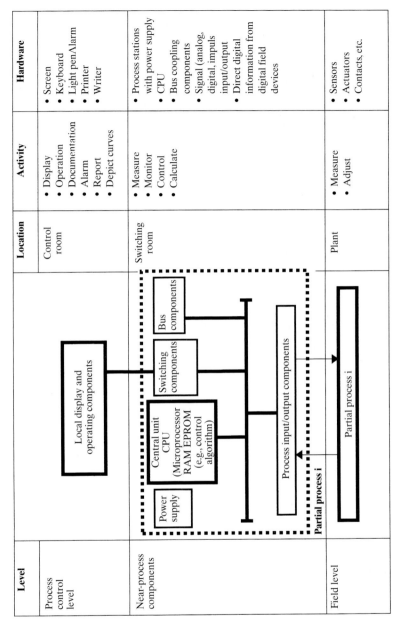

Fig. 2.8-1 Structure of a process control system.

Resistance Thermometer

In the resistance thermometer the resistance of a metal wire coil (often Pt) is measured, and the temperature is determined from the known resistance/temperature function (so-called Pt 100, i.e., 100 Ω at 0 °C). The measuring probe is inserted in a protective sleeve (Figure 2.8.2b) and can be changed while the process is running. The measurement signal is transformed into a unit signal (4–20 mA) in a transducer. The overall error in the measurement is $\pm$ 0.5 %. Depending on the required accuracy, two-, three-, or four-wire circuits are used. Common sources of error include corrosion of the clamping joint and poorly insulated protective sleeves (thermal radiation). The preferred measuring range lies between −250 and +500 °C.

Thermoelement

A thermoelement (Figure 2.8-2) consists of two wires made of different materials (e.g., Fe and CuNi) that are generally soldered together at one end. At the free wire ends (cold junction) a temperature-dependent thermoelectric voltage can be measured (e.g., for Fe and CuNi 5.26 mV/100 K at −210 to 760 °C). The measuring probe is inserted in a protective sleeve and can be changed while the process is running. The measurement signal is transformed into a unit signal (4–20 mA) in a transducer. The overall error in the measurement is $\pm$ 1 %. Common sources of error include variations in the thermoelectric voltage due to aging and drift of the cold junction. The preferred measuring range lies at higher temperatures (between 400 and +1000 °C).

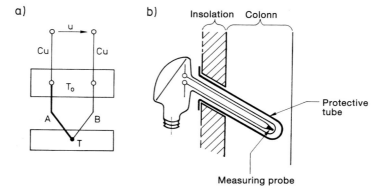

Fig. 2.8-2 a) Principle of a thermoelement circuit with three materials A, B, and Cu and the cold junction temperature T_0. The cold junction can be a thermostat (e.g., ice bath) or a correction circuit. b) Measuring probe inserted in a protective sleeve.

2.8.1.2
Pressure Measurement

The most widely used pressure unit in technical chemistry in continental Europe is the bar, while in Anglo-Saxon countries pounds per square inch (psi) is preferred. The concersion factors follow:

$$1\,\text{bar} = 10^5\,\text{Pa} = 750\,\text{Torr} = 1.02\,\text{kp cm}^{-2}\,\text{at} = 0.987\,\text{atm} = 1405\,\text{psi}.$$

Pressures can be specified as:

- Absolute pressure (P_a)
- Overpressure $P_e = P_a - 1.013\,\text{bar}$
- Pressure difference ΔP.

Three types of pressure-measuring devices can be distinguished on the basis of measurement principle:

- Pressure-measuring devices with a confining liquid (e.g., U-tube manometer).
- Pressure-measuring devices with elastic pressure sensors.
- Pressure transducers.

While U-tube manometers for routine measurements are only rarely used for safety reasons, pressure-measuring devices with elastic pressure sensors (spring-tube, spring-plate, and capsule-spring manometers) are the most widely used because they are very robust and cheap. In the case of pulsating pressures (e.g., downstream of a piston pump) liquid-filled manometers must be used (damping). The use of a transducer, which affords an electrical measurement signal, is also possible. The measuring element of a pressure transducer is a metal membrane that is slightly deformed by changing pressure (Figure 2.8-3b). This deformation can be transformed into an electrical signal by an inductive, capacitive, or piezoresistive transducer. Pressure transducers are much more expensive than manometers and are therefore preferentially used for transmitting measured data and measuring pressure differences.

2.8.1.3
Measuring Level

Precise knowledge of the filling level in columns and vessels is especially important for startup and shutdown of a plant, since the inside of the apparatus is generally not visible. Furthermore, disturbances (blockages, leaks, etc.) can often be detected at an early stage with the aid of this important measured quantity.

While simple mechanical filling-level meters such as dipsticks and inspection glasses are reliable and accurate, they are not suitable for transmitting measured values. A cheap and reliable method is bubble-through measurement (Figure 2.8-4), which in principle measures the hydrostatic pressure and thus transforms the measurement of level into the measurement of pressure difference, which can readily be

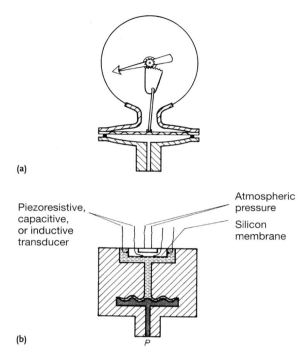

Fig. 2.8-3 Principles of pressure measurement. a) Spring-plate manometer. b) Pressure transducer.

automated, as described above. Nitrogen is generally used as bubbling gas, so that the vessel is also inerted. The level is normally indicated in per cent, whereby the limiting values of 0 and 100% should be included in the plans of the vessel.

For the level measurement of difficult media (high viscosity, foaming, high solids content, etc.), vessel balances, radiometric level measurement with a γ emitter [Kämereit 2001], capacitive techniques, and ultrasonic methods (echo depth sounder) are suitable, but they are expensive both to purchase and maintain.

2.8.1.4
Flow Measurement [CITplus 1999, Belevich 1996]

Flows are of fundamental importance for controlling a plant and for balancing and metering liquids. A large number of measurement principles are available here. Apart from the Coriolis-force meter, which directly measures the mass flux, all of them measure the volume flow.

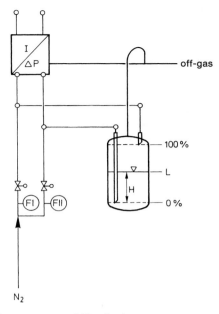

Fig. 2.8-4 Bubble-through measurement of filling level.

Throttle Devices

A mechanical resistance is introduced into the flow and the resulting pressure drop is measured. Thus, flow measurement is transformed into the measurement of pressure difference. An orifice guage or standard orifice plate is generally used (Figure 2.8-5). Other types of throttle are the standard jet, the standard Venturi jet, and the spiral tube, which have the advantage over the standard orifice plate of a lower permanent pressure drop. This measurement principle is applicable over a wide measuring range, requires

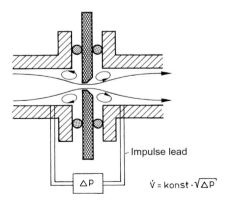

Fig. 2.8-5 Measurement principle of a standard orifice plate [Belevich 1996].

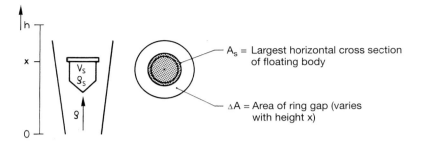

Fig. 2.8-6 Floating body flow meter [Belevich 1996].

little maintenance, is cheap, and is applicable for gases and liquids. The disadvantages are high permanent pressure drop (compression energy) and problems with difficult media (visvous liquids, media that cause fouling, multiphase flows).

Floating-Body Flow Meter
The use of these devices is restricted to the measurement of small to medium volume flows of low molecular weight liquids free of solid components. The position of the floating body can be measured inductively, so that an electrical signal is available for processing (Figure 2.8-6).

Magnetic-Induction Flow Meter
This measurement principle requires the presence of liquids with a certain minimum conductivity (at least 1 $\mu S\ cm^{-1}$; for comparison: singly distilled water has approx. 10 $\mu S\ cm^{-1}$). The volume flow to be measured flows through a magnetic field perpendicular to the direction of flow (Figure 2.8-7). The induced electrical potential is recorded by two electrodes and used as measured quantity. The measurement is independent of the pressure and viscosity of the liquid, causes no additional pressure drop, and can be used over a wide measurement range (from several liters per hour up to 100 000 $m^3\ h^{-1}$).

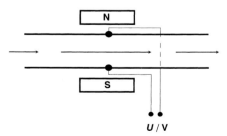

Fig. 2.8-7 Magnetic-induction flow meter [Belevich 1996].

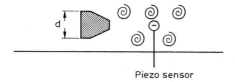

Fig. 2.8-8 Eddy current flow meter [Belevich 1996].

Eddy Current Flow Meter

In the flowing liquid eddy currents are generated behind a barrier. The eddy frequency (eddies per unit time) is measured with a suitable sensor (e.g., piezoelectric) and is proportional to the flow velocity (Figure 2.8-8). This type of flow meter is also suitable for extreme temperatures (-200 to $400\,°C$), and the measurement range is about $0.5–2000\,m^3\,h^{-1}$ in liquids, and about $10–20\,000\,m^3\,h^{-1}$ in gases.

Coriolis Flow Meter [Messtechnik 1998]

A U-tube through which the liquid stream to be measured flows is electromagnetically excited to vibrate at its natural frequency. The flowing liquid resonates with the tube and is subjected to a Coriolis force that results in twisting of the tube, which is registered optically and is a measure of the mass flux (Figure 2.8-9). Since this method measures the mass of the flowing liquid rather than its volume, it is mainly used where volume flow meters fail (e.g., two-phase flows, viscous media, foams, slurries, etc.). Another advantage is that measurements can be made practically free of pressure drops.

Thermal Flow Meters

A heated wire wound around a metal pipe generates a temperature distribution in the absence of flow. A flowing liquid changes this distribution. Measuring and reference resistances measure this changing temperature difference along the pipe. As components of a Wheatstone bridge, they control the heating current in such a way that the temperature difference remains the same for all flow conditions (Figure 2.8-10). This measurement principle is suitable for small flows ($< 500\,g\,h^{-1}$) and high pressures (up to 3000 bar), that is, in the regime of miniplants and for monitoring feed streams. Since the measurement depends on the specific heat and the thermal conductivity, the instrument must be calibrated for each medium.

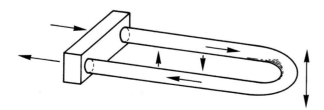

Fig. 2.8-9 Coriolis flow meter [Belevich 1996].

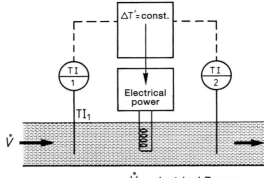

Fig. 2.8-10 Thermal flow meter [Belevich 1996].

Ultrasonic Flow Meter
Noninvasive ultrasonic flow measurement in three variants (direct measurement of time of passage, sing-around, and Doppler) has a series of advantages: low pressure drop, absence of moving mechanical parts and hence insensitivity to fouling result in ease of maintenance and prolong the working life. The main problem is measuring the rather long time of passage of the ultrasound with sufficient accuracy.

Quality
Some important process quantities follow [Melzer 1980]:

- Refractive index
- Concentration, measured by gas chromatography or special methods such as flame-ionization detectors, IR methods including URAS for CO, CO_2, NO, SO_2, paramagnetism for measuring O_2 concentration
- Density
- Rotational speed
- Electrical conductivity
- Humidity
- pH value (glass electrode)
- Viscosity.

2.8.2
Control Technology (Closed-Loop Control)

Control here means that the measured value of a process variable, the so-called control variable x, is maintained at a given setpoint w by adjusting actuators such as valves and pumps by means of a control element until the deviation between x and w is compensated. The control command (controller output y) is thus always dependent on the setpoint (feedback). Figure 2.8-11 shows the principle of a simple control circuit.

In addition to the basic components shown in Figure 2.8-11, further auxiliary elements are require for carrying out the control task (Figure 2.8-12):

Transducers
These transform the output signal of a measuring instrument into a standardized signal (e.g., 4–20 or 0–20 mA, 0.2–1 bar) [Meßumformer 1999].

Position controller
This auxiliary controller is used where the load on drive and control element is high or when the quality of control is to be improved. It mechanically measures the true valve position in situ and compares the value with the controller output y provided by the

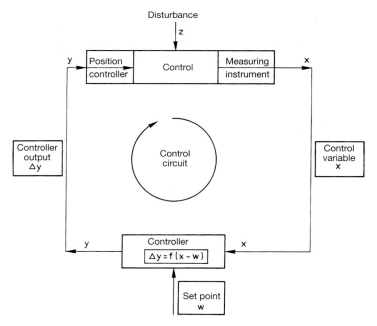

Fig. 2.8-11 Basic elements of a closed-loop control circuit. The variable x to be controlled in the control system is measured by a measuring instrument and complared with the setpoint w in the controller. The control deviation $x-w$ is converted to a controller output y by means of a control algorithm and acts on the control system via a control element that changes the control variable (feedback).

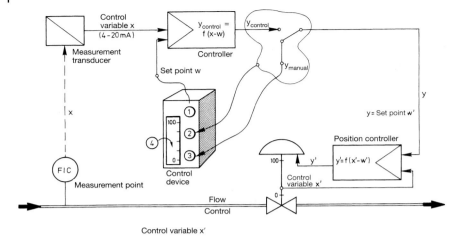

Fig. 2.8-12 Extended closed-loop control circuit for the example of flow control.

controller. In cases of control deviation, the control element is adjusted until the required position y is reached.

Time behavior of control circuits (important: controller is in manual mode)
Knowledge of the dynamic behavior of the process to be controlled (control circuit), that is the temporal behavior of the control variable $x(t)$ resulting from a sudden change in the controller output Δy is important for the design of the entire control circuit. This behavior can be analyzed by suddenly increasing the controller output y by the amount Δy (e.g., by opening the control valve from 50 to 60%) and recording the value x (Figure 2.8-13).

Time behavior of controller types (important: without control circuit)
The controller type or the parameters to be adjusted on the controller must be chosen or optimized in accordance with the expected dynamic behavior of the control circuit. An important characteristic of a controller is the temporal variation of the controller output in response to a sudden control deviation. According to this behavior, three types of controller are distinguished: P, I, and D controllers (Figure 2.8-14).

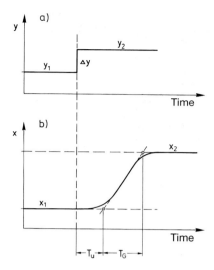

Fig. 2.8-13 Dynamic behavior of a control circuit. a) Increase in controller output y by the amount Δy. b) Response x(t). A tangent to the turning points can be used to determine the two characteristic quantities delay time T_U and compensation time T_G. When both values are zero, the control circuit is without delay or of zeroth order, and when only T_G is zero the circuit is first-order. In the general case, higher order control circuits are present with:

$$n \approx 10 \cdot \frac{T_U}{T_G} + 1 \qquad (2.8-1)$$

The following are measures of the controllability:
$T_U/T_G < 1/10$ well controllable
$T_U/T_G \approx 1/6$ just controllable
$T_U/T_G > 1/3$ difficult to control

Behavior of control circuits (interaction of the important controller types with the control circuit)

- *P controller.*
  The proportional controller reacts immediately to an x–w jump with a proportional change in y. The ratio $\Delta w/\Delta y$ is known as the proportionality range X_p. The process leads to a value of x that generally deviates from the setpoint, that is, it is not possible to use a P controller to exactly control x to the setpoint (permanent control deviation).
- *I controller.*
  The integral controller reacts to an x–w jump with a continuously increasing change in controller output $\Delta y(t)$ until the control deviation becomes zero. Pure I controllers have the disadvantage that they are either too sluggish or are prone to uncontrollable oscillations.

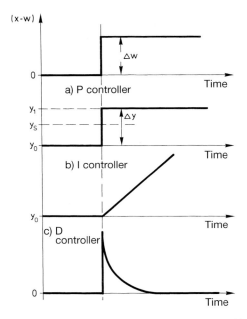

Fig. 2.8-14 Temporal variation of the controller output y in response to a sudden control deviation x–w for the three basic types of controller.

- D controller.
  The differential controller reacts to a control deviation with a y pulse, that is the controller output y rises sharply and then returns slowly to the initial value, so that no inherent control function is present.

In practice the basic types are often combined to achieve optimum control behavior. The most important combinations that are used in chemical plants are:

PI controller
Overall this controller has a favorable control behavior (no permanent control deviation). It acts immediately, but exhibits relatively slow adjustment to the setpoint. The time behavior of the PI controller is characterized by the reset time T_N (Figure 2.8-15). It is especially suitable for pressure and flow contol circuits.

PID controller
The PID controller combines the advantages of the three basic types: very fast action and fast adjustment to the setpoint without a permanent control deviation. It is especially suitable for temperature control. An ideal PID controller can be described mathematically by Equation (2.8-2).

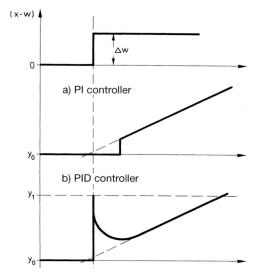

Fig. 2.8-15 Control behavior of PI and PID controllers.

$$\Delta y = \left\{ K_p \cdot \Delta x + \frac{1}{T_N} \cdot \int_{t=0}^{t} \Delta x \cdot dt' + T_V \cdot \frac{d\Delta x}{dt} \right\}. \tag{2.8 – 2}$$

The individual contributions of the three basic types to a PID controller and hence the control behavior are characterized by three characteristic values:

- X_P: proportional range, changing P behavior
  (small X_P value = large P contribution).
  The reciprocal value $K_P = 1/X_P$ is often used.
- T_N: reset time, changing I behavior
  (small T_N value = large I contribution).
- T_V: derivative action time, changing D behavior
  (large T_V value = large D contribution).

Note that the three quantities are not independent of one another.

For optimal adaption of the controller to the control circuit, the behavior of both must be known. Random adjustment of all three parameters is generally not successful. A series of methods are available for determining favorable control parameters [e.g., Ziegler–Nichols method: 1. operate the circuit with a pure P controller ($T_N \to \infty$, $T_V = 0$), 2. reduce X_P until the control circuit undergoes continuous oscillations], whereby on the basis of targeted adjustments and the resulting reaction of the control circuit, the correct setting of the controller can be found. Some other practical tips (Table 2.8-1) follow:

Tab. 2.8-1 Rough guideline for setting controller parameters.

Control quantity	Preferred controller type	$X_P/\%$	T_N/min	T_V/min	Notes
Temperature	PID	10 to 50	1 to 20	0.2 to 3	control element often sluggish with long response times
Pressure	PI	10 to 30	0.1 to 1	0	control element often very fast
Flow	PI	100 to 200	0.1 to 0.5	0	control element very fast for liquids
Level	PI	5 to 50	1 to 10	0	

Tab. 2.8-2 Abbreviations for measurement and control point. For example, PDIRCA^{++}S^{0+} denotes a differential pressure measurement that is displayed and registered in the control room and is additionally equipped with a pre-alarm *Max1* and a switch at *Max2*.

First letter	Supplementary letter	Subsequent letter
• *P* pressure	• *D* difference	• *I* indication of control variable
• *T* temperature	• *Q* sum	• *R* registration (e.g., plotter)
• *F* flow	• *F* ratio	• *C* control
• *L* level quality		• *A* alarm on exceeding boundary value
• *Q* quality (analytical value)		• *S* switching on exceeding boundary value
		• *Z* safety switch on exceeding boundary value

- If the measured quantity of the control circuit undergoes periodic oscillations, then control is unstable. The reason for this usually lies in too strong an action of the controller. Here it often helps to enlarge the proportional range (smaller P contribution) or to increase T_N. With a PID controller one must ensure that T_N is always at least four to five times T_V.

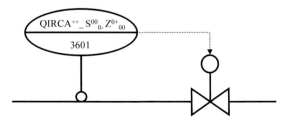

Fig. 2.8-16 Meaning of symbols on a measurement point. The combination of letters describes the type of measurement and control device (Table 2.8-2). The dash indicates that the Q value is transmitted to the control room (no dash = no transmission, processing in situ). The number is the number of the measuring point.

2.6 Measurement and Control Technology | 217

Analytical quantities such as the pH value are often relatively difficult to control, because the behavior of the control circuit is often highly complex.

The depiction and designation of measurement and control devices is standardized in DIN 19227 and DIN 19228 (Figure 2.8-16 and Table 2.8-2).

Cascade control

An auxiliary controller can be used to increase the accuracy and speed of control. Often

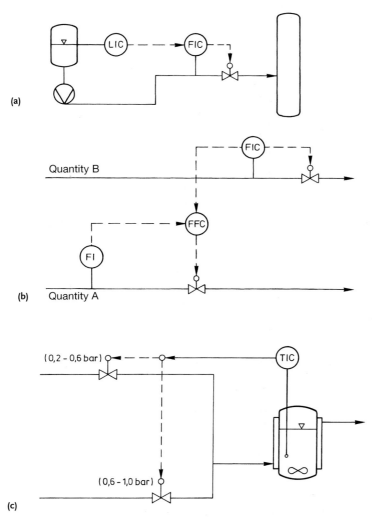

Fig. 2.8-17 Control strategies. a) Cascade control for the example "level controls quantity". b) Ratio control for the example "quantity of B controls quantity in a constant ratio". c) Split-range control for the example of "temperature controls cold water in the range of 0.2–0.6 bar, and hot water in the range of 0.6–1.0 bar".

it is more favorable not to control the flow directly (large control fluctuations), but use it as a control setpoint. If the level in the vessel increases rapidly, the setpoint of the outlet flow is slowly increased (Figure 2.8-17a).

Ratio control
Ratio control (Figure 2.8-17b) is often used in controlling constant mixing ratios (e.g., stoichiometric feeding of two starting materials A and B). The value of flow A is input via a ratio controller as the setpoint for the sequential controller (flow B).

Split range
The term split range does not refer to a control method, but means that the range of the final control element is split over two actuators (Figure 2.8-17c).

2.8.3
Control Engineering (Open-Loop Control)

Control is a process in a system in which input quantities influence output quantities on the basis of a logical operation. It is characterized by an open circuit (Figure 2.8-18).

The processing of boundary contacts and the resulting switching of actuators is an important task of control units, which can be classified as follows:

- Timed (event-controlled) control (for batch processes): Control events that are controlled by time or fulfillment of conditions and take place in a chain of steps. An actuator does not react immediately to conditions, but instead depends on the step by which it is controlled.
- Continuous control (for continuous processes): switching of an actuator (open/closed or on/off or 1/0, i.e., $A = 1$ if $A \neq 0$, $A = 0$ if $A \neq 1$) on the basis of a logical linkage of process measurments. In this logical circuit, the entire safety philosophy of a chemical plant is realised.

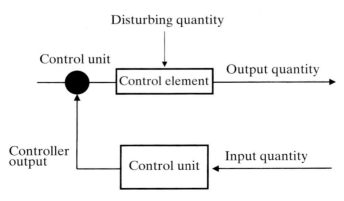

Fig. 2.8-18 Principle of open-loop control.

2.6 Measurement and Control Technology

The logical connections between the sensors and the actuators are the fundamental elements of control technology. The most important are (Figure 2.8-19):

AND operator (conjunction, logical product): the AND operation of the sensor boundary values is the strictest condition for enabling an actuator, since all requirements on the measured value must be fulfilled.

Circuit symbol	Function	a	b	x	Example with contacts
 a, b → & → x	$x = a \wedge b$ AND Conjunction	0 0 1 1	0 1 0 1	0 0 0 1	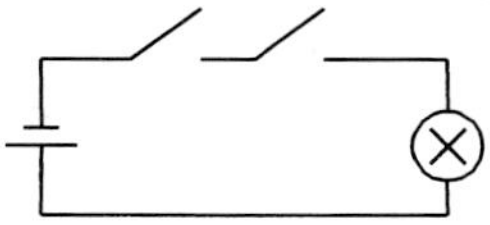
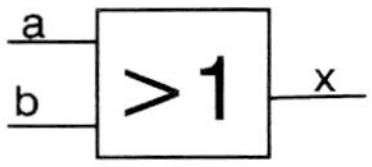 a, b → ≥1 → x	$x = a \vee b$ OR Disjunction	0 0 1 1	0 1 0 1	0 1 1 1	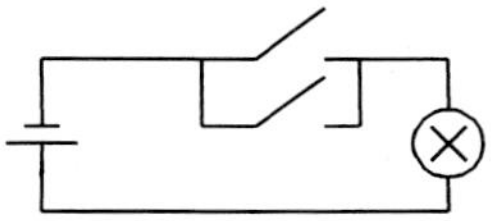
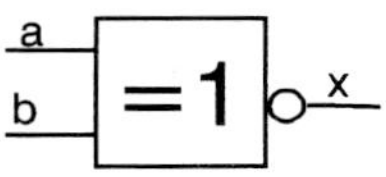 a → =1 →o x	$x = \bar{a}$ NOT Negation	0 1		1 0	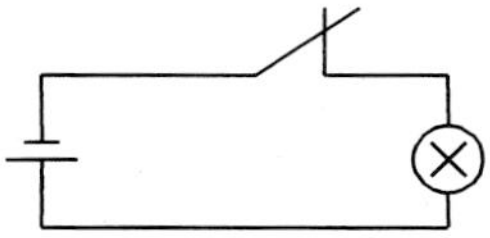
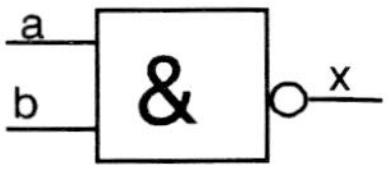 a, b → & →o x	$x = \overline{a \wedge b}$ NAND	0 0 1 1	0 1 0 1	1 1 1 0	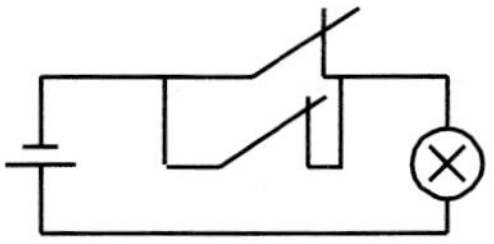
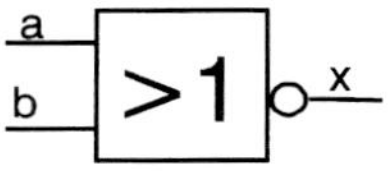 a, b → ≥1 →o x	$x = \overline{a \vee b}$ NOR	0 0 1 1	0 1 0 1	1 0 0 0	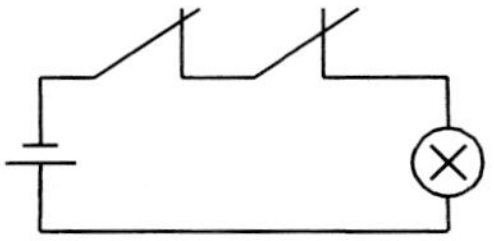
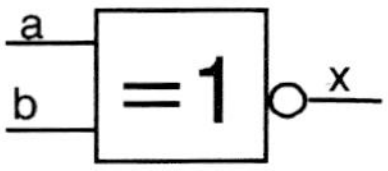 a, b → =1 →o x	$x = (a \wedge \bar{b}) \vee (\bar{a} \wedge b)$ Exclusive OR	0 0 1 1	0 1 0 1	0 1 1 0	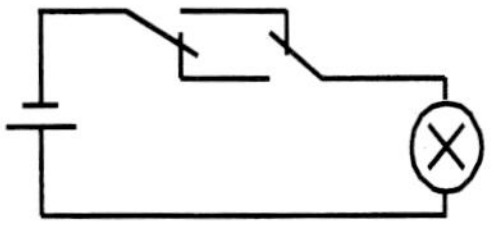

Fig. 2.8-19 Basic logical elements of open-loop control technology.

OR operator (disjunction, logical addition): the OR operator is the softest condition for enabling an actuator. One sensor value in the required range is sufficient to activate the actuator.

Regardless of this classification, all controls can be hard-wired or memory-programmed, for example, in a process control system. In control with hard-wired programming the control function is achieved by means of a wiring plan. A disadvantage is that in the case of a new control task, the wiring scheme must be changed or a new switching box must be installed. If the same problem is solved with a stored-program control (SPC), the controllers and actuators remain unchanged, but the wiring scheme is replaced by a program. When the control task is changed the program is simply changed or replaced. An SPC consists of a CPU with internal inputs and outputs or with external input and output devices. The heart of the CPU is a microprocessor that runs a previously loaded program. It takes into account the signal states of the inputs and, depending on the computational operation, assigns the outputs.

2.9
Plant Safety

Long before a test plant is built, that is, already during the construction of the laboratory apparatus and at the beginning of the planning phase, the first consideration of the safety concept must be made. The first step is to gather the relevant substance data (Section 3.2). Then the necessary auxiliary substances are considered (must an AI liquid be used, or is a B classified solvent adequate?). For the transition from a laboratory to a pilot plant, a detailed safety concept must be provided and discussed with the relevant departments (head of pilot plant, safety department, etc.). At the end of process development all safety-relevant aspects must be documented. This document is the basis of the safety studies (Section 5.2.2) in the framework of the feasibility study (Section 5.1).

The hazard potential posed by a chemical plant depends mainly on the materials used (raw materials, products, auxiliaries) and their possible chemical reactions. Therefore, knowledge of properties (Section 3.2) of all handled materials (data of pure materials) and their mixtures (binary and ternary data) are indispensible for safety considerations. In addition to physicochemical data, the ecotoxicological and safety data are required.

Many safety-relevant characteristic properties are members of the family of so-called soft material data, which are often determined in standardized test apparatus (e.g., flash point). They can be used to assess the potential hazard posed by a chemical or mixture of chemicals and thus the safety measures for their handling.

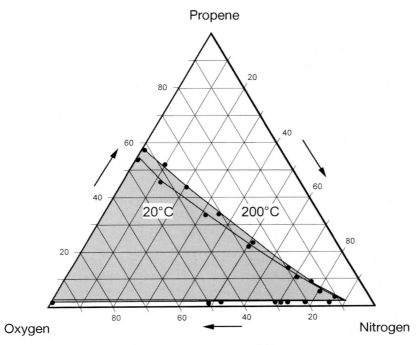

Fig. 2.9-1 Explosion diagram for propene–oxygen–nitrogen [Fehlings 1998].

Since air is ubiquitous, there is the risk of formation of ignitible mixtures when flammable substances are handled. For combustion to take place, the following prerequisites must be fulfilled:

- Flammable material (gas, vapor, dust).
- Oxygen (e.g., in the form of air or high-oxygen compounds such as peroxides, nitrates, etc.).
- Ignition source with sufficient ignition energy.
- Concentraion in the explosive range.

If one of these is not present, combustion (appearence of flames) can not occur.

Explosion diagrams (Figure 2.9.1), recorded acccording to DIN standards [Fehlings 1998, Weber 1996] give precise information on the explosion range, the explosion limits, and the limiting oxygen concentrations of gas mixtures. The limiting oxygen concentrations of hydrocarbons are usually in the range of 10–12 vol % (methane 12.1, ethane 11.0, propane, 11.4, butane 12.1, ethylene 10.0, propene 11.5). Hydrogen and carbon monoxide have values of 5.0 and 5.6 vol %, respectively.

Important characteristic quantities:

- *Ignition temperature*
  This is the lowest temperature at which an ignitible substance/air mixture just fails to spontaneously ignite (Table 2-9-1).
- *Minimum ignition energy*
  Another prerequisite for ignition is that the ignition source must have a certain minimum ignition energy. This is typically 0.01–2 mJ for gas mixtures, and 1–1000 mJ for flammable dusts. The lowest ignition energies are those for hydrogen and acetylene (ca. 0.01 mJ), and ammonia has the highest (ca. 700 mJ). Sparks from grinding (up to 100 mJ) and especially from welding (up to 10 00 mJ) generally have sufficient energy to cause a gas explosion. Even the discharge spark of an electrostatically charged person (up to 1 mJ) is generally sufficient to ignite organic/vapor air mixtures (e.g., propane 0.25 mJ). Therefore, in the design of miniplants, which are usually constructed of electrically insulating glass, measures to prevent electrostatic charging are especially important.
- *Flash point*
  The flash point is the lowest temperature of a liquid at which it generates sufficient vapor that ignition with an ignition source of higher temperature is possible. The flash point is closely related to the boiling point: gasoline < -20, ethanol 12, diesel fuel > 55, glycerol 160 °C.

The flash point is a decisive quantity in choosing safety measures for handling flammable substances. If a material is handled above its flash point, special explosion-protection measures are necessary. In Germany the handling of flammable liquids during transport, filling, and storage is governed by the *Verordnung über brennbare Flüssigkeiten* (VbF, flammable liquids regulation), which assigns them to three hazard classes according to flash point:

- *Hazard class B:* liquids with a flash point below 21 °C which are soluble in water at 15 °C (e.g., ethanol -11 °C).
- *Hazard class A:* liquids with a flash point above 100 °C which are immiscible with water. Further divided into:
  AI: flash point below 21 °C (e.g., diethyl ether).
  AII: flash point between 21 and 55 °C (e.g., 1-butanol 35 °C, cyclohexanone 43 °C).
  AIII: flash point between 55 and 100 °C (e.g., cyclohexanol 68 °C).

Tab. 2.9-1 Ignition temperatures.

Temperature class	Ignition temperature in °C	Example
T1	> 450	H_2 (560 °C), NH_3 (630 °C)
T2	> 300	1-butanol (340 °C)
T3	> 200	cyclohexane (270 °C)
T4	> 135	dimethyl ether (190 °C)
T5	> 100	
T6	> 55	carbon disulfide (95 °C)

- *Decomposition temperature*
  All organic substances and mixtures of substances can undergo decomposition in the absence of air when a ceratin temperature is exceeded. For certain substances this decomposition is exothermic. If the heat evolved is less than the heat removed, in extreme cases a thermal explosion can occur. Especially the organic, chemically undefined, high-boiling residues that are formed in many processes tend to exhibit such a behavior. Therefore, during process development appropriate investigations (DTA measurements, heat-accumulation tests under pressure) must be carried out on typical samples from the test plant.

Flammable dusts (e.g., flour) in suspended state can also form explosive mixtures with air, which due to their particularly high energy content can cause severe damage [Borho 1991]. For their safe handling, further specific safety characteristics, such as minimum ignition temperature of a dust layer, drop-hammer evaluation, smoldering point, and decomposition temperature, are necessary.

Protective measures to prevent explosions include preventive measures:

- Avoidance of explosive mixtures (e.g., by inerting with nitrogen)
- Replacing flammable by nonflammable substances
- Lowering the oxygen content to below the limiting concentration
- Avoidance of ignition sources (e.g., insulation of hot pipes).

The chosen explosion protective measures must also remain effective during startup and shutdown of the plant, as well as during disturbances. If this goal can not be achieved by these preventive measures, then constructive measures are used, for example:

- Explosion-resistant design of apparatus components
- Explosion-technological decoupling of apparatus.

It is hardly affordable to design the entire plant for the worst-case scenario. Therfore, a chemical plant is divided into different explosion-hazard zones (Table 2.9-2).

Tab. 2.9-2 Zones for explosion-hazard areas.

Flammable substance	Explosion hazard	Explosion zone	Example
Gases and vapors	• always present or present for prolonged time	0	inside evaporators, reactors, etc.
	• occasional	1	near zone 0; near filling stations or fragile apparatus
	• rare or short-term	2	areas around zone 1; flange connections in closed rooms
Dust	• prolonged or often	10	inside dryers, mills, mixers, etc.
	• occasional	11	near zone 10 apparatus

2.10
Materials Selection

The choice of a material for a plant component is not only decided by requirements for corrosion resistance, but also depends on process parameters (T, P, concentration, pH, state of flow, etc.) and product specifications. Apart from economy and good workability, there are a wide range of factors that must be taken into account in materials selection.

In the laboratory phase the question of materials is of minor imporatnce, since materials costs play no role in this stage of development. Either glass is used or a material that can be expected to resist the occurring mechanical and corrosive loading. This situation changes when an integrated miniplant is operated. At this stege it must be determined experimentally which material will be used in the technical plant. Since the investment costs are strongly dependent on the material, the goal is to find the cheapest material with the required corrosion behavior. Data on the corrosion behavior and mechanical properties of different materials as a function of temperature can be found in collections of tabulated data. Difficulties can be expected when different extreme conditions occur simultaneously. For example, when supercritical water is used as reaction medium (chemical attack, pressure > 250 bar, temperature > 400 °C [Kaul 1999]). Here an initial selection can be made on the basis of tabulated data [Dechema 1953, Hochmüller 1973]. In general, the advice of a materials specialist should be sought at as early a stage as possible.

Because of their mechanical strength, favorable corrosion behavior, and low price, materials with iron as the main alloy component (i.e., steels) are of overwhelming importance in the construction of chemical plant. Figure 2.10-1 gives an overview of the most important materials for the construction of chemical apparatus.

The most reliable tests are those in which a material sample is exposed to conditions of attack expected in the process. The test must often be carried out in a time that is a

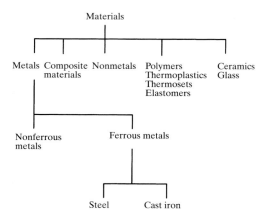

Fig. 2.10-1 Overview of the most important materials for the construction of chemical apparatus.

fraction of the service time of the material. The tests can be accelerated by using harsher conditions and by valuating less strongly attacked samples. However, in these cases only relative statements can be made.

For the final choice of material in the case of extreme operating conditions or corrosive or erosive media, plant components that are particularly at risk are constructed from the material in question and installed in the plant, monitored during operation of the test plant, and, when the test is finished, subjected to destructive testing. These are the most meaningful tests, since the conditions are closest to those of the industrial plant. Material samples that are to be tested for stress corrosion cracking are best installed in the miniplant under mechanical stress (material samples with welded seams).

These results can not always be transferred to larger plants. This is to be expected when they depend on the hydrodynamic conditions, which is the case, for example, when the rate of removal of the material increases with increasing flow rate [Johnstone 1957]. Hence, the rate of removal of the material is plotted as a function of the Re number of the fluid, and thus the expected rate of erosion in the industrial plant is obtained from the Re number of the fluid in the industrial plant. Here it is of no importance whether the material is removed by corrosion, or whether simultaneous erosion continually removes a protective layer from the surface.

With electrolytes differences in concentration or velocity at different locations in an apparatus can lead to electrochemical attack. Since the length of the field lines in the electrolyte increase with increasing size of the apparatus, while the current intensity for a given potential decreases, this type of corrosion is generally of lesser importance in a technical plant than in a model. However, if the length of the field lines is determined by the crystallite size and the mass-transfer resistance of the corroding media lies in the material itself, as in the case of stress corrosion cracking, then neither a dependence on the flow velocity nor on the plant size is to expected. Similar relationships apply when the composition of the material slowly changes due to diffusion of components. Examples are the dissolution of hydrogen in steel and the enrichment or depletion of carbon. In these cases, too, dependence on the scale of the model are not expected.

In the selection of materials it must not be forgotten that not only can the material be damaged by the medium, but the medium can also be influenced by the wall material. Surface catalysis by the wall material or homogeneous catalysis by dissolved alloy components can often not be excluded. Since both surface reactions and dissolution of wall material in chemical media increase with increasing surface area, their influence decreases in approximately inverse proportion to the apparatus volume.

2.10.1
Important Materials and their Properties

The most important materials of the chemical industry are listed in Table 2.10-1 [see also Ullmann 1992].

The choice of a suitable material is a matter of optimization between material and processing costs on the one hand, and minimum requirements with regard to mechanical properties (e.g., tensile strength), thermal stability, and corrosion resistance on the other.

Tab. 2.10-1 Materials for chemical apparatus.

Material	Example of use
1. Ferrous metals	
Soft iron	gasket material
Gray cast iron	enamelled apparatus, machine components
Cast steel	more highly loade machine components
Steel, low-alloy	apparatus scaffolds, pipelines
Steel, high-alloy • ferritic (12–17% Cr) • austenitic (0.07% C; 12–18% Cr; 9–26% Ni) • martensitic (0.075–0.9% C; 13–18% Cr) • heat-resistant (with Si, Al, etc.)	• pressureless apparatus subject to corrosion • apparatus subject to corrosion • heat-treated and hardened components • scaling-resistant apparatus
2. Nonferrous metals	
Aluminium	Storage vessels when contamination with iron must be avoided; apparatus for low temperatures
Copper • Bronze (75–99% Cu, rest Sn, Pb, Al, Mn, Ni, etc.) • Brass (56–90% Cu, rest Zn) • Monel (ca. 60% Cu, rest Ni, Fe)	heat-transfer apparatus • sea-water resistant apparatus • condensate tubes in case of risk • sea water resistant apparatus or corrosion cracking
• Red bronze (83–93% Cu, rest Sn, Zn, Pb) Nickel [Ameling 2000, Henker 1999, Crum 2000] • Inconel (45–76% Ni, 15–22% Cr, Al, Ti, Nb, Mo) • Hastelloy (ca. 60% Ni, rest Cr, Mo, Fe)	• Armatures, pumps corrosion-resistant pressure apparatus • heat- and scaling resistant apparatus • heat- and corrosion-resistant apparatus
Tantalum	rarely used due to high price; corrosion-resistant special components
Titanium	corrosion-resistent special apparatus (nitric acid, hot lower carboxylic acids)
Tungsten	heat exchangers for brackish water
Zirconium	highly acid resistant apparatus (heat exchangers for hot aqueous sulfuric acid [Donat 1989]

Tab. 2.10-1 Materials for chemical apparatus (continued).

Material	Example of use
3. Noble metals	
Gold	cladding of apparatus for highly concentrated mineral acids
Silver	small apparatus with high thermal conductivity
Platinum	claddings with highest corrosion resistance
4. Inorganic materials	
Glass	smaller corrosion-resistant apparatur for moderate temperatures and pressures; sight glasses
Quartz	corrosion-resistant small apparatus for higher temperatures
Stone and glass wool	insulating material
Asbestos	seals, when no alternative material available
Enamel	acid-resistant corrosion protection on metal surfaces
Graphite	corrosion-resistant heat exchangers up to ca. 200 °C
5. Plastics	
Thermoplastics • Polyvinyl chloride (PVC) • High-density polyethylene (HDPE) • Polypropylene (PP) • Polytetrafluorethylene (PTFE)	good chemical resistance and workability, low strength and stiffness, used for pipes, vessels, packings, fittings
Glass fiber reinforced thermosets • Phenol-formaldehyde-resin (PF) • Polyesterresin (UP) • Vinyl esterresin (VP) • Epoxide (EP)	good chemical resistance, workability, and strength, low stiffness, used for pipes, pressure vessels, watchtowers, flanges
Elastomers • Ethylen Propylene, terpolymers (EPDM) • Fluororubber	good chemical resistance and high flexibility, used for hoses, compensators, fitting internals

2.10.1.1
Mechanical Properties and Thermal Stability

A material can be loaded by tension, compression, shear, bending, and torsion. Of these, the tensile strength at the operating temperature is the most important (Table 2.10-1). The tensile strength is determined in tensile tests (Figure 2.10-2). They determine the force–deformation or stress–strain behavior under quasi-uniaxial tensile loading. In these tests a standardized specimen is stretched in a tensile testing machine at a given strain rate until it breaks. The stress is given by the ratio of the force to the intial cross-sectional area, and the strain by the ratio of the change in length to the initial length. The stress at which the limit of elastic deformation (Hooke's law) is exceeded by a predetermined tolerance is known as the yield point. For most metals a permanent strain of 0.2% is generally taken as a characteristic value for the yield strength (Table 2.10-2). The maximum achievable stress is the ultimate tensile strength. For brittle materials it is equal to the fracture strength. For example, ceramics can hardly be stretched and therefore break virtually without deformation. The initial slope of the stress–strain curve gives another important material property,

Tab. 2.10-2 Important properties of materials and classes of materials. α = linear coefficient of thermal expansion, ρ = density, λ = thermal conductivity, c_p = heat capacity.

Material class or material	yield strength (100 °C)/ N mm^{-2}	Tensile strength/ N mm^{-2}	Application range/ °C	α/ 10^{-6} K^{-1}	ρ/ g cm^{-3}	λ/ W m^{-1} K^{-1}	c_p/ kJ kg^{-1} K^{-1}
Boiler plate, HII (1.0454)	225	410–500	< 450	11.1	7.85	12	0.46
Heat-resistant steel (1.4841)	230	800	< 150	17 (400 °C)7.9		14	0.5
Hardened steel (1.7707)	1020	1300	< 450	11	8.0	19.5	0.38
Steel resistant to pitting and crevice corrosion (2.4856)	400	850	< 700	12.8	8.44	9.8	0.41
V2A-steel (1.4541)	200	700	< 550	16	7.9	15	0.5
Quarz glass			< 1000	0.59	2.2	1.4	0.73
Normal glass		7–22	< 450	8.8	2.4–3.0	0.84	0.75
Teflon (PTFE)		17–28	− 200–280	99	2.2	0.24	1.05
Fluorocarbon (Viton)		20–32	− 60–200	16	1.8	0.12–0.23	1.38
Poly(vinyl chloride) (PVC)		55	− 30–60	70	1.4	0.15	1.05

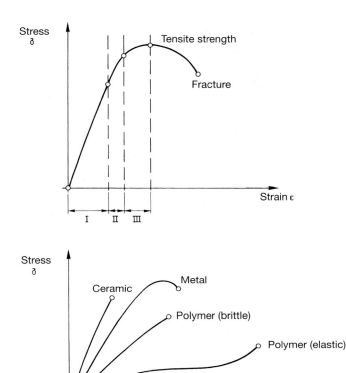

Fig. 2.10-2 Characterization of the mechanical properties of a material by measuring the stress–strain behavior in a tensile test. Stress σ = force/initial cross-sectional area of sample = $E \cdot \varepsilon$, where E is the modulus of elasticity and strain $\varepsilon = (L-L_0)/L_0$.
Region I: proportional or elastic range
Region II: elastic-plastic range
Region III: plastic range.

namely, the modulus of elasticity E, which is a measure of stiffness. A further material property is the fracture strain, that is, the elongation at breakage.

2.10.1.2
Corrosion Behavior

Corrosion is the chemical or electrochemical surface destruction of solids, and like fracture and wear it can lead to failure of the material. Corrosion can be classified as:

- Uniform surface corrosion
- Selective corrosion (e.g., intergranular corrosion)
- Localized corrosion (e.g., pitting corrosion and stress corrosion cracking).

Corrosion that occurs although all components of the apparatus are kept above the dew point of the reaction medium is purely chemical in nature. If only parts of the apparatus lie below the dew point (e.g., cold bridges on supports), electrochemical corrosion, the results of which are far more dangerous, must be expected in the condensed phase.

Uniform surface corrosion is generally the result of chemical attack by the reaction medium. It can be facilitated by erosion of a passivating layer. A well-known example is scaling of iron surfaces due to oxide formation with hot air. Studies on uniform surface corrosion require test times of 1–2 months, since the attack is more intense in the early stages. The following ranges of corrosion rates can be used for general evaluation:

- $< 2 \,\mu m \, a^{-1}$ corrosion-free
- $< 50 \,\mu m \, a^{-1}$ good
- $50 \,\mu m \, a^{-1}$ to $0.5 \, mm \, a^{-1}$ satisfactory
- $0.5–1 \, mm \, a^{-1}$ barely acceptable
- $> 1 \, mm \, a^{-1}$ unacceptable.

Selective and localized forms of corrosion are more dangerous than uniform surface corrosion, because they are unpredictable and are often only recognized after the damage has been done. They are caused by electrochemical attack by the medium, that is, electron transport to a reduction occuring at another site (formation of a local element). Metallic materials are attacked in this way when the attacking medium is liquid and has a certain minimum conductivity. The required conductivity can be very low, and the conductivity of "pure" water is already sufficient. The majority of corrosion reactions of metals can be classified as oxygen corrosion and acid corrosion (Figure 2.10-3).

Pitting corrosion often occurs in the presence of halide ions at positions where local elements can form. In alloys local elements can form between the (less noble) intergranular material and the granules and can lead to dissolution of the intergranular material starting at the surface (grain disintegration).

In stress corrosion cracking, the material breaks as the result of mechanical stress under the influence of a corrosive medium. Stress corrosion cracking is characterized by the presence of deep intergranular or intercrystalline cracks that are generally not externally evident. It can be caused by inherent stress, which can be due to cold working or arise near a welding seam.

Because of the importance of choosing the right material, it is generally unavoidable to carry out special laboratory corrosion tests [Corbett 1995] during process development. Simple and reproducible test conditions are especially important here.

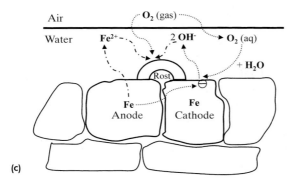

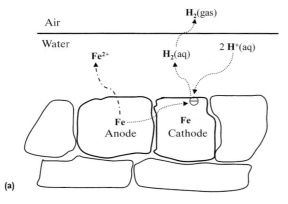

Fig. 2.10-3 Mechanism of:
a) Oxygen corrosion:
Anodic process: Me → Me^{2+} +2 e$^-$
Cathodic process, neutral/alkaline: 0.5 O_2 + H_2O + 2 e$^-$ → 2 OH$^-$
Cathodic process, acid: 0.5 O_2 + 2 H$^+$ + 2 e$^-$ → H_2O
b) Acid corrosion:
Anodic process: Me → Me^{2+} +2 e$^-$
Cathodic process: 2 H$^+$ + 2 e$^-$ → H_2

Chemical corrosion tests

In the case of uniform surface corrosion gravimetric measurements [DIN 50918] can be used to determine the linear specific rate of change of mass χ (corrosion rate, Figure 2.10-4):

$$\chi_{lin}(T,\ P,\ medium)/(\text{mg m}^{-2}\ \text{a}^{-1}) = \frac{mass\ change\ of\ probe}{area\ of\ probe\ \times\ time\ of\ action} \quad (2.10-1)$$

or the linear removal rate δ:

$$\delta_{lin}(T,\ P,\ medium)/(\mathrm{mm\ a^{-1}}) = \frac{\text{thickness decrease}}{\text{time of action}}$$

$$= \frac{\text{Mass decrease}}{\text{density of probe} \times \text{time of action} \times \text{area of probe}} \quad (2.10-2)$$

Electrochemical corrosion tests

With the aid of electrochemical tests [DIN 50918] quantitative statements can be made about the corrosion mechanism. The experimental determination of the current–potential relationship provides the most important information. It can be used to determine the limiting potentials for the occurrence of, for example, pitting corrosion and stress corrosion cracking.

Corrosion protection

An important corrosion-protection measure for metallic materials is the application of metallic, inorganic, or organic coatings. Metallic coatings can be applied by:

- Electroplating: electrochemical deposition of Zn, Cr, Ni, Ag, or Au.
- Cladding: joining of two metal plates, for example, explosive cladding of boiler plate HII with zirconium [Donat 1989].
- Melt dipping: for example, dipping pretreated steel in molten zinc.

Examples of methods for applying inorganic coating follow:

- Enamelling: A strongly adhesive glassy inorganic coating is applied to the metal surface by dipping or spraying. Enamel coatings are chemically resistant (exception: hot alkali), hard, and heat-resistant.

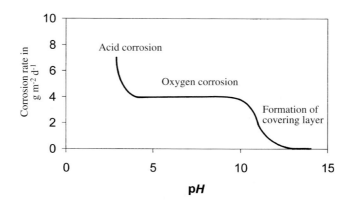

Fig. 2.10-4 Corrosion rate as function of pH for unalloyed steel in air-saturated water [Trostmann 2001].

- In phosphating or chromating the metal surface is treated with phosphoric acid, acidic phosphate solution, or acidic chromate solution by dipping or spraying at elevated temperature to give adherent phosphate or chromate layers.
- In anodization adherent protective oxide layers are produced by anodic oxidation of aluminum.

In chemical plant, the protection of metallic materials with organic coatings (paints, powder coatings, laminate coatings, plastic and rubber linings) is maninly restricted to the protection of supporting structures, and storage vessels. Because of the flammability of such protective layers and the further problems that this can cause (e.g., soot formation), their use in technical plants should be considered with care.

In addition to the above-mentioned passive protection with coatings, active corrosion-protection measures, which intervene in the corrosion reaction, are also available. Examples are:

- Electrochemical corrosion protection (cathodic corrosion protection, galvanic anodes, anodic polarization).
- Addition of corrosion inhibitors to the attacking medium.

2.10.2
Metallic Materials

Of overwhelming importance is the use of metallic materials, especially steels [Class 1982], which have a favorable combination of mechanical strength, thermal stability, corrosion behavior, and low cost. They can be roughly classified as unalloyed, low-alloy, and high-alloy steels [Heinke 1997].

Unalloyed steels contain essentially only carbon as alloying element, and the carbon content has a major influence on the properties (Table 2.10-3). The carbon is present in the form of iron carbide Fe_3C which, together with the iron matrix, forms the structural material steel.

Tab. 2.10-3 Influence of carbon content on the properties of unalloyed stells.

Carbon content	Properties	Material
3–4	brittle, hard, not forgeable	gray cast iron
0.5–1.7	hardenable, hard, difficult to weld	tool steel
0.05–0.5	dease, hard, readily weldable	- construction steels (St37, St 52, St 35.8) - boiler plate

Low-alloy steels contain less than 5% alloy components such as Cr, Co, Mn, Ni, Si, W, etc. and are used in construction and for steel vessels.

High-alloy steels (> 5% alloy components) [Henker 1999] are stainless and acid-resistant and are therefore the most important materials in chemical plant construction. Pipelines and apparatus made of these materials are resistant to all types of weather and thus do not require protective coatings. Two widely used steel types, developed by Krupp in 1914, are V2A and V4A:

V2A steel:
DIN material number[*]: 1.4541
Short name X 10 CrNiTi 18 9 [DIN 17440]
Composition: 0.08% C, 17–19% Cr, 9–12% Ni, 0.8% Ti.

V4A steel:
DIN material number: 1.4571
Short name X 10 CrNiMoTi 18 10 [DIN 17440]
Composition: 0.08% C, 16.5–18.5% Cr, 11–14% Ni, 2–2.50% Mo, 0,4% Ti.

Because of their high price relative to steel, nonferrous metal alloys are only used in special cases [Donat 1989].

2.10.3
Nonmetallic Materials

Compared to steel, other materials are of nimor importance. At low temperatures (below ca. 100 °C), organic materials are increasingly being used, and at the highest temperatures exclusively inorganic materials are used, usually as composite materials

Tab. 2.10-4 Price factors of metallic materials (carbon steel = 1) [Wilson 1971].

Material	Factor
Carbon steel	1.00
Aluminum	1.3
V2A	2.6
V4A	3.0
Titanium	8.0
Hastelloy C	14
Tantalum	55
Zirconium [Yau 1995]	> 200

[*] The first digit indicates the main group of materials (e.g. 1 for steel). The last four digites indicate the type number.

in which the components fulfill different functions (e.g., one component takes up mechanical stress, while the other is responsible for chemical and/or thermal resistance). Thus, combustion furnaces consist of heat-resistant ceramic on the inside and an outer steel mantle that provides mechanical strength; enamel (sintered glass) is resistant to strong acids, and the steel mantle absorbs mechanical forces. Cladding of steel can be used to manufacture composite materials that have the same surface properties as the cladding material but are much cheaper. An example of a metal-free composite material is glass fiber reinforced polyester resin, from which large self-supporting vessels can be constructed.

Tab. 2.10-5 Chemical resistance of important plastics in chemical plant construction.

	Thermoplastic				Thermosets			Elastomeres	
	PVC	HDPE	PP	PTFE	GF-UP	GF-EP	PF	EPDM	FPM
Water, cold	+	+	+	+	+	+	+	+	+
Water, hot	o	+	+	+	o	o	o	+	+
Weak acids	+	+	+	+	o	+	o	+	+
Strong acids	+	+	o	+	–	–	–	+	+
Oxidicing acids	o	–	–	+	–	–	–	–	+
Hydrofluoric acid	o	o	o	+	–	o	–	–	+
Weak alkalis	+	+	+	+	–	+	+	+	+
Strong alkalis	+	+	+	+	–	o	–	+	+
Inorganic salt solutions	+	+	+	+	+	+	+	+	+
Aliphatic hydrocarbons	+	+	+	+	+	+	+	–	+
Chlorinated hydrocarbons	–	o	–	+	–	o	o	o	o
Alcohols	+	+	+	+	+	o	+	+	+
Esters	–	+	o	+	o	o	o	o/–	o/–
Ketones	–	+	o	+	–	o	+	o/–	o/–
Ethers	–	o	–	+	o	+	+	–	–
Organic acids	o	+	o	+	–	+	o		o
Aromatic hydrocarbons	–	o	o	+	+	+	o	–	o
Fuels	–	+	o	+	+	+	o	–	+
Fats, oils	+	+	+	+	+	+	+	–	+
Unsaturated chlorinated hydrocarbons	–	–	–	+	–	–	–	–	–

+ resistant, o limited resistance, – not resistant

GF-UP glass fiber reinforced polyester resin
PVC Poly(vinyl chloride)
GF-EP glass fiber reinforced epoxy resin
PE-HD high-density polyethylene
PF Phenol-Formaldehyde resin
PP Polypropylene
EPDM Ethylene-propylene-terpolymers
PTFE Polytetrafluorethylene
FPM Fluororubber

Inorganic materials such as graphite and glass are growing in popularity. Complete plants can be built of glass, although here the impact and stress sensitivity of the material must be taken into account. Furthermore, the size of such plants are severely limited.

Plastics are mainly used in chemical plant when high resistance to aqueous acid, alkali, or salt solution at moderate temperature is required. Table 2.10-5 lists some important plastics and their chemical resistance (see also Table 2.10-1).

3
Process Data

When the laboratory phase has been completed and before the actual process development is started, further information must be obtained since the latter stage is normally associated with high costs.

3.1
Chemical Data

The core of a chemical plant is the reactor. Its input and output decide the structure of the entire plant built up around it. Therefore, detailed knowledge of the chemical reaction must be available at the earliest possible stage. In particular, the following questions must be clarified:

- Heats of reaction (exotherm/endotherm?)
- Thermodynamic equilibria (yes/no?)
- Kinetics of the main, secondary, and side reactions
- Dependence of selectivity and conversion on the process parameters.

3.1.1
Heat of Reaction

For the choice of a suitable reactor, the heat of reaction under standard conditions $\Delta_R H^\oslash$ is, among others, of decisive importance. Since the $\Delta_R H^\oslash$ generally can not be measured directly in a calorimeter (industrially relevant reactions do not proceed with 100% yield), they must be determined via the experimentally readily accessible heats of combustion of the reactants $\Delta_C H^\oslash{}_i$.

The enthalpies of formation of the reactants i are calculated from the heats of combustion $\Delta_C H^\oslash{}_i$ by using the Hess equation (Equation 3.1.1-1) [Yaws 1988], and the reaction enthalpy $\Delta_R H^\oslash$ calculated therefrom (Equation 3.1.1-2) is then often the difference of large numbers (see also Example 3.1.1-1), so that it is subject to large errors and should be treated with caution:

$$\Delta_f H_i^\oslash = \sum_{\text{Elements}} \Delta_C H_{\text{Elements}}^\oslash - \Delta_C H_i^\oslash. \qquad (3.1.1-1)$$

$$\Delta_R H^\oslash = \sum_i v_i \cdot \Delta_f H_i^\oslash. \qquad (3.1.1-2)$$

Thus, to minimize the risk of scale-up, it is often unavoidable to perform an (expensive) investigation of the reaction in a pilot plant under adiabatic conditions, which normally can not be achieved in small laboratory reactors (this should be checked).

Process Development. From the Initial Idea to the Chemical Production Plant. G. Herbert Vogel
Copyright © 2005 WILEY-VCH Verlag GmbH & Co. KGaA, Weinheim
ISBN: 3-527-31089-4

Example 3.1.1-1

The enthalpy of formation of benzene is required, that is, the enthalpy of reaction (3.1.1-3).

$$6\,C + 3\,H_2 \rightarrow C_6H_6. \qquad (3.1.1-3)$$

From the measured or known heats of combustion of benzene and the corresponding elements, it follows that:

$$C_6H_6 + 7.5\,O_2 \rightarrow 6\,CO_2 + 3\,H_2O;\ \Delta_c H^\varnothing = -3268\ \text{kJ mol}^{-1} \qquad (3.1.1-4)$$

$$6\,CO_2 \rightarrow 6\,C + 6\,O_2;\ 6^*\Delta_f H^\varnothing = 6^*(393{,}51\ \text{kJ mol}^{-1}) \qquad (3.1.1-5)$$

$$3\,H_2O \rightarrow 3\,H_2 + 1.5\,O_2;\ 3^*\Delta_f H^\varnothing = 3^*(285.83\ \text{kJ mol}^{-1}). \qquad (3.1.1-6)$$

Addition of Equations (3.1.1-4)–(3.1.1-6) then gives:

$$\Delta_f H^\varnothing(\text{benzene}) = -6 \times 393.51 - 3 \times 285.83 + 32868 = 49.5\ \text{kJ mol}^{-1}$$

Table 3.1.1-1 lists reaction enthalpies for some industrially important types of reaction to give some idea of the order of magnitude.

3.1.2
Thermodynamic Equilibrium

Many reactions, such as esterifications, hydrogenations, and dimerizations, proceed to chemical equilibrium. The thermodynamic equilibrium provides information about the maximum possible conversion, which can not be exceeded even with the best catalysts or kinetic tricks. To assess the potential conversion it is essential to know how far the intended reactions are from chemical equilibrium (see Example 3.1.1-2).

Example 3.1.2-1

The formation of trioxane from formaldehyde:

$$3\,CH_2O \leftrightarrow C_3H_6O_3 \qquad (3.1.2-1)$$
$$(FA) \qquad (Tri)$$

in the gas phase is a typical equilibrium reaction.
The equilibrium constant, which is given by:

$$K_p = \frac{P(Tri)}{P(FA)^3} = \frac{4 \cdot (^0P(FA) - P)}{(3P - {}^0P(FA))^3} \qquad (3.1.2-2)$$

Tab. 3.1.1-1 Reaction enthalpies for some technically important classes of reaction [Weissermel 2003] (see also Appendix 8.12).

Hydrogenation	$\Delta_R H / \text{kJ mol}^{-1}$
$CH_2 = CH_2 + H_2 \rightarrow CH_3 - CH_3$	−137
Oxidation	
$H_2 + \tfrac{1}{2} O_2 \rightarrow H_2O$	−285
$C + \tfrac{1}{2} O_2 \rightarrow CO$	−111
$C + O_2 \rightarrow CO_2$	−393
$CH_2 = CH_2 + \tfrac{1}{2} O_2 \rightarrow$ Ethylenoxid	−105
$CH_3 - CH = CH_2 + 4/2 O_2 \rightarrow 3 CO_2 + 3 H_2O$	−1920
$CH_3 - CH = CH_2 + O_2 \rightarrow CH_2 = CH - CHO + H_2O$	−340
$CH_2 = CH - CHO + \tfrac{1}{2} O_2 \rightarrow CH_2 = CH - COOH$	−250
$CH_3 CH_2 CH_3 + \tfrac{1}{2} O_2 \rightarrow CH_3 - CH = CH_2 + H_2O$	−122 at 500 °C [Watzenberger 1999]
Hydration	
$CH_2 = CH_2 + H_2O \rightarrow CH_3 CH_2 OH$	−46
$CH_3 - CH = CH_2 + H_2O \rightarrow CH_3 CH(OH) CH_3$	−50
Polymerization	
$CH_2 = CH_2 \rightarrow CH_2 - CH_2-$	−90
Neutralization	
$H^+ + OH^- \rightarrow H_2O$	−55
$NH_3(aq) + HNO_3(aq) \rightarrow NH_4 NO_3 (aq)$	−50
Chlorination	
$CH_2 = CH_2 + Cl_2 \rightarrow Cl - CH_2 CH_2 - Cl$	−180

where

P	= total pressure at equilibrium
$^0 P(FA)$	= initial partial pressure of formaldehyde
$P(Tri)$, $P(FA)$	= equilibrium partial pressure of trioxane and formaldehyde, respectively, can be expressed [Busfield 1969] as:

$$\lg(K_P / \text{bar}^{-2}) = \frac{7350}{T/K} - 19.8. \qquad (3.1.2-3)$$

It follows that if the process conditions assumed are $T = 353$ K and $^0P(FA) = 0.2$ bar, a maximum formaldehyde conversion of:

$$U = \frac{3 \cdot P(Tri)}{P(FA)} \cdot 100 = 35\ \% \qquad (3.1.2-4)$$

can be achieved in one throughput. The conversion actually achieved will be less than 35% and depends on the catalyst system and reactor type.

Chemical equilibrium is characterized by the fact that the free enthalpy of reaction is equal to zero (Equation 3.1.2-5):

$$\Delta_R G = 0 = \Delta_R G^0 + RT \cdot \ln K(T). \qquad (3.1.2-5)$$

where $\Delta_R G^0$ is the Gibbs[1] standard reaction enthalpy for the standard state of $P^0 = 1.013$ bar and the temperature T. With the aid of tabulated molar Gibbs standard enthalpies of formation $\Delta_f G^\varnothing{}_i$ for the reactants A_i, for the general reaction equation (Equation 3.1.2-6):

$$\sum_i v_i \cdot A_i = 0 \qquad (3.1.2-6)$$

v_i = stoichiometric coefficients (products +, starting materials −)

the Gibbs standard reaction enthalpy ($T^0 = 298$ K, $P^0 = 1.013$ bar; Equation 3.1.2-7):

$$\Delta_R G^\varnothing = \sum_i v_i \cdot \Delta_f G_i^\varnothing \qquad (3.1.2-7)$$

and hence the thermodynamic equilibrium constant K (Equation 3.1.2-8) can be calculated:

$$\ln K(T_0) = -\frac{\Delta_R G^\varnothing}{RT_0} = \sum_i \ln \left(\frac{f_i}{P^0}\right)^{v_i}. \qquad (3.2.1-8)$$

f_i = fugacity
P^0 = standard pressure (1.013 bar)

The fugacity f_i can be regarded as a corrected pressure, and it is only equal to the partial pressure for ideal gases, for which K is then given by Equation (3.1.2-9):

$$K_P = \Pi_i \left(\frac{P_i}{P^0}\right)^{v_i} \qquad (3.1.2-9)$$

where P_i is the partial pressure of the reactants. P_i and f_i are related by the fugacity coefficient $\phi(T,P)$ ($f_i = \phi \cdot P_i$), which, given knowledge of the critical data (T_K, P_K), can be estimated from Figure 3.1.2-1 [Atkins 2002].

---

1) Josian W. Gibbs, American physicist (1839–1903).

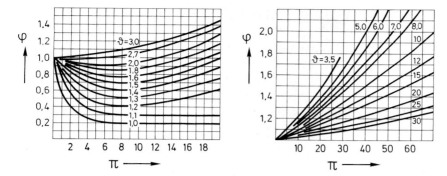

Fig. 3.1.2-1 Fugacity coefficients ϕ of a pure van der Waals gas as a function of $\Pi = P/P_K$ (abcissa) and $\vartheta = T/T_K$ (curve parameter). Example: 400 mol N_2 (T_K = 126 K, P_K = 34 bar) at 100°C in a 20 L vessel: P_{ideal} = 620 bar; from Fig. 3.1.2-1 ϕ is 1.4; from this, f_{N2} = 870 bar follows.

Owing to the exponential dependence of the equilibrium constant on the Gibbs standard enthalpy of reaction, errors in $\Delta_R G^\varnothing$ are multiplied exponentially.

$\Delta_R G^\varnothing$ values are obtained from the standard enthalpy of reaction $\Delta_R H^\varnothing$ and standard entropy of reaction $\Delta_R S^\varnothing$ according to Equation (3.1.2-10):

$$\Delta_R G^\varnothing = \Delta_R H^\varnothing - T_0 \cdot \Delta_R S^\varnothing. \qquad (3.1.2-10)$$

The $\Delta_R H^\varnothing$ values of many compounds can be calculated from tabulated standard enthalpies of formation $\Delta_f H^\varnothing{}_i$ [Landolt-Börstein] according to Equation (3.1.2-11):

$$\Delta_R H^\varnothing = \sum_i v_i \cdot \Delta_f H_i^\varnothing. \qquad (3.1.2-11)$$

If $\Delta_f H^\varnothing{}_i$ values that are not available in the literature can be calculated from the heats of combustion $\Delta_C H^\varnothing$, which are easily determined experimentally in a bomb calorimeter (Equation 3.1.1-1).

The situation is more difficult for the $\Delta_R S^\varnothing{}_i$ values. These can be calculated according to Equation (3.2.1-12) from the standard entropies of the reactants $S^\varnothing{}_i$, which, however, are often not available and can only be determined with major experimental effort form the heat capacities $c_p(T)$ and the heats of transformation $\Delta_U H$ by the schematic integration scheme of Equation (3.1.2-13):

$$\Delta_R S^\varnothing = \sum_i v_i \cdot S_i^\varnothing \qquad (3.1.2-12)$$

$$S_i^\varnothing = S(0) + \int_0^{T_1} \frac{C_{P1}(T)}{T} \cdot dT + \frac{\Delta_U H_1}{T_1} + \int_{T_1}^{\cdots} \cdots + \int_{\cdots}^{298.2} \frac{c_{Pk}(T)}{T} \cdot dT. \qquad (3.1.2-13)$$

This can be avoided by using methods of approximation (e.g., group-contribution methods) [Knapp 1987].

By using the van't Hoff[2] equation (Equation 3.1.2-14):

$$\left.\frac{\partial \ln K}{\partial T}\right|_p = \frac{\Delta_R H(T)}{RT^2} \qquad (3.1.2-14)$$

the equilibrium constants at 25°C determined with Equation (3.1.2-8) [Gmehling 1992] can be calculated for a desired temperature (Equation 3.1.2-15):

$$\ln K(T) = \ln K(T_0) + \frac{\Delta_R H^{\varnothing}}{R}\left(\frac{1}{T_0} - \frac{1}{T}\right), \qquad (3.1.2-15)$$

by assuming as a first approximation that $\Delta_R H^{\varnothing}$ is temperature-independent. If this assumption is not justified, then this influence must be taken into account with the aid of the molar heat capacities of the reactants $c_{pi}(T)$ (Equation 3.2.1-16):

$$\Delta_R H(T) = \Delta_R H^{\varnothing} + \int_{T_0}^{T} \sum_i v_i \cdot c_{pi}(T') \cdot dT'. \qquad (3.1.2-16)$$

If, as a first approximation, $\sum_i v_i \cdot c_{pi} = \Delta_R c_p(T_0) = \Delta_R c_p^{\varnothing}$ is temperature-independent, Equation (3.1.2-17) applies:

$$\frac{\partial \ln K(T)}{\partial T} = \frac{\Delta_R H^{\varnothing}(T_0) + \Delta_R c_p^{\varnothing} \cdot (T - T_0)}{RT^2} \qquad (3.1.2-17)$$

or, in integrated form, Equation (3.1.2-18):

$$\ln \frac{K(T)}{K(T_0)} = \frac{\Delta_R H^{\varnothing}}{R} \cdot \left(\frac{1}{T_0} - \frac{1}{T}\right) + \frac{\Delta_R c_p^{\varnothing}}{R} \cdot \left\{\ln \frac{T}{T_0} + \frac{T_0}{T} - 1\right\}. \qquad (3.1.2-18)$$

Example 3.1.2-2

The equilibrium constant of the water gas shift reaction ($CO + H_2O \leftrightarrow CO_2 + H_2$) *at 1000 K is required. Values of H, S, and c_p of* −41.20 kJ mol^{-1}, −42.10 J mol^{-1} K^{-1}, *and* 3.2 J mol^{-1} K^{-1}, *respectively, are obtained from Table 3.1.2-1. Hence it follows:*
from Equation (3.1.2-10): $\Delta_R G^{\varnothing}$ = −41.20 − 298(−0.0421) = −28.65 kJ mol^{-1}
from Equation (3.1.2-8): ln K(298 K) = (28.65 × 1000)/(8.313 × 298) = 11.559.
Under the assumption that $\Delta_R H^{\varnothing}$ is temperature-independent, Equation (3.1.2-15) gives K(1000 K) = 0.89.
Under the assumption that $\Delta_R c_p$ is temperature-independent, Equation (3.1.2-18) gives K(1000 K) = 1.1.

---

2) Jacobus Henricus van't Hoff (1852–1911).

Tab. 3.1.2-1 Thermodynamic quantities for some important basic and intermediate products [Fratscher 1993, Fedtke 1996, Atkins 2002, Sandler 1999] (see also Appendix 8.11.

Stoff	$\Delta_f H^i$ / kJ mol^{-1}	S^i / kJ mol^{-1} K^{-1}	$\Delta_f G^i$ / kJ mol^{-1}	$\Delta^c H^i$ / kJ mol^{-1} starting material at 25 °C H$_2$O(l), CO$_2$(g)	c_p / kJ mol^{-1} K^{-1}
H$_2$ (g)	0	130.7	0	285.8	28.8
N$_2$ (g)	0	191.6	0	/	29.1
O$_2$ (g)	0	205.0	0	/	29.4
H$_2$O (g)	$-$241.8	188.8	$-$228.6	/	33.6
H$_2$O (l)	$-$285.8	69.91	$-$237.1	/	
C (Graphit)	0	5.74	0	393.5	
CO (g)	$-$110.5	197.7	$-$137.2	283.0	29.2
CO$_2$ (g)	$-$393.5	213.7	$-$394.4	/	37.1
CH$_4$ (g)	$-$74.8	186.3	$-$50.7	890	35.8
CH$_3$OH (l)	$-$200.7	239.8	$-$162.0		
CH$_2$O (g)	$-$108.6	218.8	$-$102.5		35.3
HCOOH (l)	$-$424.7	129.0	$-$361.4		
CH$_3$CH$_3$ (g)	$-$84.7	229.6	$-$32.8	1560	52.7
CH$_2$=CH$_2$ (g)	52.3	219.6	68.15		43.6
CH$_3$CH$_2$OH (l)	$-$277.7	160.8	$-$174.8		
CH$_3$CHO (l)	$-$192.3	160.2	$-$128.1		
CH$_3$COOH (l)	$-$484.5	159.8	$-$389.9		
CH$_3$CH$_2$CH$_3$	$-$103.8	269.9	$-$23.5	2220	63.9
CH$_3$-CH=CH$_2$	20.4	267.0	62.8		
C$_6$H$_6$	82.9		129.7	3268	
C$_6$H$_5$-CH$_3$	12.2		113.6	3910	

The thermodynamic data (standard enthalpy and entropy of formation, specific heats of reactants) can be taken from collections of tabulated data or estimated by methods of approximation [Ullmann, Knapp 1987].

3.1.3
Kinetics

In order to specify the type of reactor to be used later, information must be available on the potential reaction routes to the main and secondary products and the byproducts. The rate of formation and its dependence on process parameters such as temperature, pressure, and catalyst concentration should, if possible, be known quantitatively. The elucidation of the reaction mechanism must be dispensed with in the early stages of process development for reasons of time. The necessary detailed kinetic investigations for determining the microkinetics (purely chemical kinetics without transport limitation by external mass transfer such as convection or diffusion; see also Section 2.1.3) [Forzatti 1997] are often only carried out when the industrial plant is already on-stream.

In the framework of process development it is often only possible to determine the macrokinetics (chemical kinetics superimposed by external mass- and heat-transfer processes). In this case it must be ensured that the laboratory reactor is hydrodynamically similar to the later industrial reactor (especially the length to diameter ratio), so that the transport influences are approximately the same in both. This is particularly easy to achieve in the case of tube-bundle reactors, as are often used for partial oxidations (e.g., production of phthalic acid, acrylic acid, and ethylene oxide). Here the macrokinetics can be determined in a single tube, since the subsequent hydrodynamic conditions are identical (scale-up factor = 1).

For kinetic investigations various test reactors such as differential, differential circulating (continuous stirred tank [Reisener 2000]), and integral reactors (tubular reactor [Hofe 1998], batch stirred tank) are available (Figure 3.1.3-1) [Forni 1997].

The absolute reaction rate for a general reaction $v_A A \rightarrow v_P P$ is defined as the temporal change in the number of moles of the of the starting material A or the product relative to the stoichiometric coefficients v_A and v_P (starting material negative, product positive; Equation (3.1.3-1).

$$r = \frac{+1}{v_A} \cdot \frac{dn_A}{dt} = \frac{1}{v_P} \cdot \frac{dn_P}{dt}. \qquad (3.1.3-1a)$$

$$r = \frac{+1}{v_A} \cdot r_A = \frac{1}{v_P} \cdot r_P. \qquad (3.1.3-1b)$$

Replacing the number of moles n_A by the molar concentration of the starting material A ($c_A = n_A/V = A$ mol L^{-1}) gives Equation (3.1.3-2):

$$r = \frac{+1}{v_A} \cdot \frac{d(V \cdot A)}{dt} = \frac{1}{v_A} \left[V \cdot \dot{A} + A \cdot \dot{V} \right]. \qquad (3.1.3-2)$$

Assuming a constant-volume reaction ($\dot{V} = 0$), the volumetric reaction rate is expressed as Equation (3.1.3-3a):

$$r_V = \frac{r}{V} = \frac{1}{v_A} \cdot \frac{dA}{dt}. \qquad (3.1.3-3a)$$

For heterogeneous reaction systems it is appropriate to express the reaction rate relative to the surface area or mass of the catalyst Equation (3.1.3-3b):

a) Differentialreactor

$\dot{n}_A^0 = \dot{V} \cdot c_A^0 \longrightarrow \boxtimes \longrightarrow \dot{n}_A = \dot{V} \cdot c_A$

b) Differential circulating reactor

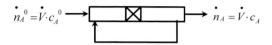

c) Integralreactor

$\dot{n}_A^0 = \dot{V} \cdot c_A^0 \longrightarrow \boxtimes \longrightarrow \dot{n}_A = \dot{V} \cdot c_A$

Fig. 3.1.3-1 Laboratory reactors for determining the kinetics of heterogeneously catalyzed reactions [Luft 1987a, Forni 1997, Cavalli 1997]:
a) Differential reactor: direct determination of the reaction rate is possible but associated with major errors if the analysis is insufficiently precise.
$r_m \approx \dfrac{1}{\nu_A} \dfrac{\dot{n}_A - \dot{n}_A^0}{m_{Kat}}$ if the conversion of A is less than about 10%.
(Caution: the difference of two large numbers is subject to large errors).
b) Differential circulating reactor: direct determination of the reaction rate is possible with high precision.
$r_m = \dfrac{1}{\nu_A} \dfrac{\dot{n}_A - \dot{n}_A^0}{m_{Kat}}$ when the recycle ratio is greater than 20.
c) Integral reactor: Only indirect determination of the reaction rate is possible. r_{Kat} must be determined from the slope of the concentration/time(position) profile and is often not unambiguously assignable to the process parameters (temperature, partial pressures, etc.). It is often difficult to ensure isothermal conditions.

$$r_m = \dfrac{r}{m_{Kat}} = \dfrac{1}{\nu_A} \cdot \dfrac{dn_A}{m_{Kat} \cdot dt}. \qquad (3.1.3 - 3b)$$

The differential circulating reactor (Figure 3.1.3-2) is especially suitable for determining the reaction rate r_V. In this type of reactor the reaction rate r_V can be directly calculated from the measured change in the concentration of component A (Equation 3.1.3-4; see also Equation 2.2-8).

$$r_V = \dfrac{1}{\nu_A} \dfrac{A_{ende} - A_0}{\tau} = \dfrac{1}{\nu_A} \cdot \dfrac{\dot{n}_{A,ende} - \dot{n}_{A,0}}{V_R} \quad \text{for homogeneous reactions}$$

$$(3.1.3 - 4a)$$

A	= concentration of component A in mol L^{-1}
τ	= residence time in reactor
V_R	= reactor volume
$\dot{n}_{A\,end}$	= moles of component A that exit the reactor per unit time in mol s^{-1}
$\dot{n}_{A\,0}$	= moles of component A that enter the reactor per unit time in mol s^{-1}

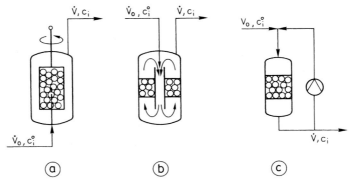

Fig. 3.1.3-2 Types of differential circulating reactors with catalyst beds [Buzzi-Ferraris 1999, Perego 1999, Forni 1997] for determining the kinetics of heterogeneously catalyzed reactions. a) Spinning-basket principle: a rotating basket containing catalyst pellets acts as a stirrer. b) Jet principle with internal recirculation [Luft 1973b, Luft 1978, Dreyer 1982]. c) External recirculation.

$$r_m = + \frac{1}{V_A} \frac{\dot{n}_{A,ende} - \dot{n}_{A,0}}{m_{Kat}} \quad \text{for heterogeneously catalyzed reactions} \quad (3.1.3-4b)$$

Thus, r_V or r_m can be unambiguously assigned to the experimental parameters (temperature, concentration, pressure), since the recycling or good mixing results in gradient-free operation [Erlwein 1998].

The integral reactor is often experimentally easier to handle and provides a faster overview of the kinetics. However, the reaction rates r_V must be derived by differentiation of the measured concentration–time curves, which is a source of errors. Furthermore, the reaction concentrations and temperatures are often not unambiguously assignable to the thus-obtained r values.

The rate data determined in the laboratory [$r_V(T)$ or A as a function of T, P] should now be fitted to suitable physical models, which may also give an indication of the reaction mechanism [Santasecaria 1999, Wang 1999]. In the first step, one checks whether the measured kinetic data can be described by a separation equation (Equation 3.1.3-5):

$$r_V = f_1(T) \cdot f_2(A_1 \ldots). \quad (3.1.3-5)$$

The T-dependent term $f_1(T)$ can often be described by a simple Arrhenius equation (Equation 3.1.3-6a) [Mezinger 1969]:

$$f_1(T) = k_A(T) = k_0 \cdot exp\left(-\frac{E_a}{RT}\right) \quad (3.1.3-6a)$$

where k_A is the rate constant, k_0 the pre-exponential factor (collision factor), and E_a the activation energy. Typical values of E_a are 50–100 kJ mol^{-1} for heterogeneously catalyzed reactions, and 200–400 kJ mol^{-1} for homogeneous gas-phase reactions. Deviations from Arrhenius behavior can be regarded as an indication of a change of mechanism (e.g., high- to low-temperature or radical to ionic mechanism). Only at very high pressures does the pressure influence reaction rate constants, via the activation volume $\Delta V_a^{\#}$ (Equation 3.1.3-6b) [Steiner 1967, Luft 1969, Luft 1989]. Typical values of $\Delta V_a^{\#}$ lie between –25 and +15 cm^3 mol^{-1}:

$$f_1(T,\ P) = k_A(T,\ P) = k_0 \cdot exp\left(-\frac{E_a + \Delta V_a^{\#} \cdot (P - P_0)}{RT}\right). \tag{3.1.3 – 6b}$$

A power law can often be used for the concentration-dependent term $f_2\ (A...)$, so that a simple expression for the reaction rate results (Equation 3.1.3-7):

$$r_V = k_A(T) \cdot A^{n_A} \cdot ..., \tag{3.1.3 – 7}$$

where n_A is the order of the reaction with respect to component A. This parameter is adjustable to the experimental data and can be obtained, for example, from a double-logarithmic plot of r_V versus A.

The above power law is widely used in industrial chemistry since it is mathematically easy to handle. However, most reactions in which a catalyst is involved have a complex reaction mechanism which can make the mathematical description of the kinetics complicated. An example of this are enzyme-catalyzed reactions according to the Michaelis–Menten mechanism (Example 3.1.3-1).

Example 3.1.3-1

Substrate A undergoes an equilibrium reaction with enzyme E to give the activated complex A*, which slowly reacts to form the product P with release of E (Equation 3.1.3-8):

$$A + E \underset{k_{-1}}{\overset{k_1}{\rightleftarrows}} A^* \overset{k_2}{\rightarrow} P + E. \tag{3.1.3 – 8}$$

Prerequisites:

- $E_0 = E + A^*$
- $A_0 = A + P + A^* \approx A + P$
- $dA^*/dt \approx 0$.

This results in the differential Equations (3.1.3-9):

$$\begin{aligned}\dot{P} &= k_2 \cdot A^* \\ \dot{A}^* &= k_1 \cdot A \cdot E - k_{-1} \cdot A^* - k_2 \cdot A^* = 0.\end{aligned} \tag{3.1.3 – 9}$$

Rearrangement gives Equation (3.1.3-10)

$$\dot{P} = -\dot{A} = k_2 \cdot E_0 \cdot \frac{A}{K_m + A}, \qquad (3.1.3-10)$$

where $K_m = (k_{-1} + k_2)/k_1$ is the Michaelis–Menten constant. Thus, the kinetics have an inhibiting term in the numerator. Depending on the boundary case these kinetics can be approximated by a first- (Equation 3.1.3-11) or zeroth-order (Equation 3.1.3-12) rate law:

$$\text{Case 1}: \; A \ll K_m \rightarrow \dot{P} = -\dot{A} = \left(\frac{k_2 \cdot E_0}{K_m}\right) \cdot A^1 \qquad (3.1.3-11)$$

$$\text{Case 2}: \; A \gg K_m \rightarrow \dot{P} = -\dot{A} = (k_2 \cdot E_0) \cdot A^0. \qquad (3.1.3-12)$$

On the basis of the elucidated kinetics the reactor can be chosen (ideal tubular reactor or continuous stirred tank (see Section 2.2.1.1).
Ideal tubular reactor (Equation 3.1.3-13):

$$0 = -u \cdot \frac{\partial A}{\partial x} - k_2 \cdot E_0 \cdot \frac{A}{K_m + A}. \qquad (3.1.3-13)$$

Integrating this differential equation and solving for the residence time τ gives Equation (3.1.3-14):

$$\tau_{\text{tube}} = \frac{1}{k_2 \cdot E_0} \cdot \left(A_0 - A - K_m \cdot \ln \frac{A}{A_0}\right). \qquad (3.1.3-14)$$

Ideal continuous stirred tank (Equation 2.2-8):

$$\dot{A} = -\frac{A_0 - A}{\tau} = -k_2 \cdot E_0 \cdot \frac{A}{K_m + A}. \qquad (3.1.3-15)$$

Solving for the residence time τ gives Equation (3.1.3-16):

$$\tau_{\text{tank}} = \frac{(A_0 - A) \cdot (K_m + A)}{k_2 \cdot E_0 \cdot A}. \qquad (3.1.3-16)$$

Since the ratio of the residence times $\tau_{\text{tube}}/\tau_{\text{tank}}$ is always smaller than one, a tubular reactor is prefereable from the viewpoint of reaction engineering (Figure 3.1.3-3).

Similar reaction kinetics to those in Example 3.1.3-1 with an inhibiting term in the numerator are also found for heterogeneously catalyzed reactions (Section 2.1.3.4) and autocatalytic reactions with an acceleration term, for example, ester hydrolysis [Krammer 1999] (see Example 3.1.3-2).

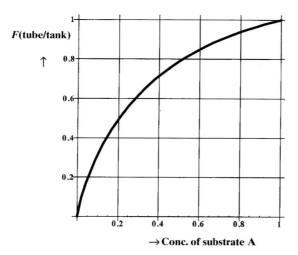

Fig. 3.1.3-3 Ratio F of the residence time of tube/tank (Equations 3.1.3-14 and 3.1.3-16) as a function of the concentration of substrate A for the example $k_1 = k_{-1} = 1$, $k_2 = 0.1$, $A_0 = 1$.

Example 3.1.3-2

The hydrolysis of ethyl acetate (E) in hot high-pressure water is a typical autocatalytic reaction, since the protons formed from the acetic acid (HOAc) catalyze the reaction. According to Krammer [Krammer 1999] the mechanism of Equation (3.1.3-17) applies:

$$E + H^+ \xrightarrow{K_2} EH^+, \quad K_2 = \frac{EH^+}{E \cdot H^+}$$

$$EH^+ + 2\,H_2O \xrightarrow{K_3} HAc + EtOH + H_3O^+ \tag{3.1.3 - 17}$$

$$HAc \xrightarrow{K_A} H^+ + Ac^-, \quad K_A = \frac{H^{+2}}{HAc}.$$

Since the rate of consumption of E is determined by the slowest step, Equation (3.1.3-18) applies to a good approximation:

$$r_V = -\frac{dE}{dt} = k_3 \cdot EH^+ \cdot H_2O^2. \tag{3.1.3 - 18}$$

Inserting the equilibrium constants (Equation 3.1.3-17) gives Equation (3.1.3-19):

$$-\frac{dE}{dt} = k_3 \cdot K_2 \cdot H_2O^2 \cdot \sqrt{K_A} \cdot E \cdot \sqrt{E_0 - E}. \tag{3.1.3 - 19}$$

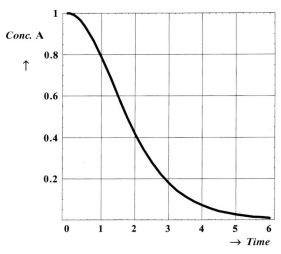

Fig. 3.1.3-4 Concentration curve for the hydrolysis of ethyl acetate in hot high-pressure water ($E_c = 1$, const. = 1) [Krammer 1989].

This differential equation can be easily solved by separation of the variables to give Equation (3.1.3-20):

$$E = E_0 \cdot \left[1 - \tanh\left\{\frac{const. \cdot \sqrt{E_0}}{2} \cdot t\right\}^2\right] \qquad (3.1.3-20)$$

where const. $= k_3 \cdot K_2 \cdot H_2O^2 \cdot \sqrt{K_A}$. *Figure 3.1.3-4 shows the typical S-shaped curve for an autocatalytic reaction.*

The situation is even more complex when not only one product but also side products are formed [Lintz 1999]. The principle for setting up the kinetic equations is described in the following:

For a general reaction according to Equation (3.1.2-6) the absolute total reaction rate is given by Equation (3.1.3-21):

$$r = \frac{1}{v_i} \cdot r_i. \qquad (3.1.3-21)$$

If the starting materials ($v_i < 0$) or products ($v_i > 0$) are involved in more than one reaction (j reactions with = 1 to N), then Equation (3.1.3-22) applies for the rate of change of the amount of substance of the ith component:

$$r_i = \sum_{j=1}^{N} v_{ij} \cdot r_{ij}, \qquad (3.1.3-22)$$

where v_{ij} is the matrix of the stoichiometric coefficients [Santacesaria 1997]. The oxidation of propylene on a mixed-oxide catalyst serves as an example of the kinetic analysis of a reaction network [König 1998].

Example 3.1.3-3

Figure 3.1.3-5 shows the assumed reaction mechanism.

There are four independent reactions ($j = 1$–4; Equations 3.1.3-23):

$$
\begin{aligned}
&1.)\ 1 \cdot C_3H_6 + 2 \cdot [O] \xrightarrow{k_{C3H6}} 1 \cdot C_3H_4O + 1 \cdot H_2O + 2 \cdot [\,] \\
&2.)\ 1 \cdot O_2 + 2 \cdot [\,] \xrightarrow{k_{O2}} 2 \cdot [O] \\
&3.)\ 1 \cdot C_3H_6 + 4{,}5 \cdot O_2 \xrightarrow{k_{1CO2}} 3 \cdot CO_2 + 3 \cdot H_2O \\
&4.)\ 1 \cdot C_3H_4O + 7 \cdot [O] \xrightarrow{k_{2CO2}} 3 \cdot CO_2 + 2 \cdot H_2O + 7 \cdot [\,]
\end{aligned}
\qquad (3.1.3-23)
$$

which can modeled by Equations (3.1.3-24) [König 1998]:

$$
\begin{aligned}
r_1 &= k_{C3H6} \cdot [O] \cdot C_3H_6 \\
r_2 &= k_{O2} \cdot (1 - [O]) \cdot O_2 \\
r_3 &= k_{1CO2} \cdot (1 - [O]) \cdot C_3H_6 \cdot O_2 \\
r_4 &= k_{2CO2} \cdot [O] \cdot \frac{C_3H_4O}{C_3H_6}
\end{aligned}
\qquad (3.1.3-24)
$$

Setting up the matrix of stoichiometric coefficients according to Equation (3.1.3-22):

j ↓	i →	C_3H_6	[O]	O_2	C_3H_4O	CO_2	H_2O	[ ]
1		−1	−2	0	1	0	1	2
2		0	2	−1	0	0	0	−2
3		−1	0	−4.5	0	3	3	0
4		0	−7	0	−1	3	2	7

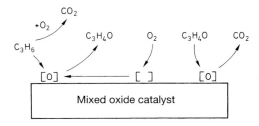

Fig. 3.1.3-5 Reaction mechanism of propylene oxidation on a mixed-oxide catalyst. [O] and [ ] denote active oxygen species and oxygen vacancies, respectively, of the mixed-oxide catalyst

from which the rate equations for all starting materials and products are obtained (Equation 3.1.3-25):

$$\dot{CO_2} = 3 \cdot r_3 + 3 \cdot r_4$$
$$\dot{O_2} = -1 \cdot r_2 - 4.5 \cdot r_3$$
$$\dot{C_3H_6} = -1 \cdot r_1 - 1 \cdot r_3$$
$$\dot{C_3H_4O} = 1 \cdot r_1 - 1 \cdot r_4.$$

(3.1.3 – 25)

This system of differential equations can only be solved numerically. The parameters (rate constants) are fitted iteratively to the concentration–time curves determined in the tubular reactor. This must be performed individually for each catalyst type. Figure 3.1.3-6 shows the simulated concentration–time curves of all reactants for a typical acrolein catalyst.

Another example is a network of parallel and sequential first-order reactions. This simple network model is often suitable for describing the technically important partial oxidation reactions.

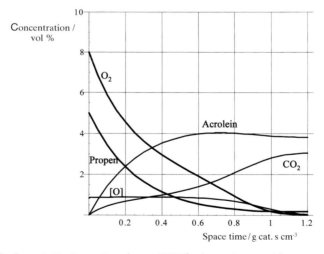

Fig. 3.1.3-6 Concentration/space time plots at 400 °C for the starting materials propene and oxygen and the products acrolein and carbon dioxide for a given set of rate constants ($k_{C3/H6}$ = 3.94 cm³ g⁻¹ s⁻¹, k_{O2} = 16.0 cm³ g⁻¹ s⁻¹, k_{1CO2} = 0.38 cm³ g⁻¹ s⁻¹, k_{2CO2} = 0.14 cm³ g⁻¹ s⁻¹), which, for the typical acrolein catalyst considered here, were individually fitted to the measurements. Instead of the residence time, the space time per unit catalyst mass [(m_{kat}/g) ÷ ($\dot{V}$/cm³ s⁻¹)] was used. This simulation was performed for a stationary degree of oxygen coverage of [O] = 0.82.

Example 3.1.3-4

The starting material A reacts not only to form product P, but also (a) to form an undesirable byproduct Y in a parallel reaction or (b) to form an undesirable secondary product X in a sequential reaction:

$$A \xrightarrow{k_1} P \xrightarrow{k_2} X$$
$$A \xrightarrow{k_3} Y \qquad (3.1.3-26)$$

a) First-order parallel reaction

For the change in concentration of A with time, expressed in the simplified form $dc_a/dt = \dot{A}$, Equation (3.1.3-27) applies:

$$\dot{A} = -(k_1 + k_3) \cdot A \qquad (3.1.3-27)$$

with $A(t=0) = A_0$, integration results in:

$$A = A_0 \cdot exp\{-(k_1 + k_3) \cdot t\}. \qquad (3.1.3-28)$$

For the useful product P:

$$\dot{P} = k_1 \cdot A. \qquad (3.1.3-29)$$

with $P(t=0) = P_0$, integration results in:

$$P = P_0 + \frac{k_1 \cdot A_0}{(k_1 + k_3)} \cdot \{1 - exp[-(k_1 + k_3) \cdot t]\}. \qquad (3.1.3-30)$$

The same applies to Y.

b) First-oder sequential reaction

This is covered by three simultaneous differential equations:

$$\begin{aligned} \dot{A} &= -k_1 \cdot A \\ \dot{P} &= k_1 \cdot A - k_2 \cdot P \\ \dot{X} &= k_2 \cdot P. \end{aligned} \qquad (3.1.3-31)$$

If $P_0 = X_0 = 0$ and $k_1 \neq k_2$, integration of these equations yields:

$$A = A_0 \cdot exp(-k_1 \cdot t) \qquad (3.1.3-32)$$

$$P = \frac{k_1 \cdot A_0}{k_1 - k_2} \cdot \{exp(-k_2 \cdot t) - exp(-k_1 \cdot t)\} \qquad (3.1.3-33)$$

$$X = A_0 \cdot \left\{1 + \frac{k_1}{k_2 - k_1} \cdot exp(-k_2 \cdot t) - \frac{k_2}{k_2 - k_1} \cdot exp(-k_1 \cdot t)\right\}. \qquad (3.1.3-34)$$

c) Combination of sequential and parallel reactions

With the system of differential equations:

$$\begin{aligned}\dot{A} &= -(k_1 + k_3) \cdot A \\ \dot{P} &= k_1 \cdot A - k_2 \cdot P \\ \dot{X} &= k_2 \cdot P \\ \dot{Y} &= k_3 \cdot A\end{aligned} \qquad (3.1.3 - 35)$$

the following solution is obtained:

$$A = A_0 \cdot exp\left\{-(k_1 + k_3) \cdot t\right\} \qquad (3.1.3 - 36)$$

$$P = \frac{k_1 \cdot A_0}{(k_1 + k_3 - k_2)} \cdot \left\{exp\left(-k_2 \cdot t\right) - exp\left(-(k_1 + k_3) \cdot t\right)\right\} \qquad (3.1.3 - 37)$$

$$X = \frac{k_1 \cdot A_0}{(k_1 + k_3)} \cdot \left\{1 - \frac{exp\left(-k_2 \cdot t\right)}{k_1 + k_3 - k_2} \cdot \left[k_1 + k_3 - k_2 \cdot exp\left(-(k_1 + k_3 - k_2) \cdot t\right)\right]\right\} \qquad (3.1.3 - 38)$$

$$Y = \frac{k_3 \cdot A_0}{k_1 + k_3} \cdot \left\{1 - exp\left(-(k_1 + k_3) \cdot t\right)\right\}. \qquad (3.1.3 - 39)$$

Figure 3.1.3-7 shows the course of the reactant concentration with time for the case of a simple sequential and parallel reaction (Equation 3.1.3-26).

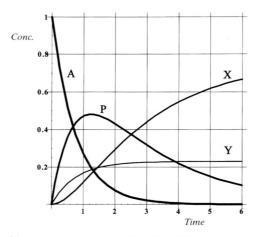

Fig. 3.1.3-7 Course of the reactant concentrations A, P, X, and Y for simple sequential and parallel reactions (Equations 3.1.3-36 to 3.1.3-39) with $k_1 = 1$, $k_2 = 0.4$, $k_3 = 0.3$, $A_0 = 1$.

3.1.4
Selectivity and Conversion as a Function of the Process Parameters

With the aid of the kinetics, dimensioning of the reactor can be performed. Other important quantities that influence the economics of the entire process are the selectivity and the conversion. If the kinetics are known, both quantities can be optimized and thus the yield (= selectivity × conversion) maximized. First we must define these quantities. Consider the reaction of starting materials A and B to give product P (Equation 3.1.4-1):

$$v_A \, A + B \text{ (in stoichiometrie excess)} \rightarrow v_P \, P \quad (3.1.4-1)$$

the conversion U_A is defined as the converted amount of substance of A (which is present in substoichiometric quantity) relative to the initial amount (Equation 3.1.4-2):

$$U_A = \frac{n_A^0 - n_A}{n_A^0} \text{ (batch) or } U_A = \frac{\dot{n}_A^0 - \dot{n}_A}{\dot{n}_A^0} \text{ (conti.).} \quad (3.1.4-2)$$

The integral selectivity $^I S_P$ with respect to P (often simply called the selectivity S) is defined as the formed amount of substance P relative to the converted amount of starting material A divided by the stoichiometric coefficients (Equation 3.1.4-3).

$$^I S_P = \frac{n_P / v_P}{(n_A^0 - n_A)/v_A}. \quad (3.1.4-3)$$

It is dependent on process conditions such as temperature, pressure, concentration, the nature of the catalyst, and the conversion (Equation 3.1.4-4):

$$^I S_P \{(T, P, conc.), (catalyst), (conversion)\}. \quad (3.1.4-4)$$

The differential (instantaneous) selectivity is defined through the ratio of the rate of formation of P and the consumption of A (Equation 3.1.4-5):

$$^D S_P = \frac{r_P}{r_A}. \quad (3.1.4-5)$$

The two selectivities can be interconverted (Equation 3.1.4-6):

$$^I S_P = \frac{1}{U_e} \cdot \int_0^{U_e} {}^D S_P(U) \cdot dU, \quad (3.1.4-6)$$

where U_e is the final conversion achieved in an integral reactor. Experiments in integral reactors provide $^I S_P$ values, and those in continuous stirred tanks $^D S_P$ values. This is now shown for a simple reaction network of sequential and parallel reactions.

Example 3.1.4-1

For the following triangular first-order scheme:

$$A \xrightarrow{k_1} P \xrightarrow{k_2} X$$
$$A \xrightarrow{k_3} X \qquad (3.1.4-7)$$

the differential equations in simplified for ($c_A = A$, dc_A, etc.) read:

$$\dot{A} = -k_1 \cdot A - k_3 \cdot A$$
$$\dot{P} = +k_1 \cdot A - k_2 \cdot P \qquad (3.1.4-8)$$
$$\dot{X} = +k_2 \cdot P + k_3 \cdot A$$

with the solutions:

$$A = A^0 \cdot \exp\{-(k_1 + k_3) \cdot t\} \qquad (3.1.4-9)$$

$$P = \frac{k_1 \cdot A_0}{(k_1 + k_3 - k_2)} \cdot \{\exp(-k_2 \cdot t) - \exp(-(k_1 + k_3] \cdot t)\} \qquad (3.1.4-10)$$

$$X = \frac{A_0}{(k_1 + k_3 - k_2)} \cdot \{k_1 \cdot \exp(-k_2 \cdot t) + (k_3 - k_2) \cdot \exp(-(k_1 + k_3) \cdot t)\}$$
$$(3.1.4-11)$$

Figure 3.1.4-1 shows the course of the reactant concentrations.

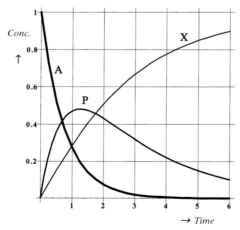

Fig. 3.1.4-1 Course of the reactant concentrations A, P, and X for a simple reaction network (Equations 3.1.4-9 to 3.1.4-11) with $k_1 = 1$, $k_2 = 0.4$, $k_3 = 0.3$, $A_0 = 1$ in the batch-mode.

Differential selectivity
From the defining Equation (3.1.4-5) it follows that:

$$^D S_P = \frac{r_1 - r_2}{r_1 + r_3} = \frac{k_1 \cdot A - k_2 \cdot P}{(k_1 + k_3) \cdot A}. \qquad (3.1.4 - 12)$$

After introducing Equations (3.1.4-9) and (3.1.4-10) and the conversion $U = (A_0-A)/A_0$ or $t = -\ln(1-U)/(k_1 + k_3)$, Equation (3.1.4-13) follows:

$$^D S_P = \frac{k_1}{(k_1 + k_3) \cdot (k_1 + k_3 - k_2)} \cdot \left\{ k_1 + k_3 - k_2 \cdot (1 - U)^{-\frac{k_1 + k_3 - k_2}{k_1 + k_3}} \right\}. \qquad (3.1.4 - 13)$$

Examining the limiting conditions shows:

a) $k_2 \to 0$ (pure parallel reaction):

$$\lim_{k_2 \to 0} {}^D S_P = \frac{k_1}{k_1 + k_3}. \qquad (3.1.4 - 14)$$

b) $k_3 \to 0$ (pure sequential reaction):

$$\lim_{k_3 \to 0} {}^D S_P = \frac{k_1 - k_2 \cdot (1 - U)^{\frac{k_2 - k_1}{k_1}}}{k_1 - k_2}. \qquad (3.1.4 - 15)$$

c) $U \to 0$ (limiting selectivity):

$$\lim_{U \to 0} {}^D S_P = \frac{k_1}{k_1 + k_3} \qquad (3.1.4 - 16)$$

Figure 3.1.4-3 shows the differential selectivity as a function of conversion.

Integral selectivity
From the definition equation (3.1.4-3), Equations (3.1.4-9) and (3.1.4-10), and the definition of conversion, Equation (3.1.4-17) results:

$$^I S_P = \frac{P_e}{A_0 - A_e} = \frac{1}{U_e} \cdot \frac{k_1}{(k_1 + k_3 - k_2)} \cdot (1 - U_e)^{\frac{k_2}{k_1 + k_3}} - (1 - U_e)\}. \qquad (3.1.4 - 17)$$

$^I S_P$ can be calculated from $^D S_P$ by using Equation (3.1.4-6):

$$^I S_P = \frac{1}{U_e} \cdot \int_0^{U_e} \frac{k_1}{(k_1 + k_3) \cdot (k_1 + k_3 - k_2)} \cdot \left\{ k_1 + k_3 - k_2 \cdot (1 - U)^{-\frac{k_1 + k_3 - k_2}{k_1 + k_3}} \right\} \cdot dU.$$

$$(3.1.4 - 18)$$

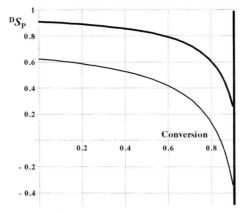

Fig. 3.1.4-2 Differential selectivity $^D S_P$ as a function of conversion U (Equation 3.1.4-13) for two cases. Thick curve: $k_1 = 1$, $k_2 = 0.1$, $k_3 = 0.1$; Thin curve: $k_1 = 1$, $k_2 = 0.4$, $k_3 = 0.6$.

Examining the limiting conditions shows:

a) $k_2 \to 0$ (pure parallel reaction):

$$\lim_{k_2 \to 0} {}^I S_P = \frac{k_1}{k_1 + k_3}. \tag{3.1.5 – 19}$$

b) $k_3 \to 0$ (pure sequential reaction):

$$\lim_{k_3 \to 0} {}^I S_P = \frac{1}{U_e} \cdot \frac{k_1}{k_1 - k_2} \cdot \left\{ (1 - U_e)^{\frac{k_2}{k_1}} - (1 - U_e) \right\}. \tag{3.1.4 – 20}$$

c) $U \to 0$ (limiting selectivity):

$$\lim_{U \to 0} {}^I S_P = \frac{k_1}{k_1 + k_3}. \tag{3.1.4 – 21}$$

Figure 3.1.4-3 *shows the integral selectivity as a function of conversion.*

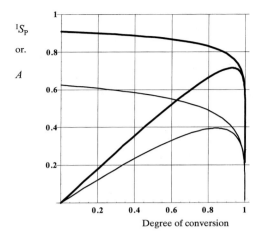

Fig. 3.1.4-3 Integral selectivity $^I S_P$ and yield A as a function of final conversion U_e (Equation 3.1.4-17). Thick curve: $k_1 = 1$, $k_2 = 0.1$, $k_3 = 0.1$; Thin curve: $k_1 = 1$, $k_2 = 0.4$, $k_3 = 0.6$.

3.2
Mass Balance

The most important pieces of information about the new process enter the mass balance. Therefore, a mass flow sheet, preferably for a unit mass of product, should be drawn up as early as possible on the basis of the first, coarse flow sheet. Together with the conversion and yield, one obtains information on the amounts of raw materials, the mass flows between reaction and separation sections, and side products.

The basis of the mass balance is the law of conservation of mass. For each section of the flow sheet (reactor, separator, mixer, etc.), the sum of the entering quantities and chemical elements is equal to the sum of the exiting amounts [Denk 1996]. The following example illustrates the basic procedure.

Example 3.2-1

Starting material A is converted to product P in reactor R (Figure 3.2-1) with 100% selectivity and 50% conversion. In the downstream separation unit P is quantitatively separated and unconverted A is recycled to the mixer, where it is mixed with fresh A and then fed to the reactor. The total balance consists of four balance regions I–IV.

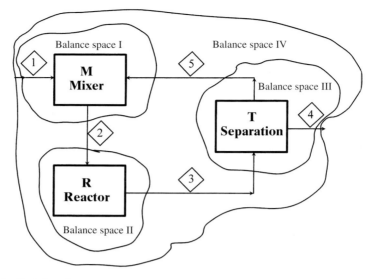

Fig. 3.2-1 Block flow diagram.

Balance space	Starting Material A	Product P
I (Mixer)	$n_A^1 + n_A^5 = n_A^2$	$n_P^1 + n_P^5 = n_P^2$
II (reactor)	$n_A^2 + n_P^2 = n_A^3 + n_P^3$	
III (separating unit)	$n_A^3 = n_A^4 + n_A^5$	$n_P^3 = n_P^4 + n_P^5$
IV (total plant)	$n_A^1 = n_P^4$	

n_A^1 means, n moles h^{-1} of A in stream 1 (see also Figure 3.2-1)

From this, six equations with ten variables (n_A^1 to n_A^5 and n_P^1 to n_P^5) result, so that 10–6 = 4 variables must be given. It is reasonable to specify:
Production: $n_P^4 = 10$ kmol h^{-1}
Product quality: $n_A^4 = 0$ kmol h–1
Separation performance: $n_P^5 = 0$ kmol h–1
Conditions ahead of reactor: $n_A^2 = 20$ kmol h^{-1}.

From this, a system of linear equations results (Equation 3.2-1):

$$\begin{pmatrix} 1 & 0 & 1 & 0 & 0 & 0 \\ 0 & 1 & 0 & 0 & -1 & 1 \\ 0 & 1 & -1 & 0 & 0 & 0 \\ 1 & 0 & 0 & 0 & 0 & 0 \\ 0 & 0 & 0 & 1 & -1 & 0 \\ 0 & 0 & 0 & 0 & 0 & 1 \end{pmatrix} \cdot \begin{pmatrix} n_A^1 \\ n_A^3 \\ n_A^5 \\ n_P^1 \\ n_P^2 \\ n_P^3 \end{pmatrix} = \begin{pmatrix} 20 \\ 20 \\ 0 \\ 10 \\ 0 \\ 10 \end{pmatrix} \qquad (3.2-1)$$

with the solution:

$$\begin{pmatrix} n_A^1 \\ n_A^3 \\ n_A^5 \\ n_P^1 \\ n_P^2 \\ n_P^3 \end{pmatrix} = \begin{pmatrix} 10 \\ 10 \\ 10 \\ 0 \\ 0 \\ 10 \end{pmatrix}. \qquad (3.2-2)$$

Thus the mass balance for this example is:

Stream-no. →	1	2	3	4	5
Substance A/kmol h^{-1}	10	20	10	0	10
Substance P/kmol h^{-1}	0	0	10	10	0
Sum/kmol h^{-1}	10	20	20	10	10

3.3
Physicochemical Data

In recent years, a knowledge of exact physicochemical data has become more important [Fratzscher 1993] for a number of reasons:

- The increasing use of simulation programs has made the requirements for exact physicochemical data increasingly important. The result of a simulation calculation can only be as good as the quality of the physicochemical data.
- To approve chemical plants, the authorities demand information on the toxicity, degradability, and safety of the materials involved [Streit 1991, Roth 1991]].
- The public are demanding more information on the effect of the materials being handled on the environment.

At the start of process development, material data files for pure substances as well as binary and ternary mixtures are started. As development advances, these grow and must be continually updated. They subsequently form a document which is passed to the planning and plant construction departments. Many companies now maintain their own material data banks. The data have been evaluated and should have been found satisfactory when used in practice. The times when a single company had ten different sets of Antoine parameters (Equation 3.3-3) for water between 0 and 100 °C ought now to be over. Ultimately, experimental determination or confirmation of the most important values on which the plant design is based is unavoidable. It is sheer negligence to rely on physicochemical data originating from a single literature reference.

3.3.1
Physicochemical Data of Pure Substances

The layout shown in Table 3.3-1 has proved useful as a model for the collection of pure-substance data. Only data which has been reliably evaluated is kept in the files, and these are the only values fed into the simulation programs. At the start of process development, the first step is to collect all available literature data on a material [Ullmann 3]. Dividing the data into physicochemical data, ecotoxicological data, and safety data, as in Table 3.3-1, has proved useful.

Tab. 3.3-1 Important pure-substance data that are often required during process development for the example of acrylic acid [Ullmann 1, BASF 1987].

Product name: acrylic acid			
CAS-no.		[79-10-7]	
Empirical Formula		$C_3H_4O_2$	
Molar mass	kg kmol^{-1}	72.06	
Melting point	°C	13.5	
Boiling point	°C	141.0	
Antoine parameters $\ln(P/\text{bar}) = A + B/(C + \langle k \, u \rangle T/°C)$	°C	A = 9.135 B = − 3245 C = 216.4	for 15 to 140 °C
Vapor pressure 20 °C	mbar	10	
Heat of evaporation	kJ kg^{-1}	633	at b.p.
Heat capacity	kJ kg^{-1} K^{-1}	1.93	liquid
Density	kg m^{-3}	1040	at 30 °C
Viscosity	mPa s	1.149	at 25 °C
Heat of formation	kJ mol^{-1}	/	
Upper heating value	kJ kg^{-1}	19 095	
Heat of transformation	kJ kg^{-1}	1075	(polymerization)
Heat of melting	kg^{-1}	154	at 13 °C
Solubility in H_2O		∞	at
Solubility of H_2O		∞	at
MAK-value		/	
Toxicity	mg kg^{-1}	LD$_{50}$ = 340	rat, oral
Water hazard class		1	slightly hazardous
Odor threshold		pongent	
Flash point	°C	54	
Ignition temperature	°C	390	
Explosion limit, lower	vol%	2.4	at 47.5 °C
, upper	vol%	16	at 88.5 °C

Tab. 3.3-2 Classification of pure-substance data into physicochemical, ecotoxicological, and safety data.

General data	• product name
	• synonyms
	• molar mass
	• CAS no.
Physicochemical data	• b.p.
	• m.p.
	• density
	• vapor pressure
	• heat of evaporation
	• heat capacity
	• viscosity
	• refractive index
	• water solubility
	• water take-up
	• dielectric contant
	• specific rotation
Safety data (see also Appendix 8.15)	• heat of transformation
	• Flash point
	• explosion limits
	• ignition temperature
	• WGK (water hazard class)
	• classification
	• R phrases
	• S phrases
Ecotoxicological	• LD_{50} (oral)
	• LD_{50} (skin)
	• LC_{50} (inhalative: gase, vapors)
	• LC_{50} (inhalative: aerosols, dusts)
	• MAK value
	• BSB_5
	• CSB

3.3.2
Data for Mixtures

Phase equilibria are a fundamental basis of many steps in chemical processes. Quantitative knowledge of these equilibria and data on the substances involved are therefore a prerequisite for developing processes and designing apparatus [Gmehling 1992].

Vapor–liquid equilibria can easily be measured experimentally and described mathematically. Equation (3.3-1) is the general relationship for the equilibrium between a liquid phase and an ideal gas phase:

$$y_i(x_i, T) = \frac{\gamma_i(x_i, T) \cdot P_i^0(T)}{P} \cdot x_i \qquad (3.3-1)$$

y_i mole fraction of component i in the gas phase
x mole fraction of component i in the liquid phase
$P^0{}_i(T)$ saturation pressure of pure component i at the system temperature T
P vapor pressure of the mixture
γ_i activity coefficient of component i in the liquid phase.

In the simplest case the vapor pressure of the pure components $P^0{}_i$ can be expressed by the Clausius–Clapeyron equation:

$$\ln P_i^0(T) = -\frac{\Delta_V H_i}{RT} + const. \qquad (3.3-2)$$

Since the heat of evaporation $\Delta_V H_i$ is not constant, in practice the three-parameter Antoine equation (3.3-3), which allows the vapor pressure to be accurately calculated even over large temperature ranges, is widely used:

$$\ln P_i^0(T) = A_i + \frac{B_i}{C_i + T}. \qquad (3.3-3)$$

The heat of evaporation can be roughly approximated by the Pictet–Trouton rule (Equation 3.3-4):

$$\Delta_V H / kJ\ kg^{-1} \approx \frac{88/kJ\ kmol^{-1} \cdot T_{boil}/K}{M/kg\ kmol^{-1}}. \qquad (3.3-4)$$

Several models are available for determining the activity coefficients γ_i, for example, the Wilson model, which is applicable to homogeneous mixtures of liquids, and the NRTL model, which is also suitable for systems with miscibility gaps [Prausnitz 1969, Gmehling 1977, Gmehling 1992].

If no material values can be found in the literature, which is often the case for binary and ternary data, they should initially be estimated by using empirical formulas or realistic values [Sandler 1999, Poling 2001].

3.4
Processing

Processing (see Section 2.3) is intimately linked to the chemical reaction. While the chemical aspects are still subject to extensive modification (for instance, in the choice of catalyst, solvents, etc.), there is little point in paying a great deal of attention to processing. It is only when the reaction mixture is being produced in a representative manner that an initial processing procedure can be devised (see Section 4.2). There is no generally accepted way of going about this. It is more an art than craft, and reliance must be placed on inspiration and experience. It has, however, always been helpful to gather together as many experienced specialists as possible and to discuss the problems regularly. Although tentative efforts are now being made to compile collective know-how in expert systems for processing strategies, these are still far from being universally applicable.

The above deliberations culminate in a separation concept which can be broken down into individual unit operations. Initially, these operations can be examined in the laboratory either on a batch basis, or in a continuous process to determine whether they are feasible in principle (e.g., are azeotropes formed?, how difficult is separation?, what is the dissolution rate?, can the phases be separated?, etc.). This gives the initial process concept, which can be used as a basis for starting the actual process development.

3.5
Patenting and Licensing Situation

A patent is the granting of exclusive rights for a limited time period (usually 20 years) by a state in exchange for the published description of an invention. The conditions for patentability are novelty, inventiveness, and economic applicability. A patent grants the right to exclude others from using the invention, but the inventor may himself require permission for its use, for example, approval of pharmaceuticals. Therefore, reliable sources of information that cover the current state of the patent literature, international markets, and economic competition are becoming of increasing importance [Boeters 1989, Cohausz 1996, Fattore 1997, Hansen 1997, Hirsch 1995, Münch 1992, Reichel 1995, Ullmann 5]. If a new idea is to be used in developing a process, a careful check should be made to determine whether any third-party proprietary rights would be infringed [Dietzsch]. The major chemical companies maintain their own patent and documentation departments for carrying out such patent searches. Smaller companies have the possibility of carrying out research in patent information centers. However, today the process developers can directly access patent information at their workplace, since about 90 % of published technical knowledge is contained in patents. Each week about 20 000 patents providing legal, economic, and technical information, are published. Some examples are [Bio World 1997]:

- Chemical Abstracts (whole field of chemistry)
- CAB International (agriculture, biotechnology, forestry, veterinary medicine)
- MEDLINE (whole field of medicine)
- PROMPT (Predicasts Overview of Markets and Technology)
- PATDPA (German Patent Office)
- INPADOC (European Patent Office)
- World Patents Index (Derwent).

The information resources are usually licensed from databank hosts such as DIMDI, ECHO, KNIGHT, RIDDER, and STN International. The folllowing research strategies are widely used [Walter 1997]:

Technical information search
- Short search to give an overview of the state of the art
- Full search to give a detailed survey of the state of the art

Statistical patent analysis of competitors
- Development trends in areas of technology
- List of applications of companies in the area in question
- Country of origin and target countries

Name search
- National or international
- Inventors, applicants

Monitoring
- According to patent class
- According to company names

The situation is most favorable when the idea is:

- Completely new
- Unpredictable
- A considerable improvement on the state of the art.

Then patents can be submitted that permit unrestricted exploitation (rare). The idea is formulated in written form and supported by the results of some laboratory experiments. The basic structure of a patent document follows:

- *Applicant*
  Name and address of the applicant (company).
- *Title*
  Short description of the invention: *"Process for manufacturing..."*
- *Patent claims*
  divided into main claims, subclaims, and side claims. The main claim (standard structure in schematic form: *"The state of the art...characterised in that...invention...process for manufacturing...characterised in that..".*
- *Discussion of the state of the art*
  Critical treatment of the literature most relevant to the idea. Here hackneyed phrases such as *"As is generally known..."* or *"As is generally known to persons skilled in the art..."* are popular.
- *Criticism of the state of the art*
  What are the disadvantages of the state of the art? Only disadvantages that are eliminated by the invention are mentioned.
- *Task and solution of the subject of the patent*
  What is new about the idea (advantages over the state of the art?)
- *Detailed description of the idea*
  Detailed description of the new process, apparatus, possibly supported by figures and tables.
- *Description of investigations.*
- *Summary.*

In the next step a patent professional translates this into patent language [Münch 1992].

In Germany this document is then submitted to the German Patent Office, and the date of submission is then the priority date of the application. During the priority deadline the applicant can decide whether the application will be registered overseas, and if so in which countries. Normally, an application is made at the European Patent Office (EPO). When this patent application names Germany (DE) as member state, the application at the German Patent Office can be dispensed with. Inventions must be patented in each country in which they are to be protected (territorial principle). Transfer of patenting processes is aided by international organizations such as the EPO in Munich and the World Intellectual Property Organization (WIPO) in Genf.

Eighteen months after the priority date the appplication is published by the EPO (patent application, EP-A1) and the examination procedure (refusal or grant, opposition, additional examples, etc.) is started. The application cannot be supplemented during the examination procedure. Argumentation during the examination must be based exclusively on documents that were submitted with the application. Then, if the examination is positive, the patent is granted (patent specification, EP-B1). In the following nine months everybody has the possibility to lodge opposition against granting of the patent.

Sometime it is advisable to make the patent application so wide-reaching that for competitors it "takes away the air that they breathe" or forces them to take countermeasures. A strong patent situation facilitates negotiations with competitors (quid pro quo). Furthermore, the value of a company depends strongly on having a strong patent portfolio, which can play a major role in takeovers. Patent litigation over industrial property rights can involve sums of tens or hundreds of million dollars (e.g., metallocene catalysts for polyolefins, acetic acid synthesis).

Apart from their protective function, patents are also a source of information for competitors [Dolder 1991, Wirth 1994]. Careful consideration must therefore be given to the question of whether it is desirable to keep an idea secret or whether it should be patented [Hayes 1994]. The risk involved in deciding against patenting in-house knowledge can be reduced by eliminating the patentable novelty of that information on a worldwide basis through publication in a suitable form, for example, in an in-house journal or a technical bulletin. If the idea is restricted by third-party proprietary rights, the following questions should be answered:

- When do those rights expire?
- How can they be rendered void?
- How can they be circumvented by modifications?

If the idea is completely covered by third-party proprietary rights, the only way out is to enter into licensing negotiations.

3.6
Development Costs

Development costs often constitute a considerable proportion of the total cost of a project. If the process is completely new, these costs are often around 50% of the investment costs for the industrial plant, and perhaps more if a pilot plant must be built. Therefore, the development costs should be estimated prior to process development according to the following scheme:

- *Man years*: This involves determining how many chemists and engineers will be employed on the project and for how long. Since the costs associated with each of the specialists working on the individual investigations are known, the man-year costs can then be calculated.
- *Setting up costs*: A rough estimate of the costs of setting up a small-scale plant (miniplant, see Section 4.4) is sufficient since this item is normally the smallest in the development costs. Once some notion of the process is available (a rough process flow diagram), these costs can be determined with reasonable accuracy.
- *Operating costs*: The operating costs can be estimated from the assumed development time and the supervisory staff (shift personnel) required. A rule of thumb is that the annual operating costs of an experimental plant will be at least equal to those required to set it up.

More difficult to estimate are the amounts required annually for modifications and repairs to the experimental plant. These depend very much on whether new or already proven technologies are being used. The development costs determined in this way do not contain any contributions arising from personnel not directly involved in development, such as management or patent specialists. Nor is account taken of unsuccessful developments. It nevertheless seems reasonable to apportion the direct development costs to the product and to allow for the other research and development costs in some other way, for example by means of a general cost provision (see Chapter 6). Although this method only accounts for the costs of successful development projects, the financial burdens on the product will be relatively high if it must bear all the development costs of the first plant.

In addition to good financial planning, effective time management is essential for competitiveness. Being six months or a year earlier on the market that a competitor can be a justification for higher development costs.

3.7
Location

The main competitive advantages such as price and availability are nowadays often dependent on the location of the production plant. Thus, there is now an increasing tendency to move the production of major basic chemicals such as methanol to the oil-producing countries. However, products whose production requires considerable know-how are less location-dependent and will continue to be produced in the industrialized countries, even in the future. Extreme cases are those in which legal requirements decide the location, as is the case in the field of gene technology. In deciding on location, consideration should be given to the following aspects:

- Proximity to the sales market
- Proximity to the sources of raw materials or precursors
- Nature and amount of the product
- Transport facilities (ship, rail, road)
- Quality of the available labor, personnel costs
- Energy situation (availability of energy sources such as natural gas, heating steam, utility and cooling water; exploitation of process heat; couplling of electrical and staeam energy).
- Specific location situation (climatic conditions, optimal coupling with a network and exploitation of synergies, for example, use of side products and process heat)
- Waste-disposal management (sewage treatment plant, oil- or gas-fired power stations that can be synergically used as incineration plants)
- Political considerations (taxes, investment aid, financial support, approval situation)
- Local society (critical attitude of the population towards new technologies, caused by "oversaturation of basic needs", readiness to strike, etc.)

3.8
Market Situation

The competitive advantages generated by process development should ultimately benefit the market and its customers. It is therefore impossible to target process development correctly without a precise knowledge of that market. The market situation is generally elucidated by a marketing department. Marketing means directing a company at the market with its customers (i.e., "marketing means thinking in the customer's head"). Thus, the company is systematically connected to its environment. The following information is required if the basic data needed to plan production (plant size, maximum production costs) are to be obtained:

- Market price (how it will develop and vary with time).
- Demand, broken down into in-house and third-party demand.
- Market growth (endangerment by new technologies?).

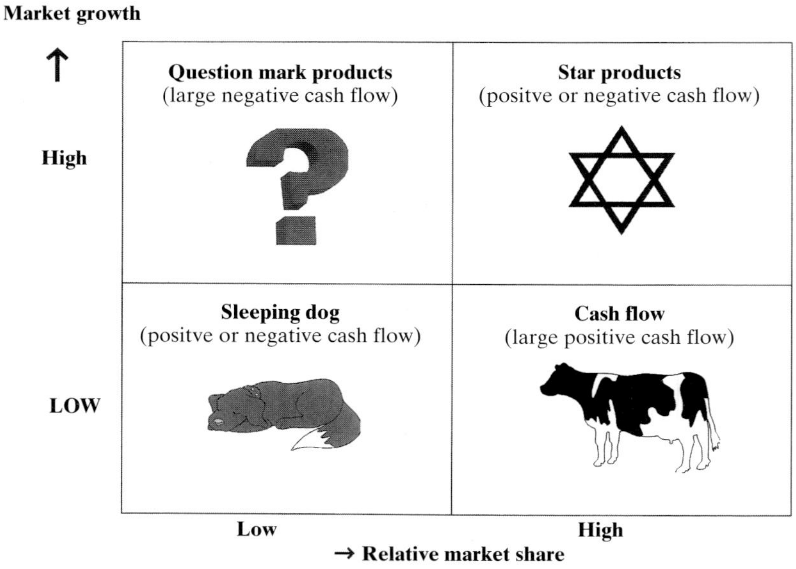

Market growth

	Question mark products (large negative cash flow)	**Star products** (positve or negative cash flow)
High		
LOW	**Sleeping dog** (positve or negative cash flow)	**Cash flow** (large positive cash flow)

Low — High
→ Relative market share

Fig. 3.2-1 Portfolio matrix [Dichtl 1987].

- Sleeping dog products: These are problem products whose competitive situation is weak because of their low market share. In such cases process development would not be worthwhile.
- Cash cow products: These are products which, although their market growth is low, have a high market share. To exploit the cost advantages to the full, there is little point in heavily investing in process improvements. Instead, these products should be "milked" and the resulting cash flow used to support the star products.
- Star products: All the efforts should be concentrated on these products since they will safeguard the survival of the company in the long term.
- Question mark products: In these cases the company must decide whether to increase the market share (turn them into star products) or to abandon them if the prospects are slim.

- Utilization of capacity in existing plants, with a breakdown in terms of location (USA, Western Europe, Japan, China, etc.) and production process.
- Competitive situation (who is the largest competitor?).
- Customer situation (are there many small customers or a few large ones?).

The question of whether a process development or improvement will be worthwhile can be analyzed by using a portfolio representation (Figure 3.8-1).

Portfolio analysis is a strategic planning tool, used to concentrate research investment on products for which market prospects appear favorable and competitive advantages can be exploited [Dichtl 1987].

3.9
Raw Materials

Raw materials, their availability, and their price structure (see Tables 6.1-3a and b) have always been crucial factors responsible for shaping the technological base of the chemical industry and, consequently, its growth and expansion. The chemical industry produces a wide variety of products from only a few inorganic and organic raw materials. Thus, raw materials such as natural gas, petroleum, air, and water, salt, and sulfur are used to produce basic chemicals such as synthesis gas (CO–H_2 mixture), acetylene, ethylene, propene, benzene, ammonia, sulfuric acid, etc. These are used in turn to manufacture intermediates such as methanol, styrene, urea, ethylene oxide, acetic acid, acrylic acid, cyclohexane, etc. (see Figure 1-2, Table 3.9-3). Any change in the price and availability of raw materials will therefore have a substantial effect on the production processes for secondary products. Since the exact purchase prices (transfer prices) are an important source of information in assessing manufacturing costs, they must always be available and kept up to date in a chemical company.

It is not only the price of potential raw materials which must be determined but also their availability and quality (purity, state as delivered). It is never possible to reach a generally applicable decision on a particular raw material since much depends on its availability at a particular location and variation with time.

The most important chemical raw material is petroleum, of which about 3.6×10^9 t/a is processed in around 600 refineries worldwide. The chemical industry only accounts for about 7% of the raw material source, the vast majority being consumed in transport

Tab. 3.9-1 Worldwide oil consumption in 10^9 t.

	1995	2000	2010
Transport	1.60	1.87	2.32
Heating	1.22	1.27	1.43
Petrochemistry	0.19 (= 6%)	0.25 (= 7%)	0.30 (= 7%)
Total	3.2	3.6	4.3

and heat generation (Table 3.9-1). Since petroleum is a complex mixture of hydrocarbons, it is characterized by sum parameters or characteristic numbers:

- Density (e.g., API density = -145.5; ρ (15.5 °C) = -131.5; API = American Petroleum Institute)
- Boiling point, boiling point analysis (true boiling point)
- S content (0.1–0.7 %)
- Wax content (n-alkanes, 5–7 %).

Petroleum is separated by atmospheric-pressure distillation into:

- Liquified petroleum gas (LPG: propane, butane) b.p. < 20 °C
- Low-density fuel (LDF[3]: light naptha) b.p. 20–75 °C
- Heavy naphtha   b.p. 75–175 °C
- Kerosene   b.p. 175–225 °C
- Gas oil   b.p. 225–350 °C
- Long residue   b.p. > 350 °C.

The long residue can be split by vacuum distillation into:

- Vacuum gas oil
- Heavy vacuum distillate
- Short residue.

The petrochemical refineries that supply the chemical industry process these cuts in various chemical, mostly catalytic, processes to produce petrochemical primary products (basic chemicals such as olefins, diolefins, aromatics, paraffins, acetylene). These raffination (i.e., shifting the C/H ratio of the products) processes can be divided into:

Carbon-out methods ($C_nH_m \rightarrow C_{n-x}H_m + x\,C$)

- Thermal cracking
- Visbreaking (internal shift of the C/H ratio; purely thermal cracking processes, e.g., $n\text{-}C_{12}H_{26} \rightarrow n$-hexene + n-hexane)
- Fluid coking
- Delayed coking (intensified thermal cracking process, still widely used in the USA)
- Flexicoking (did not become established for reason of cost; renaissance for tar sands conceivable)
- Catcracking (catalytic cracking on zeolites, today the most important process).

Hydrogen-in methods ($C_nH_m + x\,H_2 \rightarrow C_nH_{m+2x}$)

- Hydrocracking (hydrogenative cleavage at high pressure and temperature; removal of heteroatoms, especially hydrogenation of S to H_2S)
- Residue cracking
- H oil cracking.

Thus the H/C ratio (Table 3.9-2) is increased.

---

3) LDF = Low density fuel.

Tab. 3.9-2 H/C ratio of different classes of raw materials.

Natural gas	4
Aliphatics	2
Aromatics	1
Coal	0

Tab. 3.9-3a World demand for the most important basic chemicals in 10^6 t Martino 2000].

	1998	2010
Ethylene	80	120
Propylene	45	82
I-Butene	0.8	1.4
Higher α-olefins	1.0	2.2
Benzene	27	40
Toluene	13	21
p-Xylene	14	30

Tab. 3.9-3b Prices for energy sources and for selected raw materials and basic chemicals [VCI 2001].

	1997	1998	1999	2000
Energy sources				
Coal/€ t^{-1}	42	39	34	40
Heating oil/€ t^{-1}	118	101	118	191
Natural gas/€ GJ^{-1}	3.02	3.09	2.73	3.65
Electricity/€-cent/kWh				
4 MW/4000 h	7.40	7.40	7.08	5.55
10 MW/6000 h	5.84	5.84	5.41	4.23
40 MW/8000 h	4.96	4.96	4.46	3.50
Raw materials				
Crude oil/US$/bbl	19.12	12.72	17.79	28.33
Naphtha/€ t^{-1}	167	117	156	284
Olefines				
Ethylene/€ t^{-1}	508	422	422	664
Propylene/€ t^{-1}	414	290	303	548
Aromatics				
Benzoe/€ t^{-1}	270	225	236	410
o-Xylene/€ t^{-1}	351	225	236	410
p-Xylene/€ t^{-1}	411	327	343	541

3.10
Plant Capacity

Optimum plant capacity, i.e., the size of plant which yields the maximum return (Chapter 6), will be achieved if a degree of capacity utilization of 100% is reached 3 – 5 years after start-up. However, since the plant capacity must be decided about 4 – 5 years before start-up for planning and legal approval reasons, and plants which are not fully utilized over a period of time increase the specific production costs, it is essential to be able to look 7 – 10 years into the future in determining the optimum capacity. Therefore, precise knowledge of the market potential and of market growth trends is required. The relationship between production price and degree of capacity utilization is as follows:

$$\frac{PC}{PC_0} = \frac{1 + \left(\frac{FC}{VK}\right)}{1 + \left(\frac{FC}{VC}\right) \cdot \left(\frac{C_{eff}}{C_0}\right)} \qquad (3.10-1)$$

PC = production costs in € kg^{-1}
PC_0 = production costs at nominal capacity
FC = fixed costs
VC = variable costs
C_{eff} = effective plant capacity in t a^{-1}
C_0 = nominal plant capacity

Thus, as the degree of capacity utilization drops, the production price increases hyperbolically and this dependence is the greater, the higher the contribution made by the fixed costs (Figure 3.10-1).

The greater the uncertainty in the market studies and the newer the technology used, the greater will be the preference for choosing a lower capacity. Many Japanese chemical companies adopt this approach and prefer to set up several plants with smaller capacities. However, lower capacities result in higher specific plant costs and therefore in higher production costs. This is represented by the following empirical relationship:

$$I_2 = I_1 \cdot (K_2/K_1)^\kappa \qquad (3.10-2)$$

I = investment
C = capacity
κ = degression exponent.

For individual items of equipment, $\kappa = 0.6$, and for plants containing the same number of items of equipment $\kappa = 2/3$ [Schneider 1965]. However, experience shows that considerable increases in capacity can often be achieved in the first few years after startup by means of debottlenecking and continuous improvement.

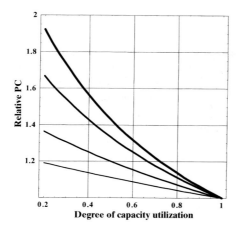

Fig. 3.10-1 Dependence of the relative production costs PC on the degree of capacity utilization C_{eff}/C_0 for different ratios of fixed costs FC and variable costs VC (from top to bottom: 0.25, 0.5, 1, 1.5, ∞).

3.11
Waste-Disposal Situation

Waste-disposal facilities (incineration, sewage treatment plant, waste-disposal site) depend on the location. However, at the beginning of process development the location will generally still be undecided and the process should therefore be designed to produce only waste products which can be readily incinerated. Processes which produce salts or large amounts of dilute aqueous solutions containing materials that are difficult to degrade biologically are inherently location-dependent. The main disposal routes are discussed in more detail in Section 2.7.

For classification and evaluation of residues, the following questions must be answered:

- General statements (responsible operation, site of formation, etc.)
- Chemical description of the known constituents
- Safety measures
- Properties of the residue
- Physical data
- Analytical data of the residue
- Test of reusability
- Test of suitability for landfill disposal.

3.12
End Product

It is essential to determine the demand for the end product as well as its specification and achievable price. A strongly competitive situation will mean particularly severe changes with time, and in this case it is necessary to adopt a particularly critical approach to forecasting.

The product specification is of particular significance for process development (Table 3.12-1). In the simplest case, it consists of a minimum purity, for example, that the end product should contain at least 99% of a particular chemical compound. The permissible impurity content may vary considerably from one product to another. Thus while it may be a few percent in the case of dyes, the limit is often only a few parts per million for monomers. However, in most cases the end product cannot be specified so simply, since its subsequent use may be affected in different ways by individual impurities. In such cases, it is not sufficient to specify an upper limit for the sum of impurities; instead, limits must be specified for important individual components. Often, the chemical analysis is not sufficient or is not sufficient on its own for specifying the end product, a case in point being, for example, polymers. In such cases chemical analysis is supplemented by processing properties which, in the simplest case, can be reduced to simple physical properties (e.g., color number).

The process development engineer endeavours to obtain as detailed a specification as possible for the end product before he starts work, but usually the specification will only be finalized during process development. Statements such as "as pure as possible" are inadequate since the removal of the final traces of impurities is time-consuming and expensive. Individual specifications which are of no significance in characterizing the end product should be avoided. If possible, a composition which can be determined analytically is preferable to a purely physical characterization since it is more closely related to the process sequence.

Tab. 3.12-1 Example of a specification.

• Purity (GC)	min. 99.0% (g g^{-1})
• Water content (Fischer method, DIN 51777)	max. 0.05% (g g^{-1})
• Acid content (as acetic acid)	max. 0.3% (g g^{-1})
• Ash content	max. 0.001% (g g^{-1})
• Color number (APHA, DIN 53409)	max. 20

4
Course of Process Development

4.1
Process Development as an Iterative Process

The development of chemical processes is a complex procedure. The first hurdle in establishing a new process is overcome when a promising synthetic route, usually with associated catalysts, is discovered, but many issues must be clarified and many problems solved before industrial implementation is feasible and all the documents necessary to design and operate a chemical plant have been assembled [Lück 1983, Harnisch 1984]. How are these planning documents arrived at? The conventional procedure is carried out in three stages (Figure 4-1):

An initial process concept is developed on the basis of an optimized laboratory synthesis. This is not usually continuous (Section 4.2), and the individual process steps are examined independently of each other in the laboratory and simulated (Section 4.3). A continuous laboratory plant (the so-called miniplant) [Behr 2000] is then designed, set up, and operated (Section 4.4-1). This is a small but complete plant handling production quantities of ca. 100 g h^{-1} and consisting of a synthesis section, workup, and all recycling streams. Once the process concept has been confirmed in the miniplant, the next step is to design and set up a trial plant with a much higher capacity, the scale of this pilot plant being between that of the miniplant and that of the industrial plant. The quantities produced are a few kilograms per hour or tonnes per annum and enable, for example, application tests to be carried out on the product or large-scale deliveries to be made to customers. Operation of the pilot plant makes it possible to complete and

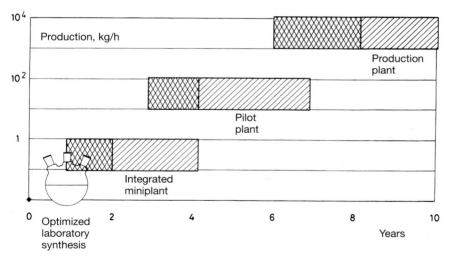

Fig. 4-1 The stages involved in developing a process. The time intervals specified for design, installation, and operation are only tentative and, in specific cases, the real values may differ greatly from those given here. The following rule of thumb applies for the plant investment: laboratory research 0.1×10^6 €, miniplant 1×10^6 €, pilot plant 10×10^6 €, production plant 100×10^6 €.

Process Development. From the Initial Idea to the Chemical Production Plant. G. Herbert Vogel
Copyright © 2005 WILEY-VCH Verlag GmbH & Co. KGaA, Weinheim
ISBN: 3-527-31089-4

verify data and documentary information obtained at an earlier stage of process development (Section 4.4.2).

The scale-up factor from one stage to the next is determined by the principle of the minimum: the process step or apparatus with the lowest possible scale-up factor determines the capacity of the next largest plant with calculable operating behavior. This offers the process developer the chance to save time and money [Hauthal 1998]. If certified documents for the planning of a production plant can be obtained directly with an external scale-up factor of 10 000 (100 g h^{-1} × 100 = 1 t h^{-1}), then the costs for the pilot plant and three to four years of development time can be saved [Hofen 1990, Jäckel 1995]. This is a major market and economic advantage. Therefore, nowadays one attempts to scale up directly from the miniplant to the production plant.

Table 4-1 summarizes some of the important process steps and typical maximum scale-up values above which reliable scale-up is no longer possible [Krekel 1985, Krekel 1992].

Larger scale-up factors are possible for gas-phase reactors such as multitubular reactors. The unit operations of distillation, rectification, and absorption can be scaled up without an intermediate stage. This explains, for example, why much effort is expended on obtaining operating conditions that allow the reaction and processing of gases and liquids. The behavior of gases and gas–liquid equilibria can be explained well in physical terms, and calculation is therefore straightforward (Chapter 3).

In solids handling unpredictable factors such as deposits, incrustations, and formation of fines lead to scale-up factors that are many orders of magnitude smaller. This applies both to synthesis (e.g., in fluidized-bed reactors) and processing (e.g., drying, sublimation, and crystallization).

Process development does not, however, take place in a one-way street. Assumptions are made for the individual development stages which are only confirmed or refused when the next stage is being worked on. It may be necessary therefore to go through

Tab. 4-1 Scale-up factors for some process steps.

Reactors	
• Tube-bundle reactor	> 10 000
• Homogeneous tube	> 10 000
• Homogeneous stirred tank	> 10 000
• Bubble column	< 1000
• Fluidized-bed reactor	50–100
Separation processes	
• Distillation and rectification	1000–50 000
• Absorption	1000–50 000
• Extraction	500–1000
• Drying	20–50
• Crystallization	20–30

the individual stages several times with modified assumptions, resulting in a cyclic pattern (Figure 4-2).

The most important task is to find the weak points and subject them to particularly close scrutiny. The entire process will then be examined again with the improved data obtained in this way, and so on. The fact that many decisions must be taken with incomplete knowledge is a fundamental but inevitable difficulty. To delay development until all uncertainties have been eliminated would be just as wrong as starting industrial development on the basis of laboratory discoveries alone. A start should be made on choosing between various processes or process variants as early as possible so

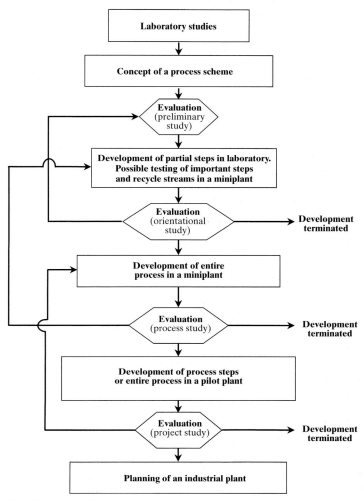

Fig. 4-2 The cyclic pattern of process development.

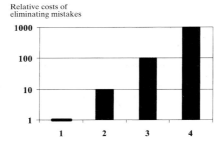

Fig. 4-3 The costs for eliminating errors and the investment costs increase from development stage to development stage by roughly a factor of ten.
1) Laboratory 3) Pilot plant
2) Miniplant 4) Production plant.

that consideration of a larger number of possibilities is restricted to the laboratory. Prolonged investigation of two variants on a trial-plant scale should be avoided.

As shown in Figure 4-2, each development stage is followed by an evaluation to decide whether development should be continued, stopped, or started again at an earlier development level (Chapter 4).

Most mistakes are made at the beginning of the activity, but it is still relatively easy and cheap to eliminate them at the miniplant stage. However, the further process development advances, the more expensive it becomes to eliminate mistakes (Figure 4-3). In the final production plant, corrections can only be made with an enormous expenditure of time and money.

4.2
Drawing up an Initial Version of the Process

Once all the information has been collected (Chapter 3), an initial version of the process is drawn up [Trotta 1997, Gilles 1998, Frey 2000, Kussi 2000a]. In general, it has been found useful to adopt the following procedure:

After the initial versions of the process, which should be the most reasonable ones possible at that stage, have been drawn up (Figure 4-4), they are discussed by the project team in an iterative problem-solving process which takes account of the general rules of process engineering, structure and system rules, but also intuition and experience with the properties of the raw-material and energy streams to be processed, the required equipment, and the economic conditions [Glanz 1999]. This discussion should take account of all the information gathered hitherto and, after all the advantages and disadvantages have been reviewed, only one version of the process should remain.

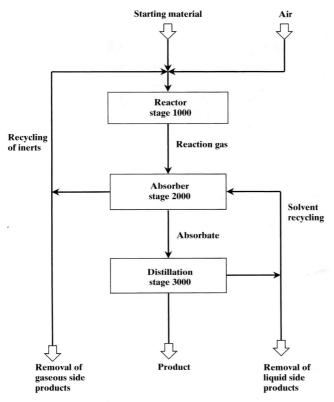

Fig. 4-4 Initial process concept for the oxidation of a starting material with air and isolation of the product by absorption and distillation.

There is at present no method which will result in the "best" version of the process being invariably chosen at the outset and there will presumably not be one in the future either [Blaß 1985, Kaibel 1989, Scholl 1995]. Finding the optimum version of a process requires not only knowledge and experience but also a considerable creativity, since the number of possible ways of carrying out a given task is almost infinite and many of the rules can still not be quantified, that is, they are not accessible to algorithms [Erdmann 1984]. This is demonstrated in the following example.

Example 4-1

The esterification of an impure organic acid with methanol yields a reaction mixture containing at least five components (N = 5): unconsumed acid, unconsumed alcohol, ester, volatiles such as H_2O, lower esters, dimethyl ether, and high-boiling components such as higher esters. According to Stephanopolous [Stephanopolous 1976] the possible number of arrangements for the separation operations can be calculated from the relationship:

$$Z = \frac{[2 \cdot (N-1)!]}{N! \cdot (N-1)!} \cdot S^{N-1} \tag{4-1}$$

Z = number of alternative arrangements
N = number of components
S = number of separation processes

*If all components can be separated by distillation, (N–1) = 4 separation columns are necessary, which can be arranged in 14 different ways. If a second separation process (e.g., an extraction if the ester, alcohol, and water form azeotropes) must be added, this already results in 224 possible arrangements. If the number of possible reaction procedures (e.g., batch [Uhlemann 1996]/continuous, gas/liquid, ion exchanger/mineral acid) is also taken into consideration, there will be an almost infinite number of ways of carrying out the process (a **so-called combinatorial explosion**).*

To find the best process among all these variants is quite impossible simply because the ideal solution is unknown [Bauer 1996]. To calculate all possible variants [Seider 1999], to compare them, and then select one is not feasible because of the large number. Perhaps this example reveals why inventing processes is still an art rather than a craft [Beßling1995]. However, the tools available for performing this art are being continuously improved [Blaß 1989, Schembecker 1996] (Section 4.2.1).

Many of the variants can of course be eliminated at the start for trivial reasons (Table 4-2). However, even the number of potential variants left is still so large that it is impossible to calculate investment and production costs for all of them. Furthermore, at this point the information available for doing this is often inadequate and there is consequently a risk of eliminating promising variants. At this stage one can further reduce the number of remaining variants (max. 3 or 4) by means of methods for the rapid estimation of costs (Section 6.1.6).

On the basis of this initial version, an industrial plant is designed. The individual unit operations (reactor, absorber, distillation, etc.) are designed on the basis of the existing information (approximate size and diameter of columns, etc.). By scaling down this hypothetical large-scale plant a trial plant is designed, which is nowadays generally a miniplant (Section 4.4.1). At the same time, the industrial plant is simulated by using a computer program and an initial complete mass and energy balance sheet is drawn up.

Tab. 4-2 K. O. criteria for reducing the number of alternatives.

Technical reasons	• high-pressure technology
	• high-vacuum technology
	• corrosion
	• catalyst lifetime, regenerability
Economic reasons	• noble metal use
	• large recycle streams
Ecological reasons	• formation of salts
	• side products that are difficult to dispose of
	• cancerogenic auxiliaries
	• side products that are difficult to remove
Legal reasons	• patent situation
	• duration of licensing procedure

4.2.1.
Tools used in Drawing up the Initial Version of the Process

Requirements which must be met for using modern EDP tools are as follows:

- Fast access to the information
- Information which can be conveniently handled at the workplace
- The information must be continuously updated

4.2.1.1
Data Banks

To carry out process development as efficiently as possible, it is important to avoid duplicate activities. This includes taking account of existing knowledge or exploiting it. For this reason, chemical companies not only build up internal data banks but also utilize external data banks [Blaß 1984]. Figure 4-5 provides an overview of the structure of an on-line data bank.

Data banks differ as to whether they contain the information itself (primary information or factual data) or whether they refer to other information sources (secondary information or literature data). Some types of data which are important for process development are [Achema 1991]:

- Material properties [Berenz 1991]
- Physicochemical data
- Ecological and toxicological data
- Costs of raw materials, intermediates, and end products
- Energy and equipment costs
- On-line literature searches [Ullmann's Encyclopedia].

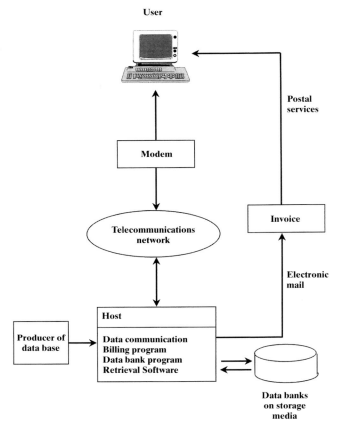

Fig. 4-5 Components of online services [Kind 1990].

4.2.1.2
Simulation Programs

Besides experiments and trial plants, simulation using computer programs furnishes the process development engineer with a very effective tool [Zeitz 1987, Schuler 1995, Messer 1998, Seider 1999]. The development of very fast, large computers and effective physical models has now made it possible to simulate individual units (e.g., distillation), networks of units of the same type (e.g., a stirred tank cascade, a heat exchanger network), and networks of dissimilar units (e.g., chemical plants). The advantages of using suitable simulation tools in process development are:

- Better understanding of the process (including safety aspects)
- Saving time
- Increasing the scale-up factor

- Reducing costs due to fewer pilot plant tests
- Optimization of process control.

In practice, the simulation program used depends on the objective; there is, after all, little point in trying to solve detailed problems with a simulation program which, although it is all-embracing, is slow if there are very much more efficient programs available which are designed for the specific requirement. Programs for solving specific problems are usually developed by the user himself and are not generally available or are available only in rare cases. Numerous programs for dealing with flow sheet problems are available from universities, industry, or the commercial market (see Tables 4-3 and 4-4) [Splanemann 2001].

Today, the term simulation is very wide ranging and accordingly imprecise [Hartmann 1980, Marquadt 1992]. Mathematically, simulation can be divided into two large groups: stationary and instationary processes (Tables 4-3 and 4-4).

Regardless of whether the task is to simulate an individual apparatus or an entire process, the following procedure is recommended [Seider 1999, Glasscock 1994]:

- Formulation of the task or problem, for example, continuous tubular reactor with recycling
- Choice of software
- Choice of hardware
- Input of the essential information
  - Plotting the flow sheet
  - Choice of the chemical components
  - Physicochemical data of the pure substances
  - Specification of the streams

Tab. 4-3 Selected simulation programs for stationary processes [Schuler 1995].

Name	Reference
ASPEN PLUS	Aspen Technology Inc.
	Cambridge (MA), USA
CHEMASIM	BASF AG
	Ludwigshafen, Germany
CHEMCAD	Chemstations GmbH Engineering Software
	www.chemstations.de, [Chemcad 1996]
DESIGN II	Chemshare
FLOWPACK	ICI
HYSYS	HYPROTECH, Düsseldorf, Germany
	www.software.aeat.com
VTPLAN (CONTI)	BAYER AG
	Leverkusen, Germany
PROCESS	Simulation Science Inc.
Pro/II	Fullerton (CA), USA

Tab. 4-4 Selected simulation programs for dynamic systems [Gilles 1986, Ingham 1994, Scheiding 1989, Wozny 1991, Daun 1999].

Name	Reference
CHEMADYN	BASF AG
	Ludwigshafen, Germany
DIVA [Kreul 1997, Weierstraß 1999]	Prof. Gilles
	Stuttgart University, Germany
SIMUSOLV	DOW Company
	Midland (Mi), USA
ASPEN Custom Modeler [Weierstraß 1999]	Prosys Technology Ltd.
	Cambridge, England, since 1991 ASPEN Technology Inc.

- Specification of the unit operations
- Linkage of the streams and units
- Choice of the thermodynamic model
- Data for binary mixtures.

There are two basic approaches to calculating stationary processes:

- Sequential modular approach
- Equation-oriented approach.

In the sequential modular approach, the solution strategy of classical calculation by hand is transferred to the flow-sheeting program. The program treats one unit operation after another. This has the following advantages and disadvantages:

Advantages

- Large flow sheets can be prepared, since the blocks are calculated individually.
- The simulation can readily be understood in terms of process engineering.
- The start-up behavior is generally well behaved.

Disadvantages

- Convergence problems in flow sheets with many recycle streams.
- Incomprehensible convergence behavior in large flow sheets.
- Slow for large flow sheets with many recycle streams and complex design specifications.

The individual unit operations are calculated on the basis of the outlet streams of the preceding unit(s). The calculation sequence corresponds roughly to the flow direction in the plant. For recycle streams, iterative solution of apparatus loops is necessary. For this, a cut stream is defined (e.g, S7 in Figure 4-6), for which the value calculated after a process run is compared with a preset value or with the value of the previous iteration.

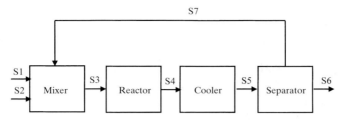

Fig. 4-6 Example of a flow sheet with recycling.

The termination criterion is a predetermined tolerance between the values.

Equation-oriented programs combine the equations for the linkage of the flow sheet and the apparatus model in a (generally weakly occupied) matrix, which is then "simultaneously" solved. However, as in the sequential modular approach, in many programs of this type, substance properties and phase equilibria are solved by subprograms. The equation-oriented approach has the following advantages and disadvantages:

Advantages

- Optimum convergence behavior in flow sheets with many recycle streams
- Certain convergence behavior
- Very fast.

Disadvantages

- Requires large storage space
- Errors difficult to find
- No insight into the units
- Specification of initial values necessary.

A modern commercial simulation program contains the following building blocks:

- User interface.
- Physicochemical data model (material systems such as conventional systems, solids, electrolytes, polymers, etc.; physicochemical data of pure substances and mixtures; equilibria etc.; see Chapter 3).
- Unit operation model (black box models such as mixers, separators, component splitters, etc.; models of phase separation and relaxation, heat-transfer model, multistage models, pumps and compressors, reactor models such as equilibrium reactor, stoichiometric reactor, tubular reactor, etc.; see Chapter 2).
- General simulation functions (sequencing, tearing, feed-forward/feedback controller, convergence, optimization, etc.).
- Additional programs.

4.2.1.3
Expert Systems [Ferrada 1990]

The difference between simulation programs and expert systems is that in the former the modeling is rigidly embodied by an algorithm, while in the latter, knowledge of the model is stored in the knowledge base independent of the deductive mechanism [Lieberam 1986, Ferrada 1990, Klar 1991]. An expert system is a computer program equipped with knowledge and ability which can carry out complex tasks by imitating human intelligence. Such systems are therefore particularly suitable for use in process development where complex problems are solved by drawing conclusions and using experimental knowledge. In this case, however, the knowledge does not only consist of facts which can be stored in data banks but is acquired by carrying out process development, experiencing failures, being successful, repeating the same process, and learning at the same time; in short, by acquiring a feel for a problem. Knowledge is gained as to when it is necessary to stick to the rules and when they can be broken; a stock of practically proven, heuristic knowledge (Table 4-5) is built up [Erdmann 1986, Hacker 1980].

Combinatorial problems such as drawing up an initial version of a process or developing a complex catalyst system result in a combinatorial explosion (Section 4.2). Time does not allow all the possible solutions to be tested. The human expert copes with such a combinatorial explosion by eliminating those possibilities which seem to him to be unfruitful, concentrating on those which seem to be feasible, and using heuristic knowledge which steers him towards the optimum solution but does not guarantee it. The main components of an expert system are as follows (Figure 4-7):

- Knowledge base: heuristic knowledge, represented in this case by a system of symbols
- Knowledge editor: a module which assists in changing, adding, or removing rules and formalizes their evaluation
- The deductive mechanism, which combines knowledge and problem data by deriving further data from the rules stored in the knowledge base
- Input–output systems.

This technique was first applied in 1969 for describing and designing composite heat exchanger systems [Masso 1969]. It was rule-oriented, i.e., the expert system involved a series of rules whose combination led to different results. The system was controlled by weighting factors, which were in turn dictated by success. The system "learned" through this method of changing weights and was therefore capable of yielding increasingly better results. The first expert systems were laboriously written in FORTRAN, but the technique acquired a new impetus in the 1980s when programming languages such as LISP (List Processor) were introduced. These are able to process symbols and symbol structures using a computer [Steele 1984]. In addition a wide range of computer tools became available. Although these have made the development and handling of expert systems easier, they have contributed almost nothing to an understanding of the phenomena [Puigjaner 1991]. It is now beyond dispute that expert systems have justified themselves, although the euphoria of the 1970s

Tab. 4-5 Examples of heuristic rules (rules of thumb) in process development.

- Maximize total yield
- Minimize number of process steps
- Favor fluid phases over solid phases
- Avoid extreme process conditions (high pressure, fine vacuum)
- Carry out each reaction or separation step only once (e.g., only one removal of low boilers).
- Remove easily isolable substances shortly after their formation; do not spread then over the entire processing train.
- Minimize recycle streams and design them so that they lead to small feedbacks.
- Preferably use separation processes with well-established technological experience (rectification over extraction, extraction over membrane technology).
- Rectification is preferred (each extraction is followed by at least three rectification columns).
- Treat separation processes that introduce foreign substances (auxiliaries) into the process as second choice (azeotropic distillation, extractive rectification, extraction).
- Auxiliaries (e.g., solvents) should preferably be substances that are already present in the reactor output. This can greatly simplify workup.

has given way to a more sober assessment. The belief that a lack of knowledge and understanding can be overcome by an expert system has been found to be completely erroneous. On the contrary, an expert system can only be used efficiently by experts [Sowell 1998].

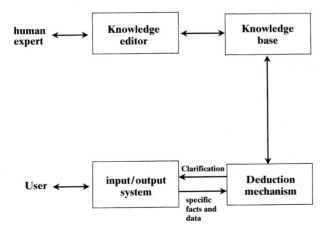

Fig. 4-7 Structure of an expert system [Trum 1986].

4.3.
Checking the Individual Steps

Like the reaction step or steps, the chosen preparatory and separation steps are first checked individually and independently of one another on a laboratory scale (Figure 4-8). Even at this stage, new requirements may be imposed on the reaction stage. For example, difficulty in removing a byproduct may make it necessary to alter the reaction conditions or modify the catalyst.

Laboratory experiments are generally carried out with pure, well-defined materials. Thus, the solvent selected for the absorption is used as received from the chemical stores. This differs, of course, from the subsequent situation, where the solvent must be recycled for cost reasons and inevitably becomes enriched with byproducts, some of which are unknown, however, at this point in time. The feedstock used is a synthetic mixture of gases prepared from pure materials which also does not contain byproducts. The experiment shows whether it is physically possible to absorb the useful product from the gas stream with reasonable amounts of solvent.

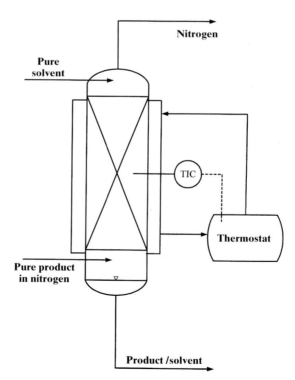

Fig. 4-8 Checking the individual steps in the laboratory using pure feedstock with the absorber stage as an example (see Fig. 4-4).

Parallel to the experimental work, for example, testing an absorber (Figure 4-8), thermodynamic simulation calculations are carried out for the unit concerned. If the absorber can be simulated with the available material data in a calculation run in parallel, work on this step will be complete. The unit is then another piece which can be fitted into the jigsaw puzzle of the entire process, which is the next item which must be considered in an integrated trial plant.

Once all the individual steps have been successfully tested, it is possible to draw up a reliable flow sheet of the entire process (Figure 4-9). However, if any subsidiary step is found not to be feasible at this stage, the process concept must be changed.

This preliminary flow sheet, which has not as yet been fully tested as a whole, can be used to draw up an initial rough cost estimate (Chapter 6).

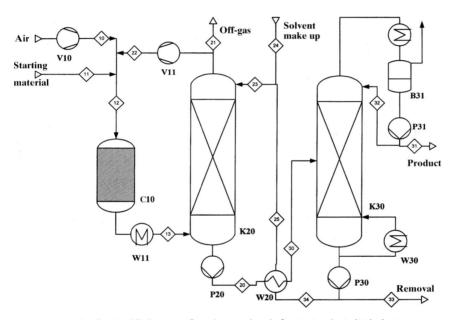

Fig. 4-9 Example of a simplified process flow chart produced after testing the individual steps.

4.4
The Microplant: The Link between the Laboratory and the Pilot Plant

Because of the high cost of pilot plants and the long times required for their construction, a newer pathway in process development attempts to introduce an intermediate step, the so-called microplant [Behr 2000], between the laboratory, in which only individual units are investigated, and the pilot plant, in which the individual units are operated in a network.

The development of a process requires, in an early stage of planning, knowledge about the substances occurring in the process and their behavior in the units. In general, not all components of a reaction product are known. Experiments that accompany process development should therefore strive to recognize the behavior of unknown components as soon as possible and to set up recycle loops, so that the accumulation of trace components and its effect on the process can be recognized and investigated.

As can be seen in Figure 4-9, when the individual steps are put together, recycling streams are created (in this case one gas and one solvent recycling path). These recycling streams or feedbacks are an economic necessity, but they raise new process engineering problems. Thus, material recycled to the reactor may drastically affect the performance of the catalyst, especially its activity and service life. In addition, the solvent circuit may become enriched in byproducts and this may lower product purity. In principle, these problems can only be solved by mathematical simulation, but since many of the quantities that are required for a mathematical description are unknown, an experimental approach must be adopted. Therefore, it is important to know at as early a stage as possible, the influence of the most important reycle streams on the overall behavior of the plant (reactor with catalyst, separation behavior) and to thus achieve further discrimination between process variants. An economical and fast solution here is a microplant.

A microplant is 100 times smaller than a miniplant (Section 4.5.1) and can be operated as simulation technology in a laboratory (Table 4-6). It is a purely experimental tool which can be used to test numerous process variants at the laboratory stage with the aim of finding the best one. Thus, a microplant should be sufficiently small and compact that it fits on a laboratory bench. A modular design allows several microplants to be constructed and operated in parallel, which considerably shortens development time and permits rapid and simple erection and dismantling. Obviously, time, costs,

Tab. 4-6 Characteristics of integrated trial plants.

	Production in $g\ h^{-1}$	Apparatus diameter in mm	Throughput in $ml\ h^{-1}$ up to
Pilot plant	10 000	500-100	1 000 000
Miniplant	100	50-10	10 000
Microplant	1	5-1	100

and laboratory space can be saved in this way. However, the construction of corresponding units such as microreactors, micro distillation columns, microextractors, and the manifolds for interconnecting them is still in the development stage. However, in ten years time, this technology can be expected to account for a cosiderable share of process development.

Example 4-2

When a plant which includes recycling streams (Figure 4-9) *is operated, the solvent circuit becomes enriched in a high-boiling substance whose existence was not previously known because it was below the analytical limit of detection in the reaction product. This high-boiling substance also has the troublesome property of depositing when it reaches a certain concentration in the solvent and forms a coating in the column.*

Since the high-boiling substance was not known before, it was not possible to include it in the mathematical simulation. Now that it has been discovered as a result of operating an integrated plant (micro- or miniplant), it can be characterized and incorporated in the simulation [Buschulte 1995]. Whereas the microplant is used to test the most important recycle streams at the laboratory stage, in the integrated miniplant all recycle streams are investigated.

4.5
Testing the Entire Process on a Small Scale

The process now leaves the research phase and enters the development phase. This is associated with a jump in costs, which makes disciplined milestoning and continuous cost and time control necessary. Today mainly miniplants are used as integrated test plants, and these are described in more detail in the following.

4.5.1
Miniplant Technology [Heimann 1998]

4.5.1.1
Introduction

Improvement in computer programs for modeling processes has resulted in the integrated miniplant technology acquiring ever increasing importance in recent years because the synergism between the miniplant technique and mathematical simulation means that the margin of safety in scaling up is as good as that obtained by setting up a pilot plant [Greß 1979, Maier 1990, Robbins, 1979, Brust 1991]. Miniplant technology has the following characteristics:

- The trial plant includes all the recycling paths (it is a so-called integrated miniplant) and it can consequently be extrapolated with a high degree of reliability [Wörz 1995].
- The components used (columns, pumps, condensers, pipelines) are often the same as those used in the laboratory.
- Where possible, reusable standardized components which have been tried and tested in continuous operation are used. Thus, investment costs are low and flexibility is high.
- The plant is operated round the clock on a shift basis for weeks and it therefore must be as fully automated as possible to keep operating costs low, but not, however, at the expense of flexibility.
- The measuring and control instruments used are standard components which are nowadays generally connected to a small process control system [Wörsdörfer, 1991], thus making it possible to carry out modifications in measurement and control rapidly.
- The entire plant is normally set up in an extracted chamber so that explosion-proof operation is possible and safety-at-work requirements [Behr 200] can be met more easily.

4.5.1.2
Construction

The following design documents are required as a minimum for constructing a miniplant:

- A process flow sheet showing all the quantity flows in the miniplant and the associated temperature and pressure conditions
- Engineering flow diagram with required information:
  - All equipment and machines
  - Internal diameters and pressure ratings of pipelines and construction material
  - The objectives of the instrumentation
  - Information on the insulation of the equipment and pipelines
- List of measuring points (measuring point number, specification, parts list, point of installation; see also Chap. 2.8.1)
- Safety concept, i.e., how the sensors (e.g., for temperature, pressure, flow, level) are linked to the actuators (e.g., valves, pumps) by the safety logic (see also Section 2.8.3)

Specification of equipment size

In a miniplant, the hold-up for a given throughput is almost always greater than in an industrial-scale plant. The size of the vessels and column bottoms should therefore be minimized, otherwise it will take too long for the miniplant to reach a steady state because of long residence times. In the column bottoms, which are particularly critical, a hold up of 100 – 200 mL is normally the lower limit. Below this level, a laboratory evaporator can usually no longer be operated efficiently (because of excessive temperature differences, foaming, etc.).

Example 4-3

A miniplant produces a high-boiling residue at a rate of $1 g h^{-1}$. This residue is removed from the bottom of a distillation column with a distillation boiler capacity of 200 mL, so that the mean residence time is 200 h. The time required to reach the steady state is consequently about 25 d (about 3 times the mean residence time). Only then will it be possible to assess this bottom residue and draw conclusions about its fouling behavior, composition, properties, and so on.

Specification of material balance

In Section 4.2 it was shown that a miniplant should be designed by scaling down a hypothetical large-scale plant. The scale-down factor is determined by finding the points in the process where the smallest and largest quantity flows occur. The levels at which these flows can be handled determine the scale-down factor. Normally only flows of at least $1 g h^{-1}$ and at most $10 kg h^{-1}$ can be handled in a miniplant. At the same time, the minimum flow is primarily determined by the time required to reach a steady state (see Example 4-3), while the upper limit of the flow is set by the maximum hydrodynamic load of the laboratory column used (max. diameter ca. 50 mm).

Example 4-4

Hypothetical large-scale plant producing $5 t h^{-1}$:

- *minimum flow: $50 kg h^{-1}$ of a high-boiling substance which must be removed and incinerated*
- *maximum flow: $100 t h^{-1}$ of the reaction throughput, which contains the useful product as a 5% aqueous solution.*

Scale-down factor:

- *for the minimum flow: $50 kg h^{-1} / 1 g h^{-1} = 50\,000$*
- *for the maximum flow: $100 t h^{-1} / 10 kg h^{-1} = 10\,000$*

Thus the miniplant should be designed to produce not less than 100 g or not more than 500 g of useful product per hour.

Specification of type of equipment

Unlike normal laboratory experiments in which the equipment operates only during the daytime, miniplants must run continuously over a period of weeks to be able to fulfil their task. This is due to the fact that the time required to reach a steady state is often long and because it is only possible to draw reliable conclusions about long-term effects such as corrosion, fouling, catalyst deactivation after a long time (1 week to 6 months). Unlike laboratory plants, the requirement imposed on the reliability of miniplant components are therefore very high and almost the same as for a large-scale plant. Occasionally this sets very tight restraints on choosing equipment and in some cases necessitates in-house development if nothing suitable is available on the market.

4.5.1.3
The Limits of Miniaturization

Table 4-7 lists the minimum dimensions for which it is possible to scale-up a process.

Columns can still be operated effectively down to 35 mm diameter when filled with an ordered packing. For plate columns having one bubble cap per plate, the limit is about 50 mm. At still smaller dimensions, wall effects are difficult to suppress.

Experimental equipment for handling solids or dealing with extreme operating conditions is often not available. This often necessitates in-house developments and co-operation with specialist firms, university institutes, and research companies.

Expenditure on instrumentation is often very much lower for a miniplant than for the corresponding large-scale plant. For this to be the case, however, most of the miniplant components must be made of glass and therefore operate at low-pressure, thus making it much easier to monitor liquid levels, deposits, foam formation, etc., than in the large-scale plant. Schulz-walz has given examples of measuring and metering techniques for miniplants [55]. In the last two decades, important advances have been made in sensor technology and, along with electronics and processor technology,

Tab. 4-7 Examples of established miniaturization limits [Maier, 1990].

Examples	Tried and tested limits
Columns	30 mm diameter, packed column
	35 mm diameter, structured ordered packings, dual-flow columns
	50 mm diameter, bubble-cap plates
Pumps	1 mL h^{-1} jet pumps
	10 mL h^{-1} piston pumps
	100 L h^{-1} rotary pumps
Belt Filters	50 mm belt width
Centrifuges	5 L h^{-1} output
Pipelines	1.5 mm outside diameter, metal or PTFE

Table 4-8. Established miniaturization limits for measurement and control components [Maier 1990].

Measured quantity	Measurement principle	Tested minimal measurement range
Gas streams	thermal	$0.02-0.6$ L h^{-1}
	volumetric	$2-200$ L h^{-1}
Liquid streams	thermal	$2-30$ g h^{-1}
	Coriolisforce	$0.07-1.5$ kg h^{-1}
	magnetically inductive	$0.6-6$ L h^{-1}
	volumetric	$0.1-1$ L h^{-1} for piston meters $2-40$ L h^{-1} for oval-wheel meters 1 mL h^{-1} for dosing pumps
	gravimetric	$2-50$ g h^{-1} (weighing range $0-5000$ g)
Temperature	resistance change	max. range $-200-600$ °C
	thermoelectric potential	max. range $0-1000$ °C
Level	conductive	> 50 mm
	capacative	> 50 mm
	hydrostatic	> 4 mbar
Pressure differences	wire strain guage	from 1.25 mbar
	piezoresistive	
	capacitive	

this has considerably simplified the miniaturization of sensors for miniplants (Table 4-8).

There are four stages in the automation of a miniplant:

Stage 1 At this stage, the electrical instrumentation of the miniplant consists of compact, self-contained individual components, for example, balances, metering controls, metering pumps, thermostats, temperature controllers, pH sensors, individual controllers, control valves, pressure sensors, and plotters. Monitoring and operation is the responsibility of the shift personnel.

Stage 2 At this stage the structure of the sensors, actuators, and individual components is the same as in Stage 1, but the individual components may be connected to a data processing computer via a data interface. Monitoring and operation is again the responsibility of the shift personnel.

Stage 3 Here the system for logging the measured data is the same as in Stage 2, but the lower-ranking self-contained individual components are controlled and provided with set points by a higher-ranking automation software, so that shift personnel are able to concentrate on monitoring tasks.

Stage 4 At this stage the miniplant is equipped with a compact process control system. While the actuator and sensor system is the same as at Stage 1, all the other functions such as controlling, measuring, regulating, calculating, processing data, optimizing, keeping records, alerting, logging, operating, and observing are performed by the process control system.

The choice of automation stage is decided during the miniplant design phase on the basis of the specific plant requirements and boundary conditions as well as on the results of trials. The higher automation Stages 3 and 4 are only justified if a miniplant is to be operated over a long period of time.

4.5.1.4
Limitations of the Miniplant Technology

A disadvantage of the miniplant approach emerges if individual steps in the process are subject to an unduly large risk in extrapolating from the miniplant to the industrial-scale plant. For example, scale-up factors of about 1000 are still not feasible for extraction and crystallization steps even today. This disadvantage can often be circumvented by isolating critical sections of the process and working on them in an intermediate stage (pilot stage). If enough of the unmodified product can then be collected from an integrated miniplant at an acceptable cost for it to be possible to operate, e.g., an extraction column or a crystallizing plant on a pilot scale with a representative feed for a sufficiently long time, the entire plant can again be extrapolated upwards with a calculable risk. This may obviate the need for constructing a cost-intensive and time-consuming pilot plant for the entire process [Berty 1979].

4.5.2
Pilot Plant

It may be necessary to construct a pilot plant if one of the following conditions applies [Palluzi 1991]:

- The scale-up risk is too large to proceed directly from the miniplant to the industrial-scale plant
- The process involves several critical stages which cannot be described by physical models (e.g., handling of solids)
- A difficult or completely novel technology is being used
- It is necessary to provide representative product quantities, for example, for the market launch, and these cannot be produced by the miniplant in a reasonable time.

The operation of the pilot plant should clarify all the issues which have not been fully dealt with in the miniplant. These may include, for example [Lowenstein 1985, Berty 1979]:

- Checking design calculations
- Solving scale-up problems
- Checking experimental results obtained with the miniplant

- Measuring the true temperature profiles in the reactors and columns under adiabatic conditions
- Gaining process know-how (dynamic behavior of the plant, start-up and shut-down procedures)
- Producing representative sample quantities in fairly large amounts
- Precisely assessing fairly small flows (e.g., residues)
- Training the personnel who are to run the plant
- Improving the estimate of the expected service life
- Carrying out material tests under realistic conditions.

A pilot plant should be designed as a scaled-down version of the industrial-scale plant and not as a larger copy of the existing miniplant. Nowadays it costs almost as much to design, construct, and operate a pilot plant as an industrial-scale plant. The decision to build a pilot plant results in considerable costs. As a rule of thumb, the initial investment is at least 10% of the investment costs of the subsequent industrial-scale plant. Moreover, if they use particularly toxic substances, pilot plants now often require approval, and this may considerably lengthen the time required to put them into operation.

Once process development on a pilot scale has been successfully concluded, the pilot plant must be kept on stand-by until the industrial-scale plant is running satisfactorily. Normally, when the larger plant is started up, the pilot plant is operated simultaneously so that any problems which occur in the former can be dealt with rapidly. Furthermore, licensing questions can also be dealt with there.

In spite of all the know-how of process engineering, very expensive flops on an industrial scale can not be ruled out, and numerous examples of these can be found in the history of the chemical industry.

5
Planning, Erection, and Start-Up of a Chemical Plant

The history of the targeted planning of chemical plants is only about 100 years old [Krätz 1990, Sattler 2000]. The construction of a complete new chemical plant was planned on the drawing board for the first time in 1898 by Duisberg. Prior to this, one simply built what one needed at the moment, and hence after a short time chemical plants developed a chaotic appearance. Today, long before a construction site is developed, concrete is poured, or steel is delivered, a plant already exists as a pile of paper or computer drives.

Chemical production plants are generally complex one-off designs that are subject to numerous rules and regulations [Wengerowski 2001]. For the planning and erection of such plants, a three-stage procedure has become established, whereby the planning process is refined in individual steps with defined targets. Especially in the early stages of the planning process, constructive thinking and creative ideas can lower the investment costs. The two main targets are:

- Achieving a high level of investment security
- Minimization of technical risk.

5.1
General Course of Project Execution

With completion of the project study (Chapter 6), proces development comes to a certain conclusion. If all of the boundary conditions are positive, then a strategic or economic decision must be made by the responsible company division or the board of directors on the basis of the available information (project study, marketing studies, patent situation, etc.). Every delay costs money, since the test plants (mini- or pilot plant) must be operated again, albeit to a lesser extent, or because a competitor could reach the market earlier. Popular excuses for further delaying a decision are:

- Have you already investigated that?
- We should wait for the detailed results!

and so on. If the decision is negative, then the consequence must be to dismantle the test plant. If it is positive, that is, application is made for preplannning permission, then an internal or external engineering department is commissioned to prepare a feasibility study. Leadership of the project is transferred from research to the corresponding engineering department. Many large chemical companies have their own development and engineering department that can carry out all necessary activities from process engineering research to plant erection (in-house engineering). However, owing to rationalization measures, in the last few years these have often become smaller or been outsourced to such an extent that only a core team is present that carries out the basic planning. Detailed planning, procurement of equipment, and supervision of construction work are performed by contractors with advice from the engineering department. In the case of new plants with a large degree of internal know-how, one attempts to carry out the project with the internal team, while standard

Process Development. From the Initial Idea to the Chemical Production Plant. G. Herbert Vogel
Copyright © 2005 WILEY-VCH Verlag GmbH & Co. KGaA, Weinheim
ISBN: 3-527-31089-4

plants are erected by external plant construction companies. The cooperation with contractors is usually specified in contracts (lump-sum contract, reimbursable contract, or a combination thereof), whereby incentive payments to contractors are playing an ever increasing role when a certain project is to be completed under budget or faster than planned. Strategic alliances between clients and contractors should also be mentioned here.

The course of the project from the end of process development in research up to start-up of the industrial plant follows the scheme shown in Figure 5-1 [Dietz 2000, Wengerowski 2001].

Phase I
The aim of this planning phase (also known as basic design phase, preproject) is to analyze at a reasonable cost whether from the viewpoint of management it is economical to realize the process described in the project study as an industrial plant. The plant must be defined; for this purpose, a planning department prepares numerous studies and analyses to facilate a decision by management. Together these are known as a feasibility study. The focus here is on evaluation of the process and its economic viability, evaluation of alternatives and preliminary clarification of the legal licensing situation. Table 5.1-1 lists important activities in the planning phase.

The time required for this phase varies quite widely from project to project and depends on many factors. Delaying tactics are not appropriate here, since capital is required (e.g., for keeping the test plants running). However, after the preplannning permission has been granted, generally by the responsible division of the company, at least 3–6 months are required to prepare the necessary documents. If possible, the potential project leader, who will later share complete responsibility for the project with the project engineer, should also be chosen in this phase.

At the end of this phase, an independent commision should examine the project application from the following viewpoints:

- Startegic importance (is the project compatible with the medium- to long-term company strategy; see also "company portfolio" in Section 3.7)
- Economic importance (return on investment; see Section 6.2)
- Technical maturity and risk (see Section 6.2)
- Official licensing situation (see Section 5.2.1).

If the decision is positive, the process passes to the basic engineering phase, possibly with the imposition of certain conditions.

Phase II
In this phase, the so-called basic engineering phase, which extends from the end of the planning phase up to licencing, detailed project documents are prepared (Table 5.1-2). These form the basic skeleton for plant plans, which subsequently enter the contract writing and ordering phase. As a result of the basic engineering phase, the plant and its functions in all components are fully defined. In this phase, too, modern engineering planning software is indispensible.

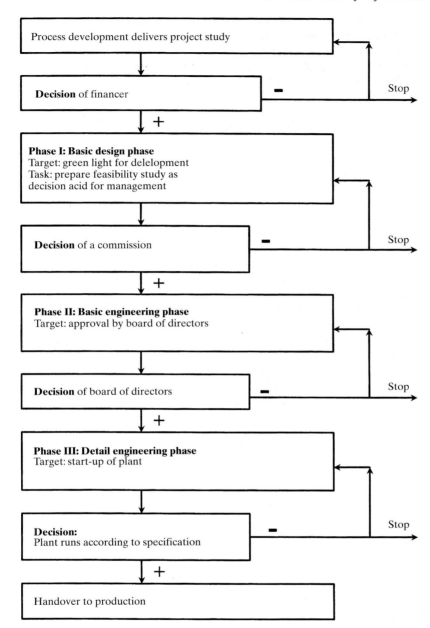

Fig. 5-1 Course of project execution.

Tab. 5.1-1 Important activities in the planning phase.

Activity	Result	Responsible persons
Determination • plant capacity • on-stream time • location	logistic requirements	marketing production
Investigation of alternatives (location costs, environment)	process concept	research, development, production
Materials recommentations	material key	research, development
Collecting material data	substance data files	research, development
Process simulation and partial optimization	process flow sheets with mass balance	development, planning
Linnhoff-analysis (Chap. 6.1.6.2)	optimized energy balance	engineering department
Rough dimensioning of machines and apparatus	technical data sheets for main equipment	development, planning
Erection study	plastic model	planning
Drawing up erection and site plans	siteplan	planning
Safety study, (1st stage)	safety requirements catalogue	safety department
Estimation of investment costs ± 15 %		engineering department
Timetable for phases II and III		engineering department
Gathering all documents for approval of project execution	project application	engineering department

Next the individual process steps are developed (design, calculations, tests, quality assurance measures). All of these activities enter the piping and instrumentation (P&I) diagram, which later forms the basis for preparing the plant model, planning construction work, technical data sheets, and the safety and measurement and control concept.

The project team, consisting of the project leader and the project engineer, must be chosen. They are responsible both for working out and implementing the project. Together with the support of experts, where necessary, they are responsible for:

- Planning
- Preparation
- Preparation and monitoring of timetables
- Supervision of plant construction
- Checking costs
- Commissioning
- Start-up.

The individual process steps and the approximate size of the required machines and apparatus are known from the project or feasibility study. If new information becomes

Tab. 5.1-2 Activities in the basic engineering phase.

Activity	Carried out by
Mass and energy balance (optimized process flow sheet)	research, development, engineering dept.
Drafting P&I schemes	project team, engineering
Safety study, 2nd stage (Formulate safety requirements for next stage)	safety dept., research
Documents for licensing authority	licensing dept.
Meeting with authority	licensing dept., responsible company section
Dimensioning of machines and apparatus	development, engineering dept.
Medium key and materials list	engineering dept.
List of electrical consumers	engineering dept.
Estimate of investment costs ± 5 %	engineering dept.
Timetable for Phase III	project team
Personal requirements	personnel dept., union
Patent and licensing situation	patent dept.
Application for licensing	

available due to safety considerations or advances in the state of the art, then the process concept is modified and redrafted. This iterative process must be carried out many times during development of the individual process steps. Therefore, documentation of the decision criteria and a record of the state of knowledge are important.

In parallel, the necessary licenses are applied for. At the end of this phase the investment cost can be estimated at reasonable expense with an accuracy of ± 10 %. Apparatuses with long delivery times are generally ordered at the end of the basic engineering phase.

Depending on the size of the project, at least 6–12 months are required to prepare the complete documents. Unexpected events such as objections from ousiders (private persons, environmental organizations, patent challenges) can greatly increase this period of time.

Phase III
After the green light from the board of directors the money required for realization of the project is available, and the detail engineering phase begins. Now the definitive design plans and documents are prepared. From this point onwards the entire project team is under pressure, since every delay means that start-up deadline has to be delayed, with the result that the invested capital cannot produce a return on investment. Thus, the most important target now is to start up the plant as soon as possible. Only with precise planning of deadlines (e.g., with the aid of block diagrams, etc.) and their "brutal" enforcement by the project leader can a smooth course of planning and construction be ensured [Hyland 1998].

After completion of the detailed planning, the results (P&I flow diagram, technical data sheets, etc.) are summarized and pictorialized. For this purpose a model of the plant is prepared (plastic or CAD model). On the basis of this model construction of the plant with all auxiliary units is begun (Table 5.1-3).

The detail engineering phase usually takes 20–24 months and generally cannot be accelerated, regardless of the efforts made by the project team, since the deadlines are detremined by delivery and worker times. Table 5.1-4 shows a typical timetable for execution of a project.

Tab. 5.1-3 Activities in the detail engineering phase.

Activity
Detailed planning:
• Process flow sheets • P&I Flow sheets • Erection plans • Media keys • Machine and apparatus data sheets
Safety study (3rd stages)
Procurement:
• Drafting of technical data sheets for machines and apparatus • Negotiations for placing of orders • Checking producer drawings and construction plans
Pipeline planning and model building
Controlling costs
Controlling deadlines
Commissioning plant
Monitoring erection
Completion of project including documentation

Tab. 5.1-4 Timetable of a project execution.

Years	Activity
0	**Start-up**
0,5	**Mechanical completion**
1	
1,5	**Start of assembly** (model on construction site)
2	**Start of construction** (ca. 80% of orders made)
2,5	License obtained Basic model complete (begin detailed model) **Approxal from board at directors** Safety study, stage III, ordering large apparatuses
3	**Discussion with authorities (if necessary)**
3,5	Publication by authorities **Discussion with authorities** Detailed P&I scheme complete **Green light for planning**
4	Safety study, stage 1 Begin to propable licensing documents **Plastic model**
4,5	**Process flow sheet** (Mengengerüst)
5	Kickoff meeting, naming of project team **Determination of capacity**
5,5	
	Project leader named

5.2
Important Aspects of Project Execution

5.2.1
Licensing

Today the licensing of a planned investment is increasingly becoming the rate-determining step in the execution of a project. Whereas the technical processes have been increasingly improved and optimized by means of automization and electronic data processing, licensing depends on external authorities. Therefore, it is important to maintain good relationships with these, to inform them as soon as possible, and to involve them in decision making. Today, a project team is faced with a myriad of laws, rules, and regulations which affect the planning, construction, and operation of chemical plants. For example, in Germany, before even the first sod is turned a legal license must be present for the erection and subsequent operation of the plant. The principle of the authorities that "anything that is not forbidden is permitted" has been switched to the opposite, that is, "only that which is permitted will be licensed". Therefore the project team requires support from experts in "authority engineering".

In Germany, materials-processing plants must comply with the Federal Antipollution Law (Bundes-Immissionsschutzgesetz, BImSchG), which in its first paragraph states: "(1) The aim of this law is to protect humans, animals, plants, soil, water, and air, as well as objects of cultural importance and other property, against harmful environmental influences. (2) In the case of plants that are subject to licensing, it also aims at:

- Integrated avoidance and reduction of harmful environmental influences due to immissions in the air, water, and soil, with inclusion of the waste-disposal industry, and thus attaining a high degree of protection for the environment.
- Protection against and prevention of risks, major disadvantages, and considerable loadings that have other causes.

Harmful environmental influences are immissions (i.e., effects of the plant on the environment) such as:

- Air pollution (e.g., smoke, soot, gases, aerosols, vapors, and odors)
- Noise
- Vibration
- Light
- Heat
- Radiation.

The most important paragraphs are summaraized in Table 5.2-1.

Tab. 5.2-1 Paragraphs of BImSchG (version of 14.5.1990) that are important for chemical plants.

Paragraph	Content
1	Purpose of the law
4	**Licensing**
5	Basic responsibilities of operators of plants subject to licensing
6	Prerequisites for licensing
7	Regulation regarding requirements for plants subject to licensing
10	Licensing process
15	**Modifications to plants subject to licensing (notification)**
16	**Major modifications to plants subject to licensing (application)**
19	**Simplified process (licensing)**
62	Infringements

Since technological development is rapid, BImSchG only formulates the target; means and measures are described in technical rules, guidelines, and handbooks; administrative regulations; and implementation regulations. These are updated according to the state of the art. The most important regulations for the licensing of chemical plant are:

- 4th BImSchV (with appandix for licensable plants)
- 12th BImSchV (Störfallverordnung: industrial emergencies regulation)
- TA Luft (clean air regulation)
- TA Lärm (low noise regulation).

The required effort (quantifiable in terms of the number of persons involved or the number of files produced) shows how complex modern-day authority engineering is in Germany, and how important coordination is for making rapid progress, since the time required for licensing is increasingly becoming the rate-determining factor in the birth of a new chemical plant. A legal distinction is made between three types of licensing process:

- Nonlicensable plants (§ 22 BImSchG, App. 1 and 2)
- Simplified process (not open to the public, no hearing, no prediscussion, unless the plant is subject to investigation for environmental safety (Umweltverträglichkeitsprüfung, UVP), in which case an application must be approved in the formal process)
- Formal process.

The formal licensing process according to BImSchG with publication is the most extensive and time-consuming process for chemical plants (Figure 5-2). After preparation of the basic concept, a preliminary discussion is held with the licensing authority and representatives of the most important specialist authorities (e.g., health and safety inspectorate, water authority) to determine the licensability, the type of application, the required application documents, and the question whether the proposal will be pub-

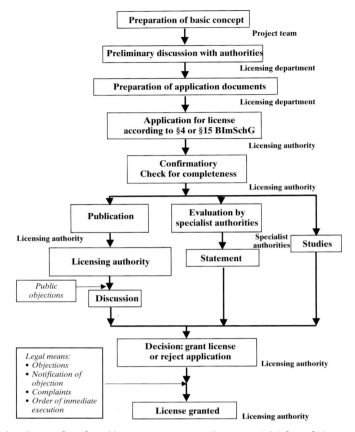

Fig. 5-2 Flow diagram for a formal licensing process according to BImSchG §4 or §10.

lished. The talks with the representatives of the authorities are repeated if necessary, also during preparation of the documents and after submission of the application, in order to discuss open questions. In the case of new plants that are classified in column 1 in the appendix to the 4th BImSchV and those subject to environmental safety tests (UVP), under certain circumstances the proposal must be published and the application documents (divided into open documents for public inspection and secret documents for protecting know-how and company secrets) must be published (Table 5.2-2). If objections are raised by affected persons (e.g., those living nearby, environmentalists), a meeting must be held to discuss these. After the discussion and the statements of the specialist authorities, the licensing authority decides on granting or refusing (generally with conditions) a license. If all prerequisites for licensing are fulfilled, then the applicant has a right to have the license granted. The applicant and the objectors have the right of appeal in court against the decision of the licensing authority, which can greatly delay the legal validity of the license.

Table 5.2-2 Contents of a licensing application according to BImSchG.

- Location of plant
- Type, size, and purpose of process
- Description of process
- Technical units of the plant; described by
  - Flow sheet
  - Apparatus list
  - Construction plans, apparatus erection plans
  - Handled substances (starting materials, intermediates, side and end products, fuels, residues)
  - Measures for off-gas purification, noise prevention, fire and explosion protection, wastewater treatment
  - type and extent of emissions
  - Occupational health and safety
  - Safety
  - Water protection

In the 1970s and early 1980s, the average time required for a decision on an application according to BImSchG was about 10 months, but with the introduction of the 5th amendment WHG (Wasserhaushaltsgesetz: water resources management act) in 1987 (so-called Sandoz effect) and the amendment (Störfallverordnung: industrial emergencies regulation) in 1988, this more than doubled by 1990. Owing to the worsening economic situation in the early 1990s, the authorities made efforts to rectify this situation, which was unpleasant for all concerned, and by the late 1990s, this time had dropped to 4–6 months following submission of the application. The deadline for examining the completeness of the applications documents is regulated in 9th BImSchV §7 (immediate, generally within one month). The deadlines for decisions by the authority and notification of the applicant are regulated in BImSchG §7 App. 6a (formal process 7 months, simplified process 3 months), whereby an extension of three months by the authority is possible under certain circumstances.

5.2.2
Safety Studies

Apart from economic aspects, important demands on chemical plants are that they function faultlessly and meet the required performance, that is, plant availability. Furthermore, they must be safe during normal operation and during disturbances to normal operation. In the ideal case, safety should be an issue at every stage of process development: laboratory, miniplant, pilotplant, and industrial operation. The most cost effective solutions are generally developed in the early stages of planning.

In comparison with other branches of industry relatively few disturbances occur in the chemical industry. However, the potential hazard involved in handling the chemical raw materials and products is relatively high. The substances must generally be handled under conditions of high pressure and temperature, and they are often toxic, caustic, flammable, explosive, and so on. Therefore, the chemical industry has recognized the importance of safety from the very beginning, and the current state of knowledge concerning specific safety-technological problems in chemical plants is accordingly advanced. The design, development and erection of a safe chemical plant is an interdisciplinary task. It can only be achieved when chemists, engineers, plant constructors, and measurement and control technicians work together. The ultimate target is the inherently safe plant.

Given today's myriad laws, it is often overlooked that the industry is itself interested in safe and environmentally friendly plants. Only safe and environmentally friendly plants operate economically. Therefore, the chemical industry often employs safety measures that go beyond the requirements of legislation. Their aim is to depict the basic safety technology concept in such a detailed form that specialists without specific knowledge of the project can evaluate it [Tröster 1985, Eichendorf 2001]. The safety concept of a chemical plant is determined, among others, by the following:

- Qualification of the employees
- Material properties
- The process
- Location
- Design of buildings and plant
- Requirements of authorites and of the board of directors

With a view to transparent project execution, it is necessary to expound the most important safety aspects from time to time in a safety study according to the state of progress of the project and to discuss them in an expert group (project team, R&D, safety department, engineering), to adopt them, and to document the results. Such a safety study is performed internally for all plants and should not be confused with that required under the extended requirements of the Störfallverordnung. In the course of plant planning, three milestones can be recorded [Henne 1994]:

1. *In the basic design phase*
   Explanation and decision that the chosen process can be carried out on safety grounds at the chosen site
2. *In the basic engineering phase*
   Development of the safety concept and its evaluation by experts without specific knowledge of the project
3. *In the detail engineering phase*
   Checking the planning documents for safety technological consistency.

Safety study in the basic design phase

The aim is to list the most important possibilities for hazards from the process and the hazard sources of the chemical plant, to formulate the tasks for the basic safety concept, and to determine whether individual safety problems must be investigated. Furthermore, it must be shown that the chosen site is suitable from the viewpoint of safety technology. The results are recorded in a safety study (basic design phase), which can have the following components:

- Process description, concentrating on safety questions (reaction control, side reactions, runaway and die-off behavior of the reaction, etc.)
- List of substances with the most important physicochemical, safety, and ecotoxicological properties of all starting and auxiliary materials, residues, intermediate and side products; approximate amounts in the whole plant, and their distribution over the individual palnt components
- Environmental load: list of all environmentally relevant influences, such as waste air, off-gases, noise, residues, and wastewater with quantities and qualities.
- Safety concept: description of all safety measures for plant components that exhibit a major hazard potential or represent potential sources of hazards (e.g., extreme operating conditions, special construction materials for corrosive media, formation of explosive mixtures, ecotoxicolgically questionable substances). The details of the basic safety concept can be drawn up later.
- Proposal of site: reasons for the choice of site from the viewpoint of safety (site plans, neighboring plants, etc.).

Safety study in the basic engineering phase

The safety concept at the proposed site is described in detail, so that it can be evaluated by specialists without specific knowledge of the project. The content of this safety study (basic engineering phase) should be structured as follows:

- *Safety concept of the construction concept*
  Locations of critical apparatus, emergency escape routes, design of the measuring room, special measures, etc.
- *Measurement and control concept*
  Tasks and function of shutdown devices and installed safety measures, redundancy and diversity of safety systems, safe positions of fittings, etc
- *Special design criteria*
  Explosion- or pressure-resistant design, etc.
- *Tightness of the plant*
  Gaskets on machines an apparatus for critical materials, etc.
- *Inerting*
  Which apparatus, pressure and its assurance, etc.
- *Explosion protection*
  Assignment of zones (Section 2.9.1), measures to avoid static charging, etc.

- *Protective measures for failures of energy supply*
  Emergency electricity supply, failure of individual energy sources, safe disposal of emissions, etc.
- *Environmental load*
  waste air, off-gases, residues, and wastewater; how are legal requirements met?, etc.
- *Emergency pressure relief, emergency emptying systems, safety valves*
  Only critcal safety valves, that is, those from which, for example, toxic substances can be released
- *Fire prevention*
  For example, special protection of important plant components
- *Special measures*
  Explosion barriers, steam barriers, etc.

Safety study in the detail engineering phase
In this phase the aim is to check the already prepared planning documents such as P&I flow sheets, and function plans for safety-technological consistency. Since this is often highly laborious, it is often restricted to plant componets that are particularly relevent to safety, which were specified in the preceding studies. Because of the large amount of work involved. This step can become rate-determining for the entire project and should therefore be carefully integrated into the overall timetable. Here documentation is especially important, that is, the considerations that led to a safety measure being implemented or omitted must later be comprehensible and legally tenable. The implementation of these safety discussions is carried out according to established methods. Here it is important to recognize all possible deviations from normal process operation, to estimate the probability that they occur and their effects, and, if necessary, to provide protective measures. A suitable method for this is a hazard and operability (HAZOP) study [Bartels 1990, Pilz 1985], that is, preventing a disturbance in a chemical plant by:

- Prognosis
- Determination of causes
- Estimation of effects
- Countermeasures.

5.2.3
German Industrial Accident Regulation (Störfallverordnung)

The term "accident" should be used carefully. Not every leak or operational disturbance can be regarded as an accident. Only when a disturbance to the normal operation of the plant endangers human life or health or could be a danger to the environment should this term be used.

The Störfallverordnung (see Appendix 8.19) applies to plants that are licensable according to BImSchG. It names substances that are to be regarded as critical. If one of these substances is to be produced in the plant or can be formed during a disturbance, the Störfallverordnung clearly specifies which safety measures must be taken in addition to those specified by BImSchG. If a plant is subject to the extended responsibilities of the Störfallverordnung, a detailed safety analysis must be prepared [Pilz 1985, Hezel 1986]. In addition to detailed descriptions of the plant and the substances, all possible deviations from normal operating behavior must be recognized and, if necessary, protective measures provided. The hazard analysis can be performed, for example, by the above-mentioned HAZOP method. This safety analysis accoring to the Störfallverordnung can be built up as follows:

- Introduction
- Description of the substances
- Description of the plant and the process
- Description of the plant components relevant to safety technology
- Effects of disturbances
- Result of the safety analysis (e.g., a maintainence and repair schedule specifies how often and at which locations the safety-relevant plant components are inspected).

This safety analysis is part of the licensing documents and must be revised in the case of new safety-technological findings.

5.2.4
P&I Flow Sheets

Before the P&I diagram (Figure 5-3) is prepared, the project leader should prepare a catalogue of guidelines for the team members, in which important conventions are recorded, for example:

- Numbering of process stages (e.g., stage 1000 (= reactor), stage 2000 (= absorber for reaction gas); see also Figure 4-9.
- Each apparatus is given a four-digit number that indicates its location and function (e.g., 20xy for all parts that can be functionally assigned to the sump of the absorber, and 21xy for all parts that can be functionally assigned to the head of the absorber).
- The spatial arrangement of the apparatus should be evident from the P&I flow sheet.
- Control stations should be drawn so that positions of valves and so on the flow sheet and on-site correspond.
- Pumping stations and so on should all be constructed with the same design philosophy.

322 | 5 Planning, Erection, and Start-Up of a Chemical Plant

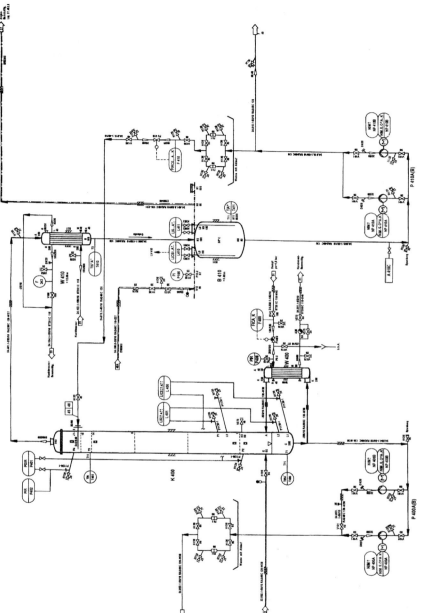

Fig. 5-3 Section of a P&I flow sheet.

5.2.5
Function Plans

Under what conditions is it permissible to switch on a pump, a valve, a stirrer, and so on? These and other questions are answered by the function plans (Figure 5-4). They also document the entire safety philosophy of the plant (Section 2.8.3). In continuous plants they couple sensor signals (T, P, L, F, Q, etc.) via logic components (AND, OR, NOT, etc.) with the actuators (pumps, valves, motors, etc.) (see Section 2.8).

5.2.6
Technical Data Sheets

On the basis of the planning documents a technical data sheet (Figure 5-5) is prepared for each required item of apparatus. It specifies exactly the process engineering and construction requirements that the device must fulfill. On the basis of the construction data contained in the technical data sheet, the appropriate technical workshop or outside supplier can make an offer with delivery date and price. This offer includes the obligation to deliver the apparatus in accordance with the requirements of the technical data sheet on time at the agreed price.

5.2.7
Construction of Models

To aid in visualizing the plans, a model-building department makes an accurate plastic model on a scale of 1:25 or 1:33 on the basis of all planning documents prepared so far. Here the medium key, which contains all information on the pipe materials, fittings, gaskets, insulation, and so on to be used for each medium, is especially important. Nowadays, CAD modeling systems with three-dimensional depiction have largely replaced conventional models. These systems offer the possibility of enlarging sections and to walk through the plant in a realistic manner. They have made huge strides since the early 1990s. The model is the basis for the following:

- Pipeline production and procurement
- Strength and stress calculations
- Isometric pipeline diagrams
- Pipe bridge occupation plans
- Material samples for procurement of pipeline materials
- Coating lists.

The following can be checked especially well at an early stage with the aid of a model:

- Arrangement of the plant
- Accessibility and operability of apparatus and fittings
- Construction sequence
- Dismantling for repairs, catalyst change, and so on

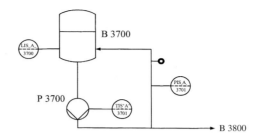
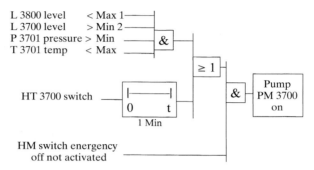

Fig. 5-4 Example of a section of a function plan. The pump P 3700 can only be switched on and remain in operation when the following conditions are fulfilled:
a) Level in vessel L 3700 > Min 2;
b) Level in vessel L 3800 < Max 2;
c) Pressure P 3701 > Min;
d) Temperature in pump head T 3701 < Max;
e) Switch EMERGENCY OFF not activated.
The timer switch HT 3700 acts as a startup bridge until p 3701 > Min.

- Feeding and pipeline routes
- Slopes of pipelines
- Studying favorable courses of pipelines (collisions?).

As in the case of the P&I flow sheets, the project leader or the subsequent plant supervisor (may or may not be the same person) should specify general guidelines for the construction of certain functional units. It has proved useful here to enlist the aid of highly experienced master craftsmen. Examples are the uniform construction of:

- Filter stations
- Pump stations
- Water coolers
- Certain units (e.g., rectification columns).

All units are to be planned according to the same criteria, and only in exceptional cases and for good reasons are deviations therefrom and the installation of special solutions permitted.

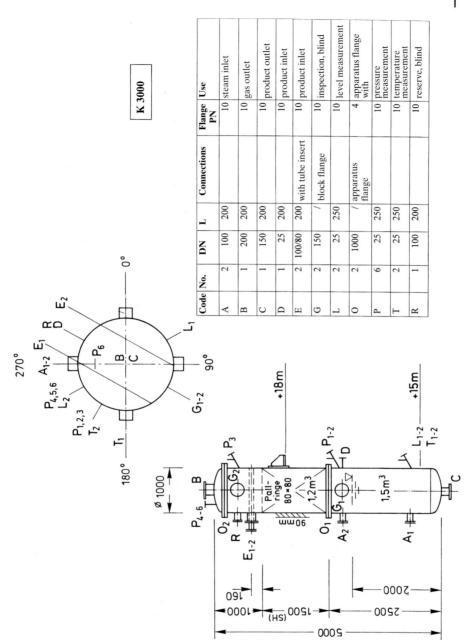

Fig. 5-5 Example of a technical data sheet.

5.2.8
Preparation of Other Documents

Analysis organization

- Develop analytical methods for all streams
- Determine sampling point
- Specify type and frequency of analyses (routine or specialized analysis, are on-line analytical instruments necessary?).

The results are recorded in an analysis file.

Prepare ecotoxicological data file (important for later encounters with authorities when reporting disturbances and accidents, etc.).

Prepare training documents for future employees.

5.3
Commissioning

The goal of the commissioning stage is to transfer the mechanically complete plant into a state ready for start-up. To save time one does not wait until mechanical completion has been achieved. Instead the project leader works closely with the commissioning engineer, so that certain process stages or connected units can preferentially be mechanical completed and tested in advance. Here high degrees of cooperation and improvisation are required. It is totally irresponsible to believe statements such as "everything's OK". The project leader or the start-up team must personally be convinced of correct operation, since practice has shown that the rate of failure is almost 100%, in accordance with Murphy's law [Henderhot 2000]. The best motto is: "everyone is lying to me, and everybody is trying to trick me". Important tasks in the commissioning phase are:

- Cleaning all pipelines
- Inspection of all columns, vessels, heat exchangers, and so on on the basis of the P&I diagram (are all bolts tight, do all internals meet their specifications?, etc.)
- Inpection of all pumps (correct operating direction?, test with water)
- Writing the operating manual
- Water test.

5.4
Start-Up

Start-up [Weber 1996] must be carefully prepared on the basis of practical experience gathered in the commissioning phase. The following preliminrary work must be done:

- Clean plant inside and out
- All insulation on pipelines must be complete
- All flammable and unnecessary material (e.g., wooden construction scaffolding) must be removed
- All temporary gaskets to be replaced by the intended gaskets
- All workmen to leave the plant.

There are no general rules for starting up new plants, so only general advice can be given here:

- Start-up of all auxiliary material systems (compressed and control air, cooling and heating water).
- Remove all blind disks and replace with perforated disks where necessary
- Start up steam system.
- Start up plant from end to beginning, that is, first activate all disposal units (flares, wastewater disposal, etc.), then workup columns, and recycle the product via auxiliary pipelines.
- The auxiliary substances are successively introduced into their storage vessels, and the solvent and starting material streams (pumps operating in recyle mode) are activated.
- Reactor is started up.

6
Process Evaluation

After each development stage (Section 4.1) the existing knowledge should be documented and the status of the process evaluated. This involves preparing study reports and is now standard in many enterprises. At this stage it is necessary to answer the question posed in Chapter 1: Introduction as to whether the process is better, cheaper and faster than that of the competitors.

6.1
Preparation of Study Reports

The study report documents knowledge about the status of the development of a process after specified time intervals or specified development stages to provide the basis for making a decision. The objective of the study report is to answer three important questions:

- Can the production process be implemented in this way in principle?
- What is the return on investment?
- How big is the risk in economic and technological terms?

At least four project study reports must be completed in the course of the development of a process:

- One during the laboratory phase (preliminary study). This is mainly concerned with the question whether the synthetic route should be followed further or not.
- One at the beginning of process development (orientational study). This is mainly concerned with questions such as: Has enough research been performed? Is the knowledge sufficient to enter the technical development phase?
- One in the course of process development (process study).
- One at the end of process development (project study.)

A study report should include the following items:

1. A summary
2. A basic flow diagram
3. A process flow diagram and a description of the process
4. A waste-disposal flow diagram
5. An estimate of the capital expenditure
6. A calculation of the production costs
7. A technology evaluation
8. The level to which the experimental details have been worked out.

Process Development. From the Initial Idea to the Chemical Production Plant. G. Herbert Vogel
Copyright © 2005 WILEY-VCH Verlag GmbH & Co. KGaA, Weinheim
ISBN: 3-527-31089-4

6.1.1
Summary

The summary must start with a precise definition of the objective. This is followed by a short description of the process and of the most important process stages.

The chemistry of the process should be represented by overall reaction equations, with the assumed conversion and selectivity being specified.

Details of the result of the study, the production costs, the capital expenditure for a specified annual production output with an indication of the accuracy of the estimate and the planned location of the plant, if known, are other important points.

The summary concludes with information on the state of knowledge and on the major risks.

6.1.2
Basic Flow Diagram

The basic flow diagram provides a quick overview of the total process. It shows the input, output, and recycling streams and identifies the individual process stages (Figure 6.1-1).

6.1.3
Process Description and Flow Diagram

All the individual steps of the process should be described as accurately as possible on the basis of all the information available at the time, reference being made to the process flow diagram (Figure 6.1-2). The latter should clearly show the process in detail and the composition of the flows. As a minimum, the following information should be included in the process flow diagram:

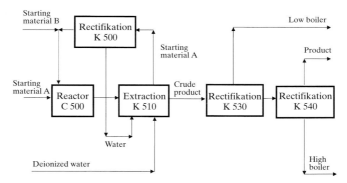

Fig. 6.1-1 Basic flow diagram.

6.1 Preparation of Study Reports

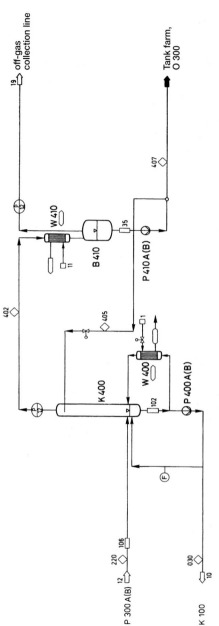

Fig. 6.1-2 Section of a process flow diagram.

Tab. 6.1-1 Example of a mass balance table.

Substance no.	Substance name	Molar mass	Stream no. 1		Stream no. 2		...
		g mol^{-1}	kg h^{-1}	% (g g^{-1})	m^3 h^{-1}	% (L L^{-1})	
1	A	100	1000	66.7			...
2	B	120	500	33.3			
3	C	32			200	9.1	
4	D	28			2000	90.9	
Sum in kg h^{-1} or m^3 h^{-1} STP			1500	100	2200	100	...
State			liquid		gaseous		...
ρ/kg m^{-3}			990		1.4		...
P/bar			4		1.2		...
T/ °C			102		220		...

- All equipment and machines (columns, vessels, heat exchangers, reactors, etc.)
- All product streams (with numbering).

Beneath the process flow diagram the material balance should be presented in a clearly laid out table (Table 6.1-1).

When development is at an advanced stage or is complete, the process flow diagram should show the following additional information:

- Identification of the energy sources (type of steam, cooling water, compressed air, electric power, nitrogen, deionized water, heating gas, refrigerants, etc.)
- Details on the construction materials, size and performance of the equipment and machines
- Typical operating conditions (e.g., pressure and temperature in columns and pipelines)
- Important fittings
- Purpose of instrumentation.

6.1.4.
Waste-Disposal Flow Diagram

The waste-disposal flow diagram illustrates the waste streams for which there is no further economic use in the plant (see Section 1.3) and indicate their flow numbers from the process flow diagram. Precise information should be given on the toxicity, degradability, water danger, flash point, ignition temperature, MAK, odor threshold, etc. of the individual components, their mass flows, and the state of aggregation of the flow (viscosity, sediment content, etc.). This information makes it possible to specify the type of waste disposal required with consideration of site-specific and economic factors, for example:

- Sewage treatment plant
- Incineration in a power station or an incineration plant
- Central hazardous waste incineration
- Landfill site

Special attention must be paid to the available capacity of existing sewage treatment plants and landfill sites. The construction of new capacities has often delayed or prevented the erection of a new production plant, since the resulting additional costs increased the production costs of the new product to too great an extent or even made it uneconomical. Here a network of production sites has major advantages.

6.1.5.
Estimation of Capital Expenditure

6.1.5.1.
Introduction

An important reason for preparing study reports is to determine the capital expenditure (investment) of a project [Muthmann, 1984, Prinzing 1985, DAdda 1997]. This is made up of the location-independent ISBL (inside battery limits) process plant costs such as investment in the production plant and the associated control rooms, laboratories, employee facilities, tank farms, and loading and unloading stations, as well as the location-dependent OSBL (outside battery limits) costs of all the service facilities,

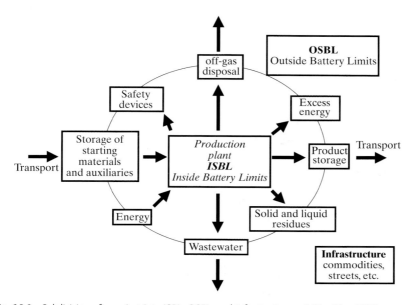

Fig. 6.1-3   Subdivision of a project into ISBL, OSBL, and infrastructure activities [Frey 1990].

such as investment in storage buildings, cooling towers, and waste-disposal facilities. Infrastructure costs are also to be included (Figure 6.1-3).

The accuracy of the investment cost estimate depends critically on the level of maturity the process has reached [Jung 1983].

6.1.5.2
ISBL Investment Costs

Since direct plant costs are always estimated for some date in the future, it is necessary to determine the effect of increases in price. Use is therefore made of price indexes which are published for chemical plants in the individual countries (Table 6.1-2). The price index i determined by extrapolation is divided by the index i_0 at the date of the cost determination. The ratio i/i_0 is then a measure of the increase in investment costs to be expected.

Of the many methods for determining the capital required to set up a plant, five are mentioned in the following:

Degression method
This very simple method can only be used for the same or very similar plants. With the aid of the degression coefficient χ the investment I_1 of a plant given capacity Cap_1 can be extrapolated to a plant with a different capacity Cap_2 [Schembra 1993]:

$$I_2 = I_1 \cdot \left(\frac{Cap_2}{Cap_1}\right)^{\chi}. \tag{6-1}$$

The degression coefficient χ has a value of about 2/3.

Tab. 6.1-2 Price indices i of chemical plants in Germany [VCI "chemie PRODUCTION", "EUROPA CHEMIE" 2001].

Year	i
1988	76.4
1989	79.2
1990	83.1
1991	88.7
1992	93.1
1993	96.1
1994	97.7
1995	100.0
1996	101.5
1997	102.6
1998	101.7
1999	101.1
2000	102.7

Additional cost calculation method

This method derives the costs of a new plant from the known costs of existing plants with an accuracy of ± 30%. Such a method is only possible because the structure of chemical plants is always very similar, that is, they are made up of interlinked pieces of equipment of similar construction whose number does not vary widely. Therefore, the overall investment cost of the entire plant when installed exceeds the average value of the machines and equipment (= total machine and equipment costs divided by the number of machines and pieces of equipment) by a factor which depends on the capacity. The total machine and equipment costs are determined from the parts list, while the costs of individual units can be obtained from in-house data banks or directly from the manufacturers.

Example 6-1 [Prinzing 1985]

The overall factor for the chemical plant construction projects handled by BASF in 1984 was on average 3.9; i.e., the direct plant costs of an average chemical plant is 3.9 times the machine and equipment costs.

The overall factor is essentially determined by the following items:

- Operating conditions (pressure, temperature)
- Type of construction materials used (steel, stainless steel, special materials)
- Size of machines and pieces of equipment.

According to Miller [Miller 1965], the effect these items have on the overall factor can be determined from the average value of the machines and equipment (see Figure 6.1-4).

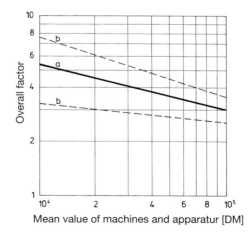

Fig. 6.1-4 Overall factor for the direct plant costs as a function of the average value of the machines and pieces of equipment [Prinzing 1985].
a) Best-fit curve;
b) Confidence limits for a statistical confidence interval of 95%.

This results in the following simple relationship:

$$\text{direct plant costs} = \text{overall factor} \\ (< \text{average machine and equipment valve} >). \qquad (6-2) \\ \sum \text{machines and equipment}$$

Since direct plant costs are always estimated for some date in the future, it is necessary to determine the effect of increases in price. Use is therefore made of price indexes which are published for chemical plants in the individual countries (see Table 6.1-2). The direct plant costs make no allowance for the cost of engineers (usually 10–20%) or for contingencies (ca. 5–10%). This method of estimation has the disadvantage that while the machine and equipment costs can be determined fairly precisely, the average value of these items is multiplied by a comparatively inexact overall factor. The accuracy can be considerably increased by breaking the overall factor down into individual factors.

Method of specific units

For plants with similar structures (petrochemical plants, biochemical plants, etc.), the ratio Q of the investment costs to the capacity K and the number of units N is approximately constant:

$$Q = \frac{I}{Kap \cdot N} \approx konstant. \qquad (6-3)$$

Here the term unit refers to a complete, functioning reactor or separation device, for example, a rectification column with all pumps, sump and head heat exchangers, and so on. The ISBL cost are then given by:

$$ISBL/Euro = Q/(\text{Euro t}^{-1}\,\text{a}^{-1}\,\text{Unit}^{-1}) \cdot Kap/(\text{t a}^{-1}) \cdot N. \qquad (6-4)$$

Any chemical company can calculate Q by analysis of earlier project calculations. This simple and fast method (accuracy $\pm 60\%$) is mainly suitable for use in the stage of preliminary studies, when knowledge about the complete process is still very incomplete. For major petrochemical basic and intermediate products the factor is about 60 € $(t/a)^{-1}$ $(unit)^{-1}$ (basic: 1995).

Method of specific units taking into account plant complexity

An improved version of this method was reported by DuPont [Zevnik 1963]. Here not only the capacity and the number of units is taken into account, but also the complexity (maximum process temperature, type of material). The following quantities are required as input data for calculating the total investment:

- Capacity Cap
- Cost index i/i_0

- Number of units N
- Complexity factor CF.

The total investment is then given by:

$$I = 1.33 \cdot \frac{i}{i_0} \cdot N \cdot CPF(CF), \tag{6-5}$$

where the factor of 1.33 takes the OSBL costs into account and the quantity CPF represents the cost per unit, which is function of CF and *Cap* [Zevnik 1963].

6.1.5.3
OSBL Investment Costs

The OSBL investment depends on the location of the plant and are required solely for operating the plant, for example, the cost of storage buildings, cooling towers, waste-disposal facilities, linking the OSBL facilities to the location in question (e.g., water and electricity supply), and so on. Since the question of location is generally still unresolved at an early stage in development, it is not possible to specify the OSBL costs explicitly, but they can be tentatively taken as 20–30% of the ISBL costs. For small plants the percentage tends to lie near the upper limit, while for larger plants it is likely to be close to the lower limit. If the location is known, the OSBL costs can be estimated very accurately.

6.1.5.4
Infrastructure Costs

The infrastructure costs depend on the location of the plant. They are not, however, related exclusively to the project since the infrastructure, such as workshops, the central store, and the canal system, is also used by other production facilities. Obviously, this type of cost can only be specified if the precise position of the plant at a location is known.

6.1.6
Calculation of Production Costs

The cost effectiveness of a new process depends on the production costs of a product (Section 6.2). The following information is required to determine these costs:

- Material balance
- Waste-disposal flow diagram
- Utilities
- Investment.

The production costs can be calculated by adding up the following items (see also Chap. 1):

- Feedstock costs
  - Raw materials
  - Auxiliaries
  - Catalysts
- Manufacturing costs, which can be subdivided into:
  - Energy costs
  - Waste-disposal costs
  - Staff costs (wages and salaries)
  - Maintainance
  - Other costs (company overheads)
  - Capital-dependent costs (depreciation, interest, taxes).

6.1.6.1
Feedstock Costs

The raw material requirements are given by the material balance for the chosen capacity, but to determine the feedstock costs the prices of the raw materials are required (Table 6.1-3). The change in raw material prices while the process is being developed can be estimated, but it is more difficult to allow for the fact that the erection of the

Tab. 6.1-3 Prices of some selected raw materials, basic chemicals, and intermediate products based on fossil and renewable resources in Germany (as of December 1999) [Chemische Rundschau 1996, Eggersdorfer 2000].

	Fossil basis	**Price/€ kg^{-1}**	**Renewable basis**	**Price/€ kg^{-1}**
Raw materials	crude oil[1]	0.17	Corn	0.90
	Natural gas	0.27	Wheat	1.05
	Naphtha	0.21	Soy beans	0.22
Basic chemicals	Benzolene	0.31	Rase-seed oil	0.40
	Ethylene	0.35	Palm oil	0.37
	Propylene	0.27	Melasses	0.14
	Methanol	0.18	Sugar	0.41
	Ammonia	0.12	Starch	0.30
Intermediates	Ethylene oxide	0.61	Sorbit	0.63
	Propylene oxide	1.03	Glycerol	1.30
	1,2-Propane diol	0.73	Furfural	0.80
	1,4-Butane diol	1.60	Citriacid	1.60
	Acrylicacid	1.15	Fatty alcohols	0.90

[1] Crude oil prices in $ barrel: 23.69 (*1990*)/17.05 (*1995*)/20.45 (*1996*)/19.12 (*1997*)/12.72 (*1998*)/17.79 (*1999*)/28.23 (*2000*) [VCI 2001].

plant will have an impact on the raw material market and consequently alter the price structure. Therefore, in some cases it is necessary to enter into negotiations with the raw material suppliers at an early stage. If the raw materials originate from another plant in the same company, market prices are replaced by transfer prices [Lunde, 1985], but this does not overcome the problem of the impact of the proposed plant on the market, it only shifts it. The significance of raw material costs in cost estimation varies widely. Thus, in processes involving a high level of material refinement (as is the case for some pharmaceuticals) or in those with high specific energy costs (e.g., chlorine production), precise determination is not as important as in the case of cheap, mass-produced products, such as petrochemicals, manufactured in large plants.

Credits for byproducts can be set against the raw material costs from the outset.

6.1.6.2.
Energy Costs

For large-scale processes, the cost effectiveness depends appreciably on the cost of the utilities. The combination of utilities required in the plant therefore acquires considerable importance and consideration must be given to the ideal thermal coupling even at an early stage in the process development since it can have a considerable impact on the configuration of the process [Körner, 1988]. For example, whether the energy derived from condensing vapors can be utilized depends on the behavior of the materials. Clean and stable condensates usually present no problem in using the heat of condensation to raise the temperature of cold streams, but in the case of products which tend to form deposits (by decomposition, polymerization, etc.) optimum thermal utilization of this type of energy is not possible since the vapors must be quenched to prevent deposits, and this results in loss of energy. This type of question should be resolved in the miniplant phase.

A suitable aid for systematizing such considerations is Linnhoff analysis (pinch technology) [1981, Linnhoff 1983, Ullmann]. The Linnhoff analysis determines the process configuration for which the energy supplied from or lost to the surroundings is a minimum. The problem which must be solved is how to couple the heat sinks and heat sources within the plant such that minimum energy consumption is achieved. The initial step in the calculation is to determine the so-called composite curves [Σ(hot streams) = hot composite curve, Σ(cold streams) = cold composite curve) and pinch temperature, which is the temperature above which heat can be lost and below which heat can be absorbed (Figure 6.1-5).

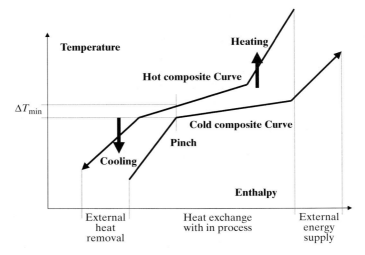

Fig. 6.1-5 Temperature–enthalpy diagram.

Example 6-2

The hot composite curve of two hot streams that are to be cooled is to be calculated:

Hot stream 1:
  from 200 to 100 °C, $\dot{m} \cdot cp = 1$ kW K^{-1}

Hot stream 2:
  from 150 to 50 °C, $\dot{m} \cdot cp = 2$ kW K^{-1}

These two streams are added as shown in Figure 6.1-6 and thus give the hot composite curve.

For a given interconnection system, the Linnhoff analysis shows how far the heat utilization is from being ideal. However, there are often reasons for not choosing

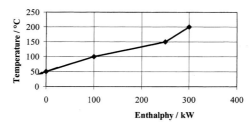

Fig. 6.1-6 Hot composite curve for Example 6-2.

an ideal interconnection:

- Start-up operations often require start-up heat exchangers, and this means higher investment costs.
- Energy utilization only becomes possible as a result of increasing the column pressure, and this means higher investment costs and may result in material problems (decomposition, side reactions, etc.).
- Problems associated with the formation of deposits during heat transfer, which necessitates direct heat removal (quenching).

Generally, as the plant capacity increases, reducing the operating utilities (i.e., optimizing the thermal combination) rather than increasing the investment will have a more beneficial effect on production costs. The energy analysis is used to find the amounts of energy required or released and to calculate the energy per unit quantity. The determination of energy costs requires information on the prices of the different types of energy, which depend on the quantities required and the location. Important types of energy are presented in the following.

Steam (see Section 2.5.1)

An initial estimate of the steam requirement of a column is given by:

$$\text{Steam quantity}/(t\ h^{-1}) \approx \frac{(\text{reflux} + \text{head production})/(t\ h^{-1})}{5}. \qquad (6-6)$$

This is based on the assumption that:

- The thermal combination is optimum (e.g., the hot discharge from the bottom is used to heat the feed)
- Normal organic liquids are involved (heats of evaporation ca. 110 kW h t^{-1}; see Table 6.1-4)

For a steam price of 10–20 € t^{-1} (central Europe, as of 1995) the steam costs in € kg^{-1} for N rectification columns are given by:

$$\text{Steam costs}/(\text{Euro kg}^{-1}) =$$

$$\frac{\sum_{i=1}^{N} \text{Steam quantity of column } i/(t\ h^{-1}) \cdot \text{steam price}/(\text{Euro t}^{-1})}{\text{nominal capacity}/(\text{kg h}^{-1})} \qquad (6-7)$$

Electrical energy (see chapter 2.5.2)

Electrical energy is more than twice as expensive as steam, and because of the explosion hazard it is normally not suitable for heating columns. Its most important application is in driving machinery, for example, pumps, impellers, and compressors for feeding material (Chapter 2.4) and in carrying out operations such as size reduction, mixing, and separation.

Tab. 6.1-4 Heats of evaporation of some organic substances and the ratio of the heats of evaporation of water and the substance in question [Handbook 1978].

Substance	$\Delta_V H$/kJ kg-1	$\Delta_V H_{H2O}/\Delta_V H$
Methane	557	4.1
Ethane	522	4.3
Propane	433	5.2
Water	2257	1
Methanol	1227	1.8
n-Propanol	788	2.8
Acetone	552	4.1
Formic acid	901	2.5
Acetic acid	695	3.2
Propionic acid	705	3.2
Benzene	437	5.1
Toluene	436	5.2

The pump delivery N required can be estimated using the following formula [Henglein 1963]:

$$N/\text{kW} = \frac{\dot{V} \cdot H \cdot \rho \cdot g}{3600 \cdot 10^3 \cdot \eta_{tot}} = \frac{\dot{V} \cdot \Delta P}{36 \cdot \eta_{tot}} \qquad (6-8a)$$

$\dot{V}$ = the useful feed flow in m³ h⁻¹
H = delivery head in m
ρ = density in kg m⁻³
g = acceleration due to gravity (ca. 10 m s⁻²),
η_{tot} = overall efficiency
ΔP = feed pressure in bar.

The overall efficiencya is made up of the mechanical and the electrical efficiencies:

$$\eta_{ges} = \eta_{mech} \cdot \eta_{eletr}. \qquad (6.1.1-8b)$$

For rotary pumps, the overall efficiency is ca. 0.7–0.8, while for piston pumps it is 0.8–0.9. For small pumps ($< 5\,\text{m}^3\,\text{h}^{-1}$), the overall efficiency may, however, be much lower [Hirschberg 1999]. An initial rough estimate of the electricity costs in euros per kilogram for all pumps in a plant consisting of N units is given by Equation (6-9):

$$\text{Electricity costs/Euro kg}^{-1} = \frac{CN}{1000} \cdot \text{electricity price/(Euro kWh}^{-1}) \cdot N. \qquad (6-9)$$

The characteristic number CN (ca. 4 kW per unit per t h^{-1}) results from the empirical fact that, in a large-scale plant, each unit with a product throughput of 1 t h^{-1} requires four pumps, each of which must deliver about 4 m^3 h^{-1}.

Another application for electrical energy in chemical plants is in auxiliary electrical heating systems (e.g., to prevent freezing of pipelines during plant disturbances, especially in winter).

Fuels

Fuels are only used in chemical plants if streams must be heated to such high temperatures that steam is no longer economical (the limit is ca. 100 bar of steam, i.e., 310 °C). However, the use of fuels in chemical plants always entails a risk from the ignition source. Only the net calorific value of the fuel can be used, since the steam formed can generally not be condensed. A few typical values of the standard enthalpy of combustion are given in Table 6.1-5.

Cooling water

The amount of cooling water required to remove a given quantity of heat depends on the temperature of the water and its quality. If more detailed information is not available, it is sufficient as a first approximation to determine the amount of cooling water required by assuming its temperature rises by 10 °C. For high product temperatures, an alternative to water cooling is to use air coolers. Although these require a higher investment than water coolers, they are often cheaper to operate and are less prone to

Tab. 6.1-5 Standard enthalpies of combustion $\Delta_c H^0$ at 25 °C [Chemikerkalender 1984, Handbook 1991].

Substance	$\Delta_c H^0$ / kWh t^{-1}
Carbon	9110
Carbon monoxide	2810
Hydrogen	39 740
Methane	15 460
Methanol	6310
Formaldehyd	5190
Formic acid	1590
Ethane	14 440
Ethanol	8250
Acetaldehyd	7360
Acetic acid	4040

fouling. However, unless there is a shortage of cooling water at a known location, it is generally sufficient to assume water cooling in an initial cost effectiveness calculation.

The cost in euros per kilogram of the required river water is obtained by assuming that all required steam energy is removed by river water with a temperature rise from 25 to 30 °C:

$$\begin{aligned} &\text{River water costs}/(\text{Euro kg}^{-1}) = \\ &\text{river water price}/(\text{Euro m}^{-3}) \cdot \\ &\text{steam consumption}/(\text{t steam/t product}). \end{aligned} \qquad (6-10)$$

The price of river water depends on location; in central Europe it is about € 8 per 1000 m³.

Refrigeration energy

Refrigeration is generally achieved by compressing, cooling, and adiabatically expanding coolants in the plant itself, that is, the refrigeration requirement can be expressed as a requirement for electrical energy and cooling water. To remove a given quantity of heat, about 20–50% of that quantity, depending on the temperature level required, must additionally be expended in the form of electrical energy, thereby increasing the amount of heat which must be removed to about 120–150%.

Compressed air

Small quantities of compressed air are drawn from pipelines. An example is the control air required by closed-loop control instruments (the rule of thumb is $1 \text{ m}^3 \text{ h}^{-1}$ of control air per instrument). Larger amounts are produced by an air compressor in the plant, and the requirement can then be expressed in electrical energy [Coulson 1990]. The power N_{theo} expended on compressing gases is given by (see Section 2.4):

$$N_{theo} = P_1 \cdot \dot{V}_1 \cdot \frac{\chi}{\chi - 1} \cdot \left\{ \left(\frac{P_2}{P_1} \right)^{\frac{\chi-1}{\chi}} - 1 \right\} \qquad (6-11)$$

χ = c_p/c_v
P_1 or P_2 = initial or final pressure, respectively
c_p or c_v = heat capacity of the gas at constant pressure or constant volume
$\dot{V}_1$ = initial volumetric flow.

For compression of air (χ = 1.4) with an overall efficiency of $\eta \approx 0.7$):

$$N/\text{kW} = 0.14 \cdot \dot{V}_1/(\text{m}^3/\text{h}) \cdot \left\{ (P_2/\text{bar})^{0.286} - 1 \right\}. \qquad (6-12)$$

6.1.6.3
Waste-Disposal Costs

Since the 1980s, the waste-disposal situation has become much more difficult for chemical plants, a trend which continues today. The delivery of effluent and wastes to central sewage plants, waste incineration plants, or waste-disposal sites is restricted by official regulations, and as a result waste-disposal costs have risen dramatically in countries such as Germany in recent years. This poses a challenge to process development. The slogan "integrated rather than end of price environmental protection" expresses the current demand for environmental pollution to be reduced in the manufacturing process itself and for efforts to be made to put even the byproducts of a chemical process to good use if possible [Swodenk 1984, Bakay 1989, Lenz 1989, Lipphardt 1989, Schierbaum 1999]. Such demands have a major influence on the development of new chemical processes. It is important to make use of synergies that are offered by the plant and site network. The key concept here is sustainable development. Important contributions to sustainable development are provided by the development of new, improved catalysts and processes. In particular, the increasing use of catalysts in industrial processes is leading to higher yields, reduced consumption of resources, less waste, and easier recycling.

Sewage treatment plant

Effluents discharged into a sewage treatment plant must be readily capable of biological degradation and must not be toxic to the bacteria in the plant. In addition, each substance in an effluent stream should have a sufficiently high biological degradation rate, regardless of its total content in the stream. This requirement prevents individual undegradable substances being lost in an otherwise readily degradable mixture. To assess whether disposal of a stream via a sewage treatment plant is possible, the following information must be collected for each substances it contains:

- BOD_5 and COD value [DIN 38 409]
- Zahn–Wellens test [OECD 1981]
- Bacteria and fish toxicity [Juhnke 1978, Bringmann 1977, DIN 38 412, Kühn 1996, DFG 1983]
- Water hazard class [Kühn 1996]

An initial rough estimate of the upper limit of the cost in euros per kilogram of product of the sewage-treatment plant can be obtained by assuming that all side products are discharged into it:

$$Sewage\ plant\ costs/(Euro\ kg^{-1}) = \frac{M\ (Starting\ material)}{M\ (product)} \cdot \frac{(1-S)}{S}$$
$$\cdot\ C - content\ (starting\ material) \cdot C - price$$
$$(6-13)$$

M (starting material) = molar mass of starting material
M (product) = molar mass of product
S = selectivity with respect to the product
1−S = selectivity with respect to the byproduct
C content = mass fraction of carbon in the starting material
C cost = € per kg organically bound carbon (operators of sewage treatment plants charge ca. € 1 per kg C)

Incineration plants [Jahrbuch 1991, Geiger 2000]
Waste streams which consist only of compounds containing carbon, hydrogen, and oxygen as aqueous solutions with a concentration >10% are ideally suitable for disposal in an incineration plant or a power station (if the content is >10%, the heat of incineration is roughly equal to the heat of evaporation of the water). Disposing of waste by incineration is more difficult if other elements such as nitrogen, chlorine, sulfur, and metals are present. Such wastes can only be disposed of economically in a central waste incineration plant equipped with suitable absorption systems for NO_x, HCl, SO_2, and so on [Hüning 1984, Hüning 1989].

Waste-disposal site
The restrictions imposed on wastes which can be dumped are continuously increasing, and new dumping space is no longer available. Consequently, this waste-disposal option is virtually no longer available for new processes.

6.1.6.4
Staff Costs

Operating staff costs in chemical plants can vary widely. For large automated plants, they may be less than 3% but for small batch processes they may amount to 25% of the manufacturing costs. Often they are 5–10% of the manufacturing costs. In new plants, whether batch-operated or of the single-train type, the trend is towards the fullest possible automation. The introduction of process control engineering has made it possible to operate and monitor virtually every aspect of a plant using pictorial displays [Jahrbuch 1991]. Almost regardless of the size of the plant, a well-planned and automated single-train installation requires the following minimum staff:

- One works manager
- One shift foreman or his deputy
- Two shift employees for the control room
- One shift employee for outside operations
- One shift employee as a reserve.

This minimum staff of five for 24 h operation should be multiplied by a factor, specific to the company, which reflects the type of shift model, the sickness statistics, and vacations to give the total number of shift employees required. An initial rough estimate of the staff costs euros per kilogram of product is given by the following rule of thumb (Germany as of 1992):

$$\text{Personel costs}/(\text{Euro kg}^{-1}) = \frac{2000}{\text{nominal capacity}/(\text{t a}^{-1})}. \tag{6-14}$$

Special operating requirements such as frequent changes in production schedule, periodic start-up and shut-down of the plant, loading and unloading of tankers may make additional staff necessary.

6.1.6.5
Maintenance Costs

Annual repair and maintenance costs generally amount to 3–6% of the invested capital [Wagener 1972]. About half of this amount is material costs and the rest wages. Their long-term increase can therefore be assumed to be between that of the investment costs (Table 6.1-2) and that of the wage rates (Table 6.1-6). Experience shows that high repair costs are incurred shortly after a plant is put into operation, in eliminating start-up difficulties, and after a prolonged operation because of the need to carry out major repairs. Depending on how new a process is, annual costs of 5–10% of the estimated total investment should be allowed for in an initial estimate.

6.1.6.6
Overheads

The overheads of the company consist of the total costs of the central facilities (research, personnel, energy-supply department), works management, staff facilities, internal plant traffic, etc. These costs are very dependent on the structure of the company. As a rule they are ca. 5–10% of the combined energy, staff, and depreciation costs.

6.1.6.7
Capital-Dependent Costs (Depreciation)

The plant capital can be depreciated by various methods [Kölbel 1967]. For a cost efficiency calculation it is generally sufficient to assume linear depreciation over a ten-year utilization period. The annual depreciation is then 10% of the plant capital:

$$\text{Depreciation}/(\text{Euro kg}^{-1}) = \frac{\text{total investment}/\text{Euro}}{10 \text{ a} \cdot \text{nominal capacity}/(\text{kg a}^{-1})}. \tag{6-15}$$

Tab. 6.1-6 Gross hourly earnings of an average worker in the German chemical industry [VCI 2001, Jahrbuch 1991].

Year	€	Inflation rate/%	Index 1995 = 100	Inflation rate/%
1985	8.99	+ 3.1	64.8	+ 3.2
1986	9.30	+ 3.4	66.9	+ 3.2
1987	9.69	+ 4.2	69.7	+ 4.2
1988	10.06	+ 3.8	72.4	+ 3.9
1989	10.42	+ 4.2	75.0	+ 3.6
1990	11.11	+ 6.7	80.0	+ 6.7
1991	11.75	+ 5.9	84.6	+ 5.8
1992	12.51	+ 6.3	89.8	+ 6.1
1993	13.20	+ 5.5	94.6	+ 5.3
1994	13.62	+ 3.2	97.6	+ 3.2
1995	14.01	+ 2.9	100.0	+ 2.5
1996	14.41	+ 2.8	101.6	+ 1.6
1997	14.54	+ 0.9	102.7	+ 1.1
1998	14.65	+ 0.8	103.2	+ 0.5
1999	14.93	+ 1.9	105.1	+ 1.8
2000	15.46	+ 3.5	108.9	+ 3.6

6.1.7
Technology Evaluation

Technology evaluation should provide information on the technical risk associated with a process. This risk is equivalent to the economic damage which would occur in the extreme case if the process did not work at all or could only be rendered operational by carrying out subsequent improvements. Technology evaluation should also reveal whether the individual unit operations or the equipment and machines used are of technically established types and whether particular risks are incurred by breaking new ground (size, construction, material, scale-up factor, etc.). It is also necessary to include a statement as to whether the process is based on a technology with which the company is familiar (e.g., high pressure, gas-phase oxidation, phosgenation technology, etc.) or whether fundamental innovations are involved. If limits are exceeded (e.g., if the dimensions of the largest extraction column previously operated by the company are considerably exceeded in the new project), the limits which have been tested should be specified. This information should be elaborated in detail as early as possible in order to recognize, consider, and evaluate potential risks and causes of failure. During operation of the miniplant, close attention should be paid to deposits, frequently occurring malfunctions, etc., since such problems often subsequently lead to insuperable difficulties in the industrial-scale plant (e.g., foaming, emulsion formation, aerosol formation, etc.).

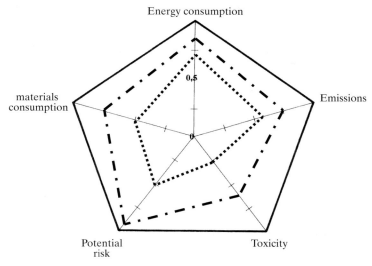

Fig. 6.1-7 Ecological fingerprint, a tool for determining the influence on the environment.

Besides cost aspects, the environmental aspects of the process should be analyzed at as early a stage as possible. For example, eco-efficiency analysis (Figure 6.1-7) can be used to analyze the process from the viewpoints of cost and the environment. The aim is to obtain a clear depiction of the economic and ecological influences during the entire lifetime of a product, so that investment decisions can choose the alternative with the best long-term perspective [Kircherer 2001, Gärtner 2000].

Example 6-3

The result of an eco-efficiency analysis [BASF 2001, Gärtner 2000] in which BASF investigated different chemicals for bleaching magazine paper showed that hydrosulfite bleaching earth is cheaper and less polluting than peroxide bleach. Therefore, where moderate bleaching quality is sufficient, paper producers should use hydrosulfite, and additional hydrogen peroxide should only be used when very high bleaching quality is required. The major advantage of hydrosulfite is that it requires no auxiliary substances, in contrast to hydrogen peroxide, which only exhibits its bleaching action in the presence of additives such as caustic soda or water glass. These increase costs and energy consumption, which, over the complete lifetime, are twice as high for hydrogen peroxide than for hydrosulfite. Another advantage of hydrosulfite is the mild reductive bleaching process. The agressive oxidizing agent hydrogen peroxide partially destroys the pulp fibers, and this results in a high TOC load in the wastewater from the paper mill.

Characteristic quantities for evaluating the sustainability of chemical processes are [Eissen 2002]:

- Cumulative energy consumption (process energy per unit amount of product).
- Atom exploitation factor (molar mass of product divided sum of molar masses of all starting materials).
- Environmental factor (mass of waste, i.e., all used substances that are not product, per unit amount of product).

The technology evaluation gives a list of weak points arranged in order of technical risk (e.g., catalyst service life, corrosion, etc.). This list can then be used to take steps which reduce such risks. In principle, two measures are possible:

- Expenditure on R & D can be increased at the weak point (e.g., the unit concerned may be checked on a pilot scale or a well-established solution may be sought). This option should always be chosen if, in an extreme case, failure of the unit concerned would be accompanied by a total investment loss.
- Failure scenarios can be developed, that is, what can be done later if problems occur or if a unit fails completely?

Normally, these two options have to be weighed against each other, that is, it is necessary to clarify whether it is more economic to minimize the risk by increasing R & D expenditure (e.g., by building a pilot plant), or to reduce or eliminate it at a later date by technical measures (e.g., additional instrumentation or backup units). Some general aspects which may be useful in assessing the technical reliability are summarized below. From the estimate of the risk and the measures it suggests, the economic effect can be quantified very exactly (surcharges on the development costs or on the investment and repair costs). This transforms technical reliability into cost effectiveness considerations and results in the possibility of expressing the risk of a development project in monetary units.

6.1.8
Measures for Improving Technical Reliability

Decoupling process steps
The susceptibility to failure of a process carried out in a single-train plant increases as the number of process steps and the probability of failure of individual steps increase: a single-train plant can only function if each step is operational. If partial streams are fed back to earlier stages, the coupling between them is particularly strong. This is also true of the energy cross-links within a process (Section 6.1.6.2). The increase in risk as the number of process steps increases is independent of whether the process is continuous or batch. In practice, various options are now adopted in resolving this dilemma (Figures 6.1-8 and 6.1-9):

- Decoupling by inserting tanks for intermediates (the larger the buffer tank, the greater the independence). For larger quantities of product or gaseous intermediates, this approach is virtually impracticable.

- Decoupling by installing parallel arrangements of equipment which is particularly susceptible. For example, filters and fittings that are susceptible to failure should be provided with a bypass if safety considerations permit (e.g., heat exchangers in the main flow in which severe fouling is to be expected).

Each measure which is aimed at increasing the uptime of a single-train plant results in an increase in investment costs. In practice it is often the case that such steps are only taken after the plant is put into operation for the first time.

Choice of process steps which can be safely scaled-up

An example of the dependence of technical reliability on the operational method employed is mass transfer. While simple standard equipment can be used for rectification, a wide variety of equipment in which phases are mixed in several stages and then separated has been developed for extractions. Such equipment generally requires more expenditure for safe operation than distillation columns. The complexity of an operation increases as the number of substances and phases it involves increases and as the operating conditions become more extreme. Thus, the presence of two

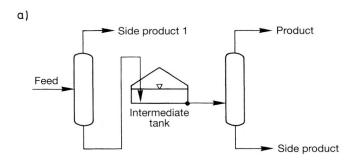

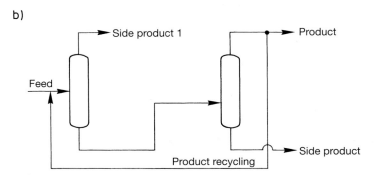

Fig. 6.1-8 Example of decoupling a single-train plant.
a) Intermediate tank to smooth out variations in feed flow.
b) Recycling to smooth out variations in feed flow.

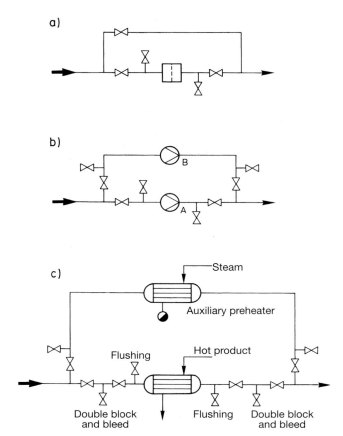

Fig. 6.1-9 Decoupling of a single-train plant by including redundant equipment.
a) Filter station with single bypass.
b) Pumping station with A and B pumps.
c) Preheater with flushing facility.

liquid phases requires special precautions to be taken. The same applies if particularly high or low temperatures or pressures are involved. If solids are processed, the process must be assessed with particular care. The scaling-up of equipment for separating solids should be subjected to particularly careful scrutiny. Even the bunkering and transportation of solids present difficulties which are often specific to the materials.

If new technology is employed, the operational safety can be expected to be lower than that of well-established processes. It can be increased if the types of equipment used in the experimental plant and in the industrial-scale plant are similar. This requires scaling-up to be possible, but in the case of reaction equipment, this is not always the case. One way out of this dilemma could be the introduction of microreactors [Srinivasan 1997], which, however, are still in the phase of research and development.

Choice of the materials to be processed
If toxic, inflammable, caustic, or otherwise hazardous substances are to be handled, a higher degree of operational safety is required from the outset than in the case of harmless substances.

Choice of construction materials
Corrosion is an ever-present risk and necessitates the use of special materials. In making an initial selection, use may be made of published tables [Ullmann], but in most cases it is necessary to carry out corrosion tests. Trace impurities can lead to considerable difficulties if they accumulate at particular points in the equipment. The most reliable information in this connection is provided by operating an integrated miniplant. If the flow rate of mixed phases is high, erosion of the wall material can be expected. The extent of erosion can often be lowered by the choice of material and by effective design. Increases in wall thickness are generally of no help because the erosion usually takes the form of pits. Miniplant experiments are of no assistance in this connection since they are usually not representative in hydrodynamic terms. In the case of moving solids, not only the walls of the equipment but also the solids themselves are subject to abrasion. If a process is intended to give a certain particle size, the abrasion loss may have a considerable effect; moreover, an increase in this abrasion is always expected on scaling up a process.

Solids deposits
Solids often deposit inside equipment and result in fouling. Valuable information about this is provided by miniplant results. In heat exchangers, allowance is made for fouling by increasing the heat transfer resistance by empirically determined amounts (fouling factors; Figure 6.1-10).

If the expected operating time is inadequate, a back-up heat exchanger must be installed in parallel. In larger plants, heat exchangers are usually subdivided into several units for design reasons. Thus, if suitable fittings (double block and bleed) are provided to ensure that each heat exchanger can be cleaned while the plant is operating, the effect on the availability of the entire installation is small (see Figure 6.1-9c). However, this is only true provided the cleaning frequency is technically feasible. The design of heat exchangers and the cleaning operations should be carefully matched to the type of fouling, and extensive automation is always worthwhile in the case of large pieces of equipment.

If fouling is expected in columns, larger columns should be designed for whole-body access to each plate, while in small columns a manual-access hole should be provided for each plate to allow cleaning. Furthermore, designs should be used whose operation is relatively insensitive to fouling and which can readily be cleaned (e.g., polished double-flow column plates).

The choice of reaction equipment can be considerably influenced by fouling. As fouling increases, the pressure drop in fixed-bed reactors increases, whereas fluidized-bed reactors are virtually unaffected. Moreover, fixed-bed reactors must be opened and emptied in order to regenerate the contaminated catalyst, but fluidized-bed catalysts can usually be regenerated in the reactor. Often it is also possible

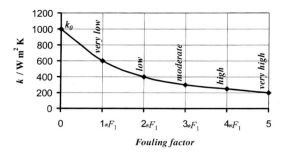

Fig. 6.1-10 Effect of the fouling factor F_i on the heat transfer coefficient k, where
$$\frac{1}{k} = \frac{1}{k_0} + F_i$$
$F_1 = 7 \cdot 10^{-4}$ m² K W^{-1} (typical value for river water), and
$k_0 = 1000$ W m^{-2} K^{-1}.

to remove, regenerate, and replace the catalyst in a sidestream while the plant is operating. Fouling that arises both from the material being processed and from the material of which the equipment is made is more troublesome. For example, in many polymerization reactors, wall deposits can be suppressed by suitable choice of the wall material and processing method. In such cases, steps should be taken to ensure that the same wall material is used in the trial plant and in the industrial-scale plant. If wall erosion or deposit growth occurs at a constant rate, the difficulties can be expected to decrease in bulk-storage equipment and pipelines with increasing scale-up but to remain unchanged in columns and heat exchangers (equipment with large surface areas).

Machines

Since machines have moving parts, they are always a potential source of malfunction and the reliability of a process decreases as the proportion of machines it employs increases. In general, machines having reciprocating parts are regarded as less reliable than those employing only rotating parts. For this reason it is particularly important to provide a reasonable level of backup.

Because of their relatively low cost, duplicate full-capacity pumps are usually installed, but occasionally two pumps having different functions can be provided with a common backup. In the case of reciprocating machines, in addition to a complete backup (2×100% capacity), 2 × 66% capacity, or in exceptional cases 3 × 50% capacity, may also be considered. Turbo machines are usually installed at 1 × 100% capacity, but provision is made for rapid replaceability of parts which are subject to wear. Alternatively, two machines with 50% capacity are often installed.

Measurement and control devices

The narrower the limits in which quantity flows and operating conditions have to be kept for reliable operation, the higher will be the requirements imposed on the instrumentation. To increase plant availability, provision is made for n measurements instead of one measurement in the case of critical circuits and control loops. The complete unit will operate as specified provided m of the n measurements ($m < n$) are in the "good" range (m of n circuits). The greater m is, the higher will be the safety margin and the lower the plant availability. In the case of chemical plants, "two-out-of-three" circuits have been found to be satisfactory, thus providing a compromise between safety margin and availability.

Multiple-train design

According to Equation (6-1) the increase in investment costs on scaling up the plant is approximately proportional to the capacity to the power 2/3. Consequently, the specific investment costs for a single-train plant are approximately 20% lower than for a two-train plant of the same total capacity. In addition, the personnel costs are lower for a single-train plant than for several individual plants of the same total capacity. However, certain plant components can often not be scaled up to the required total capacity with an adequate safety margin. Thus, the maximum capacity of polymerization vessels may be limited because a breakdown of the cooling system must be insufficient to bring about dangerous operating conditions, or it may be impossible to increase the size of an oxidation reactor for cost reasons. In such cases, the total capacity is provided by using the requisite number of individual pieces of equipment.

A significant advantage of multiple-train design is that it is easier to cope with a temporary reduction in load, which would normally mean shutting down and then starting up again. If the plant is underloaded, the specific consumption figures of multiple-train plants are lower. Furthermore, the capacity of such plants can more easily be increased at a later date, particularly if this has been planned from the outset. The advantages of multiple-train design can, of course, only be exploited if it is possible to link any line in a process stage with any of the lines of the subsequent stage. For this purpose manifolds will be needed.

6.1.9
Assessment of the Experimental Work

This part of the project study should contain information on whether the existing experimental results have been obtained in-house or are derived from third parties. If in-house knowledge is available, the test units in which the experiments were carried out (e.g., laboratory, integrated miniplant) should be specified and the duration and conditions of the experiments given. The aim is to arrive at as an objective assessment as possible of the quality of the data obtained since this is of major importance for the accuracy of the project cost estimate.

6.2
Return on Investment

After the individual cost factors have been estimated, a rough cost effectiveness calculation is possible [Solinas 1997]. This should provide information on the return on investment (= ratio of profit to capital employed) for the planned project. Investment in the process under consideration will be profitable if the sum of the revenues exceeds the total outlay and the profit (revenue–outlay) makes it possible to amortize and pay reasonable interest on the capital invested. The return on investment provided by a process can be increased by minimizing the production costs. Exposing the main cost factors (raw material, energy, waste disposal, personnel costs, depreciation) will therefore indicate the direction the development should take in order to improve the process (Table 6.2-1).

Minimizing the production costs does not, however, maximize the return on investment if the full capacity of the plant is not reached. In such cases, a plant with lower capacity and higher production costs might provide a larger return on investment despite being less economical.

Some of the indicators frequently used in practice to assess return on investment are described below.

6.2.1
Static Return on Investment

In its simplest form, the return on investment calculation widely adopted in practice relates the expected annual profit to the capital invested:

$$R = \frac{(U - H)}{I} \cdot 100 \qquad (6 - 16)$$

R = percentage return in % a^{-1}
U = annual sales proceeds (= amount sold multiplied by price) in € a^{-1}
H = annual production costs in € a^{-1}
I = the capital expenditure in €
$U-H$ = annual profit in € a^{-1}.

The method can be made more informative by taking the sales into consideration separately:

$$R = \frac{(U - H)}{U} \cdot \frac{U}{I} \cdot 100. \qquad (6 - 17)$$

The first factor indicates the sales success and the second the capital turnover.

Instead of R, use may also be made of the reciprocal value, that is, the payback time $100/R$. This indicates the time required to recover the invested capital so that the plant begins to make a profit.

Tab. 6.2-1 Examples of measures that can lead to a reduction in the major cost factors.

Major cost factor	Measure
Raw material	• Increase selectivity at catalyst
	• Minimize workup losses
	• Use lower quality raw materials
Energy	• Optimize energy network
	• Use multistage evagoration
	• Extraction instead of distillation
	• River water cooler instead of air cooler
Disposal	• Integrated waste exploitation
	• Choice of location
Personnel	• Increase degree of automation of plant
	• Process control system
Depreciation	• Simplify reaction and processing
	• Check choice of material
	• Check dimensioning of pipelines
	• Exploit synergies of location

The return on capital from the process being studied should be compared with that of known processes. Above a threshold value, which varies from case to case, the process will provide a return on investment. A minimum return R of 20% is often employed, but because of the considerable vulnerability to technical risk and market certainty, this is only a guideline. While the value may be lower if the risk is low, it is prudent to impose a minimum value up to a factor of two higher in the case of new high-risk products.

Generally the process development engineer is not faced with the question of how profitable a project is but of whether a particular process engineering approach is more profitable than an-other, alternative solution within the framework of the process being studied or not [Frey 1990]. In this case, the sales are, of course, unchanged but the production and investment costs are:

$$H_2 = H_1 - \Delta H,$$
$$I_2 = I_1 - \Delta I. \tag{6-18}$$

The return offered by a particular approach is then given by the ratio of the changes:

$$R_{12} = \frac{\Delta H}{\Delta I}. \tag{6-19}$$

The change in the overall return resulting from adopting this particular approach is then:

$$R_2 = \frac{(R_1 \cdot I_1 + \Delta H)}{(I_1 + \Delta I)}. \qquad (6-20)$$

The weakness of static return-on-investment indicators is that they take a short-term view and, since they do not allow for future changes, they favor short-lived investments. Their advantage is that they can easily be determined.

6.2.2
Dynamic Return on Investment

The dynamic return on investment method is superior to the static method because allowance is made for the timing of the cash movements. Since only contemporaneous monies are comparable, the investment outlay and the earnings from the profit must be discounted to a fixed point in time, usually the start of production. If a certain rate of interest p is assumed, all the capital movements can be referred to that point to obtain the so-called net value of a capital movement:

$$(E - A)_{t=0} = (E - A)_t \cdot (1 + p)^{-t} \qquad (6-21)$$

E = expected earnings (e.g., from product sales) in € a^{-1}
A = expected outlay (e.g., investment costs, production costs € a^{-1})
p = interest rate
t = number of years (negative before start-up and positive after start-up).

For example, when started up, a plant has to bear double the cost of an outlay incurred five years before start-up if the rate of interest is 15 %. Conversely, a revenue five years after start-up will only carry half the weight, and one ten years after start-up will be virtually insignificant. The capital value K of a project is therefore the sum of the net values:

$$K = \sum_{t=-m}^{t=+n} (E - A)_t \cdot (1 + p)^{-t} \qquad (6-22)$$

m = duration of process development in years
n = useful life of project in years (usually 10 years).

To compare different projects, a useful life of ten years is usually assumed for chemical plants. An investment is only beneficial if its capital value is equal to zero or is positive. The internal rate of return method (discounted cash flow, DCF) aims to find the rate of interest which will result in a capital value of zero, that is, the rate at which the net values of the incoming and outgoing payments are equally large (the internal rate of return). The internal rate of return p_0 is determined by equating the above capital value function with zero:

$$\sum_{z=-m}^{t=+n}(E-A)_t \cdot (1+p)^{-t} = 0. \tag{6-23}$$

Projects are compared by comparing their calculated internal rates of return, the project with the highest internal rate of return being the most profitable.

6.3
Economic Risk

The return on capital calculation indicates whether or not a project is economically worthwhile. To calculate the return, assumptions must be made about the most probable product prices, the production outlay, the plant loading (actual production/nominal capacity), investment costs, etc. Time will see whether these assumptions are correct. The impact of these uncertainties on the cost effectiveness of the project can be quantified by using the following variables.

6.3.1
Sensitivity Analysis

Sensitivity analysis yields information about the sensitivity of the process to variations in individual influencing variables and therefore about the reliability of the result of the return on capital analysis. It also provides pointers to the direction in which the process should be developed further. The development is particularly worthwhile if it alters influencing variables to which the return on capital is highly sensitive. The procedure is to vary the different influencing variables in Equation (6-23) one after the other in the unfavorable direction (while keeping all the others constant) by a certain

Tab. 6.2-2 Relative change in the internal interest rate p_0 with varying influencing factor x. For successive changes in x in the unfavorable direction, $\Delta p_0/p_0$ can be calculated from Equation (6-23).

influencing factor x	$\Delta x/x$
Development costs	+ 10 %
Investment costs	+ 10 %
Product price	− 10 %
Degree of plant use	− 10 %
Net raw material costs	+ 10 %
Costs for operating materials	+ 10 %
Personnel costs	+ 10 %

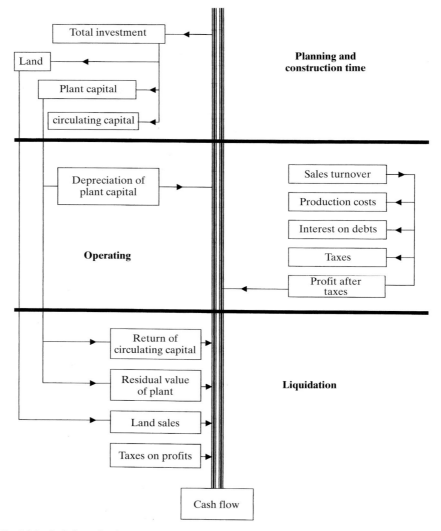

Fig. 6.3-1 Cash flow of a plant investment.

fraction (e.g. 10%) and then determine how much the internal rate of return p_0 differs from the original value (see Table 6.2-2).

6.3.2
Amortization Time

The risk as a function of time can be quantified by means of the amortization time (payback time) t_w. This indicates the time which will be necessary for the cost of the capital employed to be recovered. As an initial estimate, t_w is the ratio of the investment costs I to the annual surplus income (E–A) and the depreciation:

$$t_w/\text{a} \approx \frac{I/\text{Euro}}{(E-A) + depreciation/(\text{Euro a}^{-1})} \qquad (6-24)$$

The reciprocal value is known as return on investment.

6.3.3
Cash Flow [Hirschberg 1999]

The cash flow is the difference between the sales turnover and the monies expended therefor. Figure 6.3-1 shows the schematic cash flow of an investment.

7
Trends in Process Development

Increasingly intense international competition is forcing the chemical industry to continually develop and use new methods of process development. Apart from economic aspects such as higher reaction yields [Eigenberger 1991, Emig 1993, Misonon 1999], improved catalysts, factors such as safety, sustainability, careful use of resources, and environmental compatibility also play an important role [Standort 1995, Zlokarnik 1989]. Process development can also make a contribution here [Onken 1997], by means of a course of process development designed so that competitive advantages arise, that is, a process must be developed quickly and cheaply. Some trends that aid in achieving this target are discussed in the following [Bogenstätter 1985, Frey 1998].

Nowadays only a team made up of members from all the scientific disciplines is capable of coping with the complex task of developing a process. For this reason, a project team, usually consisting of a chemical process engineer, a preparative chemist, an engineer, and possibly the future works manager should be brought together at the beginning [Sowa 1997]. If necessary, this group may recruit or consult other specialists (Table 7-1).

Team work and team spirit are essential to the success of the activity, the synergies resulting from this type of activity being exploited to the full. The team is capable of more than the sum of its members. This way of working is generally not familar to persons new to the job, who are often only used to individual competition at university and thus must make a major readjustment in their first year at work [Frey 1995]. When process development is complete, it is advantageous for at least one member of the project team to continue working on the project during planning, construction, commissioning, and start-up, since otherwise a great deal of detailed knowledge would be lost.

Tab. 7-1. Example of a project team for the development of a chemical process.

Temporary project team
• Preparative chemist with experience of laboratory synthesis
• Engineer, subsequently plant engineer
• Process chemist with operational experience, subsequently project
Specialist from special departments
• Analysis
• Disposal
• Dept. for unit operations
• Licensing process
• Marketing
• Patent dept.
• Safety
• Toxicology
• Materials technology

Process Development. From the Initial Idea to the Chemical Production Plant. G. Herbert Vogel
Copyright © 2005 WILEY-VCH Verlag GmbH & Co. KGaA, Weinheim
ISBN: 3-527-31089-4

In the case of short-lived consumer-related products (dyes, active substances, dispersions, etc.), which are preferably produced in batch processes, the trend is towards equipment which is as flexible as possible (multipurpose production plants [Schuch 1992], multipurpose small-scale plants [Rauch 2003]). Innovations in such products are usually due to changes in formulation, and much less a result of modifications to the equipment. The ideal is to produce a wide variety of products in a batch plant over a long period of time.

If operating a pilot plant is unavoidable, nowadays it is often cheaper to have the plant built by another company (custom manufacturing) [Sowa 1997].

In the case of basic products and intermediates produced in continuous plants, process development can make a contribution to competitiveness by reducing the time and the cost required to develop improved manufacturing processes. This objective can be achieved by increasing the scale-up factors (Section 3.1). In order to limit the resulting increase in technical risk, use is now made of the synergy between an integrated miniplant and a mathematical simulation. This makes it possible to dispense with the pilot plant stage, that is, the synergism between an integrated miniplant technique and a mathematical simulation results in scale-up safety margins of similar quality to those obtained by using a pilot plant. As a result it is possible to shorten the time required for process development by ca. 30%, which can have a noticeable effect on the cost effectiveness of a new process.

Other current trends are the increased use of process simulations [Sowa 1997] and the development and use of expert systems. These are already available for simple, limited systems such as the optimization of individual unit operations [Göttert 1991]. For much more complex systems such as catalyst development or the optimization of entire plants, expert systems are at present still a long way from general practical application. However, since expert systems can only be used efficiently by experts, a responsible user must also be an expert in his own field.

Future tasks in process development will be in the following fields [Arntz 1993, Plotkin 1999, Klos 2000, Meyer-Galow 2000, Shinnar 1991, Villermaux 1995, Wiesner 1990]:

- Production with careful use of resources in the case of chemicals with a worldwide production volume of more than 10^6 t/a, for example, propylene oxide.
- Improvement of plants and processes with a view to integrated environmental protection [Swodenk 1984, Christmann 1985, Zlokarnik 1989, Scharfe 1991, Dechema 1990, Christ 1999, 2000]:
    - *Residue reduction:*
      Reduction in the amounts of byproducts by improving the catalysts and the reaction control.
    - *Integrated residue utilization:*
      Integrated workup of unavoidable waste streams within the main plant, that is, exploitation of on-site synergies (thermal and material coupling) and effective reuse of worked-up wastes.

- Development of new processes for the basic products and intermediates on the basis of:
  - Natural gas (C_1 chemistry)
  - Renewable raw materials
  - Waste streams as raw materials
  - Direct functionalization of alkanes by selective partial oxidation to oxygenates [Baerns 2000, Petzny 1999].
- Replacement of chemical reactions with stoichiometric consumption of reagents by catalytic processes and transfer of these processes to industrial scale [Metivier 2000]. There are several prerequisites for this. First, a catalyst with sufficient performance must be available. Therefore, an important R&D task is the development and improvement of catalysts:
  - By rational catalyst design (direct insight on the atomic scale)
  - By combinatorial chemistry with the aid of high-throughput processes
  - By following the example of nature (copying enzymes with structures that have a similar mode of action (zeozymes)

  Second, the required technology should be sufficiently well known. Third, a major potential for economic and ecological improvement should be present. Examples are:
  - Friedel–Crafts reactions: Replacement of stoichiometric quantities of Lewis acids ($AlCl_3$, $FeCl_3$, BF_3, $ZnCl_2$, $TiCl_4$) or Brønsted acids (H_3PO_4, HF) by heterogeneous catalysts (zeolites, ion exchangers, heteropolyacids, etc.).
  - Oxidation of alcohols and ketones: Replacement of stoichiometric quantities of oxidizing agents such as permanganate and hypochlorite by molecular oxygen activated by supported metal catalysts.
  - Reduction of carboxylic acids to aldehydes: Replacement of the Rosenmund reaction (conversion of carboxylic acids to acid chlorides followed by catalytic hydrogenation) or the reaction of Grignard reagents with formic esters by molecular hydrogen and heterogeneous catalysts (Ru_3Sn_7 alloy).
- Catalytic functionalization of renewable raw materials [Eissen 2002].
- Use of biotechnology in chemical production [Eissen 2002], that is, use of biocatalysts for the production of fine chemicals.
- Acceleration of process development [Faz 1999] by using integrated miniplant technology (laboratory plant → microplant → miniplant → industrial plant).
- Development of processes that result in smaller, more economical, and cheaper plants, for example, by using multifunctional reactors [Westerterp 1992, Eigenberger 1991] and the use of unconventional conditions (chemistry under supercritical conditions [Vogel 1999], sonochemistry, etc.).
- Cooperation of chemical, refinery, and chemical engineering companies in new processes, whereby the partners use their specific know-how to accelerate process development, for example, Geminox (BP/Lurgi: fluidized-bed oxidation and selective hydrogenation), Cyclar (BP/UOP: fluidized-bed and regeneration technology).
- Improved training [Gillett 2001, Molzahn 2001].

In the future, the most important technology will probably be biotechnology, which offers fantastic possibilities. Its interdisciplinary character influences many other branches of technology, such as microelectronics, medical technology, energy and environmental technology, agricultural and foodstuffs technology, as well as chemical technology, apparatus technology, and metrology. Materials science and chemical technology, including catalysis, will continue to be of major importance.

8
Appendix

Appendix 8.1
Mathematical Formulas

Mathematical constants

$\pi = 3.14159$
$e = 2.71828$
$\ln 2 = 0.693147$
$\ln 10 = 2.302585$

Solution of a quadratic equation

$$a \cdot x^2 + b \cdot x + c = 0$$

$$x_{1/2} = \frac{-b \pm \sqrt{b^2 - 4 \cdot a \cdot c}}{2 \cdot a}$$

Taylor[1] series

$$f(x_0 + x) = f(x_0) + f'(x_0) \cdot x + \frac{f''(x_0)}{2!} \cdot x^2 + \ldots$$

Example 8.1-1

Transformation of the Arrhenius equation:

$$k = k_0 \cdot exp\left(-\frac{E_a}{R} \cdot T^{-1}\right)$$

through a Taylor expansion of T^{-1} around a reference temperature T_0:

$$\frac{1}{T} \approx \frac{2}{T_0} - \frac{T}{T_0^2} = \frac{1}{T_0^2} \cdot (2 \cdot T_0 - T)$$

into an integrable form:

$$k = \left[k_o \cdot \exp\left(-\frac{2 \cdot E_a}{RT_0}\right)\right] \cdot \left[\exp\left(\frac{E_a}{RT_0^2} \cdot T\right)\right] \propto b \cdot \exp(a \cdot T)$$

---

1) Brook Taylor, English mathematician (1685–1731).

Process Development. From the Initial Idea to the Chemical Production Plant. G. Herbert Vogel
Copyright © 2005 WILEY-VCH Verlag GmbH & Co. KGaA, Weinheim
ISBN: 3-527-31089-4

Trigonometry

$$\sin(a) = \sin(a + n \cdot 360°)$$
$$\cos(a) = \cos(a + n \cdot 360°)$$
$$\tan(a) = \tan(a + n \cdot 180°)$$
$$\cot(a) = \cot(a + n \cdot 180°)$$

$$\sin(a) = -\sin(-a)$$
$$\cos(a) = \cos(-a)$$
$$\tan(a) = -\tan(-a)$$
$$\cot(a) = -\cot(-a)$$

$$\sin^2 a + \cos^2 a = (\sin a)^2 + (\cos a)^2 = 1$$
$$\sin(a)/\cos(a) = \tan(a)$$

Graphical depicition of $y = sin(x)$ (thin line) and $y = cos(x)$ (thick line)

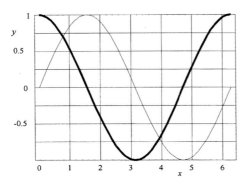

Graphical depicition of $y = tan(x)$ (thin line) and $y = cot(x)$ (thick line)

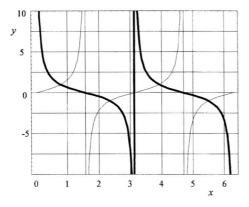

Inverse trigometric functions

$$y = \sin(x), \quad x = \text{arc } \sin(y)$$
$$y = \cos(x), \quad x = \text{arc } \cos(y)$$
$$y = \tan(x), \quad x = \text{arc } \tan(y)$$
$$y = \cot(x), \quad x = \text{arc } \cot(y)$$

Hyperbolic functions

$$\sinh(x) = \frac{e^x - e^{-x}}{2}$$
$$\cosh(x) = \frac{e^x + e^{-x}}{2}$$
$$\tanh(x) = \frac{e^x - e^{-x}}{e^x + e^{-x}}$$
$$(\cosh(x))^2 = 1 + (\sinh(x))^2$$

Inverse hyperbolic functions

$$y = \sinh(x), \quad x = \text{arc } \sinh(y)$$
$$y = \cosh(x), \quad x = \text{arc } \cosh(y)$$
$$y = \tanh(x), \quad x = \text{arc } \tanh(y)$$
$$y = \coth(x), \quad x = \text{arc } \coth(y)$$

Important series

$$\frac{1}{1 \pm x} \approx 1 \mp x \text{ for } x < 1$$

$$\frac{1}{a \pm x} \approx \frac{1}{a^2}(a \mp x) \text{ for } x < a$$

$$\sqrt{1+x} \approx 1 + \frac{x}{2}$$

$$\exp(x) \approx 1 + x$$

$$\ln(1+x) \approx x - \frac{x^2}{2} + \frac{x^3}{3} - \ldots \text{ for } -1 < x < 1$$

$$\sin(x) \approx x - \frac{x^3}{3!} + \frac{x^5}{5!} - \ldots$$

$$cox(x) \approx 1 - \frac{x^2}{2} + \frac{x^4}{4!} - \ldots$$

$$\tan(x) \approx x + \frac{x^3}{3}$$

Important formulas of differential calculus

Function	Derivative
x^n	$n \cdot x^{n-1}$
$a^{f(x)}$	$a^{f(x)} \cdot \ln a \cdot f'(x)$
$\exp[f(x)]$	$\exp[f(x)] \cdot f'(x)$
$\ln[f(x)]$	$\dfrac{f'(x)}{f(x)}$
$\sin(x)$	$\cos(x)$
$\tan(x)$	$1 + \tan^2(x) = \dfrac{1}{\cos^2(x)}$
$\cot(x)$	$-(1 + \cot^2(x)) = -\dfrac{1}{\sin^2(x)}$
$\dfrac{1}{f(x)}$	$\dfrac{f'(x)}{(f(x))^2}$

Important formulas of integral calculus

Function	Integral
x^n	$\dfrac{x^{n+1}}{n+1}$
$\exp(a \cdot x)$	$\dfrac{\exp(a \cdot x)}{a}$
$\dfrac{1}{a + b \cdot x}$	$\dfrac{\ln(a + b \cdot x)}{b}$
$\ln(x)$	$x \cdot \ln(x) - x$
$(a + b \cdot x)^n$	$\dfrac{(a + b \cdot x)^{n+1}}{b \cdot (n+1)}$ for $n \neq 1$
$\sin x$	$-\cos x$
$\cos x$	$\sin x$
$\tan x$	$-\ln(\cos x)$
$\cot x$	$\ln(\sin x)$

Cartesian, cyllindrical, and spherical coordinate

Cylindrical coordinates (r, θ, z)

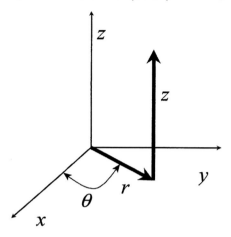

$$x = r \cdot \cos \theta$$
$$y = r \cdot \cos \theta$$
$$z = z$$
$$r = \sqrt{x^2 + y^2}$$
$$\theta = \arctan \frac{y}{x}$$
$$z = z$$

$$dV = r \cdot dr \cdot d\theta \cdot dz$$

Spherical coordinates (r, θ, ϕ or ϕ')

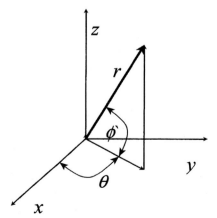

$$x = r \cdot \cos\theta \cdot \cos\phi'$$
$$y = r \cdot \sin\theta \cdot \cos\phi'$$
$$z = r \cdot \sin\phi'$$
$$\text{mit } \phi' + \phi = \frac{\pi}{2} = 90°$$
$$r = \sqrt{x^2 + y^2 + z^2}$$
$$\theta = \arctan\frac{y}{x}$$
$$\phi' = \arcsin\frac{z}{\sqrt{x^2 + y^2 + z^2}}$$

$$dV = r^2 \cdot \cos\phi' \cdot dr \cdot d\theta \cdot d\phi'$$

Transformation of a vector $\vec{u}(x, y, z)$ in *Cartesian* coordinates into a vector in *cylilndrical* coordinates and vice versa:

$$\begin{pmatrix} u_r \\ u_\theta \\ u_z \end{pmatrix} = \begin{bmatrix} \cos\theta & \sin\theta & 0 \\ -\sin\theta & \cos\theta & 0 \\ 0 & 0 & 1 \end{bmatrix} \begin{pmatrix} u_x \\ u_y \\ u_z \end{pmatrix}$$

$$\begin{pmatrix} u_x \\ u_y \\ u_z \end{pmatrix} = \begin{bmatrix} \cos\theta & -\sin\theta & 0 \\ \sin\theta & \cos\theta & 0 \\ 0 & 0 & 1 \end{bmatrix} \begin{pmatrix} u_r \\ u_\theta \\ u_z \end{pmatrix}$$

Transformation of a vector $\vec{u}(x, y, z)$ in *Cartesian* coordinates into a vector in *spherical* coordinates and vice versa:

$$\begin{pmatrix} u_r \\ u_\theta \\ u_{\phi'} \end{pmatrix} = \begin{bmatrix} \cos\theta \cdot \cos\phi' & \sin\theta \cdot \cos\phi' & \sin\phi' \\ -\sin\theta & \cos\theta & 0 \\ -\cos\theta \cdot \sin\phi' & -\sin\theta \cdot \sin\phi' & \cos\phi' \end{bmatrix} \begin{pmatrix} u_x \\ u_y \\ u_z \end{pmatrix}$$

$$\begin{pmatrix} u_x \\ u_y \\ u_z \end{pmatrix} = \begin{bmatrix} \dfrac{x}{\sqrt{x^2+y^2+z^2}} & -\dfrac{y}{\sqrt{x^2+y^2}} & \dfrac{x \cdot z}{\sqrt{x^2+y^2} \cdot \sqrt{x^2+y^2+z^2}} \\ \dfrac{y}{\sqrt{x^2+y^2+z^2}} & \dfrac{x}{\sqrt{x^2+y^2}} & -\dfrac{y \cdot z}{\sqrt{x^2+y^2} \cdot \sqrt{x^2+y^2+z^2}} \\ \dfrac{z}{\sqrt{x^2+y^2+z^2}} & 0 & \dfrac{\sqrt{x^2+y^2}}{\sqrt{x^2+y^2+z^2}} \end{bmatrix} \begin{pmatrix} u_r \\ u_\theta \\ u_{\phi'} \end{pmatrix}$$

Vector analysis

The Gradient
The gradient of a scalar field $A(x, y, z)$ is a vector field, each point of which gives the magnitude and direction of the largest change in the scalar field:

$$\text{grad } A(x, y, z) = \nabla \cdot A = \begin{pmatrix} \frac{\partial}{\partial x} \\ \frac{\partial}{\partial y} \\ \frac{\partial}{\partial z} \end{pmatrix} \cdot A = \begin{pmatrix} \frac{\partial A}{\partial x} \\ \frac{\partial A}{\partial x} \\ \frac{\partial A}{\partial z} \end{pmatrix}$$

where ∇ is the Nabla operator.

In cylindrical coordinates, the gradient reads:

$$\text{grad } A(r, \theta, z) = \begin{pmatrix} \frac{\partial A}{\partial r} \\ \frac{1}{r} \cdot \frac{\partial A}{\partial \theta} \\ \frac{\partial A}{\partial z} \end{pmatrix}$$

and in spherical coordinates:

$$\text{grad } A(r, \theta, \phi') = \begin{pmatrix} \frac{\partial A}{\partial r} \\ \frac{1}{r \cdot \cos \phi'} \cdot \frac{\partial A}{\partial \theta} \\ \frac{1}{r} \cdot \frac{\partial A}{\partial \phi'} \end{pmatrix}$$

The Divergence
The divergence of a vector field $\vec{u}(x, y, z)$ is a scalar field which for each point of the vector field gives the inflow or outflow per unit volume in the immediate vicinity of this point:

$$\text{div } \vec{u}(x, y, z) = \nabla \cdot \vec{u} = \begin{pmatrix} \frac{\partial}{\partial x} \\ \frac{\partial}{\partial y} \\ \frac{\partial}{\partial z} \end{pmatrix} \cdot \begin{pmatrix} u_x \\ u_y \\ u_z \end{pmatrix} = \frac{\partial u_x}{\partial x} + \frac{\partial u_y}{\partial y} + \frac{\partial u_z}{\partial z}$$

In cylindrical coordinates the divergence reads:

$$\text{div } \vec{u} \, (r, \theta, z) = \frac{1}{r} \cdot \frac{\partial (r \cdot u_r)}{\partial r} + \frac{1}{r} \cdot \frac{\partial u_\theta}{\partial \theta} + \frac{\partial u_z}{\partial z}$$

and in spherical coordinates:

$$\text{div } \vec{u} \, (r, \theta, \phi') = \frac{1}{r^2} \cdot \frac{\partial (r^2 \cdot u_r)}{\partial r} + \frac{1}{r \cdot \cos \phi'} \cdot \frac{\partial u_\theta}{\partial \theta} + \frac{1}{r \cdot \cos \phi'} \cdot \frac{\partial u_{\phi'} \cdot \cos \phi'}{\partial \phi'}$$

Example of application in chemical reaction technology (Chap. 2.2):

$$\frac{\partial A}{\partial t} = -\nabla \cdot (A \cdot \vec{u}) + \nabla \cdot [D \cdot \nabla \cdot A] + \nu_A \cdot r_{V,A}$$

$$\frac{\partial A}{\partial t} = -\text{div}\,(A \cdot \vec{u}) + \text{div}\,[D \cdot \text{grad}\,A] + \nu_A \cdot r_{V,A} \, .$$

For $D = $ const. it follows that:

$$\frac{\partial A}{\partial t} = -\nabla \cdot (A \cdot \vec{u}) + D \cdot \nabla \cdot [\nabla \cdot A] + \nu_A \cdot r_V$$

$$\frac{\partial A}{\partial t} = -\vec{u} \cdot \nabla \cdot A - A \cdot \nabla \cdot \vec{u} + D \cdot \nabla^2 A + \nu_A \cdot r_V \, .$$

with $\vec{u} = $ constant it follows that:

$$\frac{\partial A}{\partial t} = -\vec{u} \cdot \nabla \cdot A + D \cdot \nabla^2 A + \nu_A \cdot r_V$$

or written differently:

$$\frac{\partial A}{\partial t} = -\vec{u} \cdot \text{grad}\,A + D \cdot \text{div}\,(\text{grad}\,A) + \nu_A \cdot r_V.$$

The design equations for chemical reactors are coupled partial differential equations that can still only be solved numerically. Nevertheless, for the numerical solution one requires a series of initial conditions, so that a basic understanding of the theory of partial differential equations is necessary.

The fundamental problem of the theory of partial differential equations can be formulated as follows [Vvedensky 1994]. A function:

$$F(x, y, ..., A, A_x, A_y, ..., A_{xx}, A_{xy}, A_{yy}, ...) = 0 \tag{1}$$

of the variables $x, y, ..., A, A_x, A_y, ..., A_{xx}$, (where $A_x = \partial A/\partial x$ etc.) is given. We seek a function $A(x, y, ...)$ of the independent variables $x, y, ...$ that, together with their derivatives, fulfills Equation (1) in a certain certain region of the independent variables. Such a function is known as a solution of (1). The aim is not only to find particular solutions of (1), but also to gain an overview of the total collective of solutions and to characterize individual solutions through additional conditions.

The total collective of solutions of conventional differential equations of nth order is given by a function of the independent variables x, which in turn depend on n arbitrary integration constant. In the case of partial differential equations the relationships are more complicated. Here we can also seek the total collective of solutions or the general integral. However, instead of arbitrary inegration constants, arbitrary functions arise, the number of which is generally equal to the order of the differential equation, that is, equal to the degree of the highest derivative occurring in the partial differential equation. These arbitrary functions depend on one variable fewer than the solution A.

In general, each family of functions that depends on arbitrary functions can be assigned a partial differential equation in which the arbitrary functions no longer occur.

Numerous physical problems can be reduced to the solution of second-order partial differential equations:

- The diffusion or thermal conduction equation:

$$A_t - D \cdot A_{xx} = 0 \tag{2}$$

- The wave equation:

$$A_{tt} - c^2 \cdot A_{xx} = 0 \tag{3}$$

- The Laplaceian or potential equation:

$$A_{xx} - A_{yy} = 0. \tag{4}$$

The total collective of solutions of Equations (2)–(4) must contain the above-mentioned arbitrary functions. For Equations (3) and (4) such solution collectives can readily be obtained; for (2), which is of special importance for chemical reaction technology, this is not the case. A large class of solutions $A(x,t)$ can be found for Equation (2) with the aid of the product expression:

$$A(x,t) = X(x) \cdot T(t) \tag{5}$$

which leads to two conventional linear differential equations.

Appendix 8.2
Constants.

Constants	Symbol	value	Unit
Atomic mass unit	m_u	1.660540×10^{-27}	kg
Avogadro constant	N_A	6.022137×10^{23}	mol^{-1}
Bohr magneton	μ_B	9.274015×10^{-24}	$J\ T^{-1}$
Electric field constant in vacuum	ε_0	8.854188×10^{-12}	$F\ m^{-1}$
Electron radius	r_e	2.817941×10^{-15}	m
Elementarly charge	e	1.602177×10^{-19}	C
Acceleration due to gravity	g	9.80665	$m\ s^{-2}$
Faraday constant	F	9.648531×10^4	$C\ mol^{-1}$
Gas constant	R	8.31451	$J\ mol^{-1}\ K^{-1}$
Gravitational constant	f	6.672×10^{-11}	$N\ m^2\ kg^{-2}$
Velocity of light in vacuum	c	2.99792×10^8	$m\ s^{-1}$
Magnetic field constant in vacuum	μ_0	1.256637×10^{-7}	$H\ m^{-1}$
Molar volume of ideal gas at 298 K and 1.01325 bar	V	24.47	$L\ mol^{-1}$
Planck constant	h	6.626×10^{-34}	J s
Rest mass of electron	m_e	9.109390×10^{-31}	kg
Rest mass of neutron	m_n	1.674929×10^{-27}	kg
Rest mass of proton	m_p	1.672623×10^{-27}	kg
Stefan-Boltzmann constant	σ	5.6705×10^{-8}	$W\ m^{-2}\ K^{-4}$

Appendix 8.3
List of elements with relative atomic masses bonding radis and melting and boiling points.

Atomic number	Chem. Symbol	Element	Atomic-masses	r(atom)/ pm	r(coval)/ pm	m. p./°C	b. p./°C
1	H	Hydrogen	1.008	37.3	30	− 259	− 253
2	He	Helium	4.003	/	99	− 270	− 269
3	Li	Lithium	6.941	152.0	123	180	1330
4	Be	Beryllium	9.012	111.3	89	1284	2480
5	B	Boron	10.811	79.5	81	2150	3900
6	C	Carbon	12.011	77.3	77	3730	4830
7	N	Nitrogen	14.007	54.9	70	− 210	− 196
8	O	Oxygen	16.000	60.4	66	− 219	− 183
9	F	Fluorine	18.998	70.9	64	− 220	− 188

Appendix 8.3 List of elements with relative atomic masses bonding radis and melting and boiling points

Continue

Atomic number	Chem. Symbol	Element	Atomic-masses	r(atom)/pm	r(coval)/pm	m. p./°C	b. p./°C
10	Ne	Neon	20.180	/	160	−249	−246
11	Na	Sodium	22.990	185.8	157	98	892
12	Mg	Magnesium	24.305	160.0	157	649	1103
13	Al	Aluminum	26.981	143.2	125	660	2450
14	Si	Silicon	28.085	117.6	117	1412	2680
15	P	**Phosphorus**	30.974	110.5	110	44	280
16	S	**Sulfur**	32.066	103.5	102	113	445
17	Cl	**Chlorine**	35.4527	99.4	99	−101	−35
18	Ar	Argon	39.948	/	192	−189	−186
19	K	Potassium	39.098	227.2	203	64	760
20	Ca	Calcium	40.078	197.4	174	838	1482
21	Sc	Scandium	44.956	160.6	144	1530	2730
22	Ti	Titanium	47.88	144.8	132	1668	3270
23	V	**Vanadium**	50.941	131.1	122	1905	3450
24	Cr	Chromium	51.996	125.0	118	1903	2642
25	Mn	Manganese	54.9380	136.7	118	1243	2097
26	Fe	Iron	55.847	124.1	116	1534	3000
27	Co	Cobalt	58.933	125.3	116	1495	2900
28	Ni	Nickel	58.69	124.6	115	1453	2730
29	Cu	**Copper**	63.546	127.8	117	1083	2600
30	Zn	Zinc	65.390	133.5	125	419	907
31	Ga	Gallium	69.723	122.1	125	29.8	2227
32	Ge	Germanium	72.610	122.5	122	937	2830
33	As	Arsenic	74.921	124.5	121	814	616
34	Se	Selenium	78.960	116.0	117	217	685
35	Br	Bromine	79.904	114.5	114	−7	58
36	Kr	Krypton	83.800	197.0	/	−157	−152
37	Rb	Rubidium	85.468	247.5	216	39	688
38	Sr	Strontium	87.620	215.1	191	770	1360
39	Y	Yttrium	88.906	177.6	162	1502	2930

Continue

Atomic number	Chem. Symbol	Element	Atomic-masses	r(atom)/pm	r(coval)/pm	m. p./°C	b. p./°C
40	Zr	Zirconium	91.224	159.0	145	1852	3600
41	Nb	Niobium	92.906	142.9	134	2420	5127
42	Mo	Molybdenum	95.940	136.3	130	2620	5560
43	Tc	Technetium	98.906	135.2	127	2200	4900
44	Ru	Ruthenium	101.07	132.5	125	2280	4900
45	Rh	Rhodium	102.90	134.5	125	1966	7300
46	Pd	Palladium	106.42	137.6	128	1552	3125
47	Ag	*Silver*	107.87	144.5	134	960	2210
48	Cd	Cadmium	112.41	148.9	141	321	765
49	In	Indium	114.82	162.6	150	156	2000
50	Sn	Tin	118.71	140.5	140	232	2270
51	Sb	Antimony	121.75	145.0	141	630	1380
52	Te	Tellurium	127.60	143.2	137	450	1390
53	I	Iodine	126.90	133.1	133	114	183
54	Xe	Xenon	131.29	/	217	−112	−108
55	Cs	Caesium	132.90	265.5	253	28.65	685
56	Ba	Barium	137.33	217.4	198	714	1640
57	La	Lanthanum	138.91	187.0	169	920	3470
58	Ce	Cerium	140.11	182.5	165	797	3470
59	Pr	Praseodymium	140.91	182.0	165	935	3130
60	Nd	Neodymium	144.24	181.4	164	1024	3030
61	Pm	Promethium	146.91	/	163	1030	2730
62	Sm	Samarium	150.36	179.0	162	1072	1900
63	Eu	Europium	151.96	199.5	185	826	1430
64	Gd	Gadolinium	157.25	178.5	161	1312	2800
65	Tb	Terbium	158.92	176.3	159	1364	2800
66	Dy	Dysprosium	162.50	175.2	159	1407	2600
67	Ho	Holmium	164.93	174.3	158	1461	2490
68	Er	Erbium	167.26	173.0	/	1497	2390
69	Tm	Thulium	168.93	172.4	156	1545	1720

Continue

Atomic number	Chem. Symbol	Element	Atomic-masses	r(atom)/pm	r(coval)/pm	m. p./°C	b. p./°C
70	Yb	Ytterbium	173.04	194.0	174	824	1430
71	Lu	Lutetium	174.97	171.8	156	1652	3330
72	Hf	Hafnium	178.49	156.4	144	2220	5400
73	Ta	Tantalum	180.95	143.0	134	3000	6030
74	W	Tungsten	183.85	237.0	130	3410	5930
75	Re	Rhenium	186.21	137.1	128	3160	5642
76	Os	Osmium	190.20	133.8	126	3027	5500
77	Ir	Iridium	192.22	135.7	127	2450	4500
78	Pt	Platinum	195.08	137.3	130	1769	3825
79	Au	Gold	196.97	144.2	134	1064	2970
80	Hg	Mercurry	200.59	150.3	144	−38	357
81	Tl	Thallium	204.38	170.0	155	303	1460
82	Pb	Lead	207.20	175.0	154	327	1740
83	Bi	Bismuth	208.98	154.5	146	271	1560
84	Po	Polonium	208.98	167.3	146	254	962
85	At	Astatine	209.99	/	145	302	335
86	Rn	Radon	222.02	/	/	−71	−62
87	Fr	Francium	223.08	/	/	27	680
88	Ra	Radium	226.02	/	/	700	1530
89	Ac	Actinium	227.03	187.8	/	1050	3200
90	Th	Thorium	232.04	179.8	165	1700	4200
91	Pa	Protactinium	231.03	156.1	/	1575	4000
92	U	Uranium	238.03	138.5	142	1133	3818

Appendix 8.4
Conversion of various units to SI units

Length

1 Å (Ångström)	$= 1 \times 10^{-10}$	m
1 in	$= 2.5400 \times 10^{-2}$	m
1 ft = 12 in	$= 3.0480 \times 10^{-1}$	m
1 yd = 3 ft = 36 in	$= 9.1440 \times 10^{-1}$	m
1 thou	$= 2.5400 \times 10^{-5}$	m
1 mile (statute)	$= 1.6094 \times 10^{3}$	m
1 mile (nautical)	$= 1.8533 \times 10^{5}$	m
1 rod = 1 perch = 5.5 yd	$= 5.292$	m
1 chain	$= 2.0117$	m
1 furlong	$= 2.0117 \times 10^{2}$	m

Area

1 in²	$= 6.4516 \cdot 10^{-4}$	m²
1 ft²	$= 9.2903 \cdot 10^{-2}$	m²
1 yd²	$= 8.3613 \cdot 10^{-1}$	m²
1 acre	$= 4.0469 \cdot 10^{3}$	m²
1 mile²	$= 2.5900 \cdot 10^{6}$	m²

Volumen

1 in³	$= 1.6387 \times 10^{-5}$	m³
1 ft³	$= 2.8317 \times 10^{-2}$	m³
1 yd³	$= 7.6455 \times 10^{-1}$	m³
1 US gal	$= 4.5460 \times 10^{-3}$	m³
1 UK gal	$= 4.5460 \times 10^{-3}$	m³
1 US bushel (dry)	$= 3.5239 \times 10^{-2}$	m³
1 UK bushel (dry)	$= 3.6369 \times 10^{-2}$	m³
1 barrel (petroleum US)	$= 1.5898 \times 10^{-1}$	m³
1 lübe oil barrel	$= 2.0819 \times 10^{-1}$	m³
1 gill	$= 1.1829 \times 10^{-4}$	m³
1 register ton = 100 ft³	$= 2.8317$	m³
1 quart	$= 2.9095 \times 10^{-1}$	m³

    = 8 UK bushels
    = 32 pecks
    = 64 UK gallons
    = 256 quarts
    = 512 pints

Mass

1 kp s²/m	$= 9.80665$	kg
1 grain	$= 6.4800 \times 10^{-5}$	kg
1 lb	$= 4.5359 \times 10^{-1}$	kg
1 ton (short) = 20 cwt US	$= 9.0718 \times 10^{2}$	kg
1 ton (long) = 20 cwt UK	$= 1.0160 \times 10^{3}$	kg

Density

1 grain/ft³	$= 2.2884 \times 10^{-3}$	kg m⁻³
1 lb/ft³	$= 1.6018 \times 10$	kg m⁻³
1 lb/UK gal	$= 9.9779 \times 10$	kg m⁻³
1 lb/US gal	$= 1.1983 \times 10^{2}$	kg m⁻³

Velocity

1 ft/hr	$= 8.4667 \times 10^{-5}$	m s^{-1}
1 ft/min	$= 5.0300 \times 10^{-3}$	m s^{-1}
1 ft/s	$= 3.0480 \times 10^{-1}$	m s^{-1}
1 mile/hr	$= 4.4704 \times 10^{-1}$	m s^{-1}

Force

1 kp	$= 9.8067$	N
1 dyn	$= 1.0000 \times 10^{-5}$	N
1 Dyn	$= 1.3825 \times 10^{-1}$	N
1 lbf	$= 4.4482$	N
1 ton f	$= 9.9640 \times 10^{3}$	N

Pressure, mechanical stress

1 bar	$= 1.0000 \times 10^{5}$	Pa
1 at	$= 9.8067 \times 10^{4}$	Pa
1 kp/cm^2	$= 9.8067 \times 10^{4}$	Pa
1 atm	$= 1.0133 \times 10^{5}$	Pa
1 Torr	$= 1.3332 \times 10^{2}$	Pa
1 mmHg (1 mm QS)	$= 1.3332 \times 10^{2}$	Pa
1 mm/WS	$= 9.8067$	Pa
1 dyn/cm^2	$= 1.0000 \times 10^{-1}$	Pa
1 pdl/ft^2	$= 1.4881$	Pa
1 lbf/ft^2 (psf)	$= 4.7880 \times 10^{1}$	Pa
1 pdl/in^2	$= 2.1429 \times 10^{2}$	Pa
1 in water	$= 2.4909 \times 10^{2}$	Pa
1 ft water	$= 2.9891 \times 10^{3}$	Pa
1 in Hg (l in mercury)	$= 3.3866 \times 10^{3}$	Pa
1 lbf/in^2 (oder psi)	$= 6.8948 \times 10^{3}$	Pa
1 ton f/in^2	$= 1.5444 \times 10^{7}$	Pa

Energy

1 N m	= 1.0000	J
1 W s	= 1.0000	J
1 dyn cm	= 1.0000×10^{-7}	J
1 erg	= 1.0000×10^{-7}	J
1 Dyn m	= 1.0000	J
1 kpm	= 9.8067	J
1 kcal	= 4.1868×10^3	J
1 kW h	= 3.6000×10^6	J
1 Ps h	= 2.6478×10^6	J
1 Btu	= $1.0551\ 10^3$	J
1 Chu	= $1.8991\ 10^3$	J
1 ft pdl	= $4.2139\ 10^{-2}$	J
1 ft lbf	= 1.3558	J
1 hp hr (british)	= $2.6845\ 10^6$	J
1 therm	= $1.0551\ 10^8$	J
1 eV	= $1.602\ 10^{-19}$	J
1 SKE (Steinkohleneinheit)	= $2.9308\ 10^7$	J

Power, thermal flux

1 mkp/s	= 9.80665	W
1 kcal/h	= 1.1630	W
1 erg/s	= 1.0000×10^{-7}	W
1 PS	= 7.3548×10^2	W
1 m^3 atm/h	= 2.8150×10	W
1 ft lbf/min	= 2.2597×10^{-2}	W
1 ft lbf/s	= 1.3558	W
1 ft pdl/s	= 4.2139×10^{-2}	W
1 Btu/hr	= 2.9308×10^{-1}	W
1 Chu/hr	= 5.2754×10^{-1}	W
1 hp (british)	= 7.4570×10^2	W
1 ton refrigeration	= 3.5169×10^3	W
1 therm/hr	= 2.9308×10^4	W

Caloric, quantities, volumetric

1 kcal/m^3	= 4.1868×10^3	J m^{-3}
1 Btu/ft^3	= 3.7260×10^4	J m^{-3}
1 Chu/ft^3	= 6.7067×10^4	J m^{-3}
1 therm/ft^3	= 3.7260×10^9	J m^{-3}

Heat-transfer coefficient

1 kcal/m² h °C	= 1.1630	W m^{-2} K^{-1}
1 cal/m² s °C	= 4.1868 × 10^4	W m^{-2} K^{-1}
1 kcal/ft² hr °C	= 1.2518 × 10	W m^{-2} K^{-1}
1 Btu/ft² hr °C	= 5.6785	W m^{-2} K^{-1}
1 Chu/ft² hr °C	= 5.6783	W m^{-2} K^{-1}

Thermal conductivity

1 kcal/m h °C	= 1.1630	W m^{-1} K^{-1}
1 cal/cams °C	= 4.1868 × 10^2	W m^{-1} K^{-1}
1 Btu/ft² hr (°F/in)	= 1.4423 × 10^{-1}	W m^{-1} K^{-1}
1 Btu/ft hr °F	= 1.7308	W m^{-1} K^{-1}
1 Chu/ft hr °C	= 1.7308	W m^{-1} K^{-1}

Heat capacity, specific

1 kcal/kg °C	= 4.1868 × 10^3	J kg^{-1} K^{-1}
1 cal/g °C	= 4.1868 × 10^3	J kg^{-1} K^{-1}
1 Btu/lb °F	= 4.1868 × 10^3	J kg^{-1} K^{-1}
1 Chu/lb °C	= 4.1868 × 10^3	J kg^{-1} K^{-1}

Viscosity, dynamic

1 kps/m²	= 9.80665	Pa s
1 kph/m²	= 3.532 × 10^{-4}	Pa s
1 Poise = 1 g/cm s	= 1.0000 × 10^{-1}	Pa s
1 lb/ft hr	= 4.1338 × 10^{-4}	Pa s
1 kg/ft hr	= 9.1134 × 10^{-4}	Pa s
1 lb/ft s	= 1.4882	Pa s

Viscosity, kinematic

1 Stoke = 1 cm²/s	= 1.0000 × 10^{-4}	m² s^{-1}
1 dm³/hr in	= 1.0936 × 10^{-5}	m² s^{-1}
1 ft²/hr	= 2.5806 × 10^{-5}	m² s^{-1}
1 ft²/s	= 9.2903 × 10^{-2}	m² s^{-1}

Appendix 8.5
Relationships between derived and base units.

1 N (Newton), force	1 kg m s^{-2}
1 Pa (Pascal), pressure	1 kg m^{-1} s^{-2}
1 J (Joule), energy	1 kg m^2 s^{-2}
1 W (Watt), power	1 kg m^2 s^{-3}
1 V (Volt), electric potential	1 kg m^2 A^{-1} s^{-3}
1 Ω (Ohm), electrical resistance	1 kg m^2 A^{-2} s^{-3}
1 T (Tesla), magnetic flux density	1 V s m^{-2}
1 Wb (Weber), magnetic flux	1 V s

Appendix 8.6
Conversion of concentrations for binary mixtures of dissolved component A in solvent B.

$w_A/\%$ (g g^{-1}): mass per cent
x_A = mole fraction
c_A/mol L^{-1}: molarity
M_A/g mol^{-1}: molar mass of component A (ditto for B).
ρ_M/g cm^{-3}: Density of mixture at temperature T

	$w_A/\%$	x_A	c_A/mol L^{-1}
$w_A =$	/	$\dfrac{100}{\left(1 + \dfrac{(1 - x_A)}{x_A} \cdot \dfrac{M_B}{M_A}\right)}$	$\dfrac{M_A}{10 \cdot \rho_M} \cdot c_A$
$x_A =$	$\dfrac{1}{1 + \dfrac{(100 - w_A)}{w_A} \cdot \dfrac{M_A}{M_B}}$	/	$\dfrac{c_A}{c_A \cdot \left(1 - \dfrac{M_A}{M_B}\right) + \dfrac{1000 \cdot \rho_M}{M_B}}$ $\approx \dfrac{M_B}{1000 \cdot \rho_M} \cdot c_A$ at very small c_A or $M_A \approx M_B$
$c_A =$	$\dfrac{10 \cdot \rho_M}{M_A} \cdot w_A$	$\dfrac{\rho_M}{M_B} \cdot \dfrac{1000 \cdot x_A}{1 - x_A \cdot \left(1 - \dfrac{M_A}{M_B}\right)}$	/

Appendix 8.7
van der Waals constants *a* and *b* and critical values for some gases.

$$\left(P + \frac{a}{\bar{V}^2}\right) \cdot (\bar{V} - b) = RT$$

Compound	a/0.1 Pa m^6 mol^{-2}	b/10^{-5} m^3 mol^{-1}	T_{crit}/K	P_{crit}/bar
Ammonia	4.46	5.15	405.6	112.8
Argon	1.37	3.23	150.8	48.7
Benzene	18.24	11.50	562.1	48.9
Chlorine	6.59	5.64	417.0	77.0
Hydrogen chloride	3.73	4.09	324.6	83.0
Hydrogen cyanide	10.90	8.25	456.8	53.9
Dinitrogen monoxid	3.84	4.43	309.6	72.4
Acetic acid	17.59	10.68	594.4	57.9
Ethane	5.53	6.47	305.4	48.8
Ethene	4.53	5.73	282.4	50.4
Ethyne	4.46	5.15	308.3	61.4
Ethylene oxide	17.63	13.50	466.7	36.4
Helium	0.03	2.38	5.2	2.27
Carbon dioxide	3.65	4.28	304.2	73.8
Carbon monoxide	1.51	4.00	132.9	35.0
Methane	2.29	4.30	190.0	46.0
Methanol	9.67	6.71	512.6	81.0
Chloromethane	7.57	6.51	416.3	66.8
n-Butane	14.69	12.30	425.2	38.0
Propane	8.77	8.47	369.8	42.5
Propene	8.49	3.30	365.0	46.2
Oxygen	1.38	3.19	154.6	50.5
Sulfur dioxide	6.81	5.65	430.8	78.8
Hydrogen sulfide	4.49	4.30	373.2	89.4
Nitrogen	1.41	3.92	126.2	33.9
Nitrogen monoxide	1.36	2.80	1.80	65.0
Water	5.53	3.60	647.3	220.5
Hydrogen	0.25	2.67	33.2	13.0

Appendix 8.8
Heat capacities of some substances and their temperature dependance.

The temperature dependence of the molar heat capacity is approximated by:
$c_p / \text{J mol}^{-1} \text{ K}^{-1} = a + b \cdot T + c \cdot T^2 + d \cdot T^3$ (T in K).

Gas	$c_p/$ kJ kg^{-1} K^{-1}	$c_p/$ J mol^{-1} K^{-1}	a	$b/\ 10^{-3}$	$c/\ 10^{-6}$	$d/\ 10^{-9}$
Hydrogen	14.3	29.1	27.124	9.267	− 13.799	7.640
Oxygen	0.94	20.2	28.087	− 0.004	17.447	− 10.644
Steam	1.94	34.9	32.220	1.923	10.548	3.594
Nitrogen	1.10	30.7	31.128	− 13.556	26.777	11.673
Dinitrogen monoxide	0.87	40.2	24.216	48.324	− 20.794	0.293
Ammonia	2.27	38.7	27.296	23.815	17.062	− 11.840
Sulfur dioxide	0.68	43.5	23.836	66.940	− 49.580	13.270
Carbon monoxide	1.05	29.3	30.848	− 12.840	27.870	− 12.710
Carbon dioxide	0.94	41.2	19.780	73.390	− 55.980	17.140
Methane	2.57	41.2	19.238	52.090	11.966	− 11.309
Methanol	1.61	51.6	21.127	70.880	25.820	− 28.500
Dichlorodifluoromethane	0.67	81.3	31.577	178.110	− 150.750	43.388
Ethane	2.17	65.2	5.406	177.980	− 69.330	− 1.916
Ethene	1.89	53.0	2.803	156.48	− 83.430	17.540
Ethanol	1.76	81.0	9.008	213.900	− 83.846	1.372
Ethylene oxide	1.42	62.6	− 7.524	222.100	− 125.560	25.900
Acetic acid	1.36	81.6	4.837	254.680	− 175.180	49.454
Propane	2.14	94.5	− 4.222	306.050	− 158.530	32.120
Propene	1.90	80.0	3.707	234.380	− 115.940	22.030
Butane	2.73	158.7	9.481	331.100	110.750	− 2.820
Hexane	2.11	181.9	− 4.41	581.570	− 311.660	64.890
Benzene	1.43	111.6	− 33.890	474.040	− 301.490	71.250
2-Methylpropane	2.14	124.3	− 1.389	384.500	− 184.470	28.932

Appendix 8.9
Thermodynamic data of selected organic compounds.

($\varnothing$ = Standard temperature and pressure [Sandler 1999].

Compound	μ / 10^{-30} C m	$\Delta_f H^\varnothing$ / kJ mol^{-1}	$S^\varnothing$ / J mol^{-1} K^{-1}	c_p / J mol^{-1} K^{-1}	$\Delta_V H^\varnothing$ / kJ mol^{-1}
Alkanes					
Methane	0.0	−75	186	36	
Ethane	0.0	−85	230	53	
Propane	0.0	−104	270	74	
Butane	0.0	−126	310	97	22
Pentane	0.0	−146	349	120	27
Hexane	0.0	−167	388	143	32
Cyclohexane		−123	298	106	33
Alkenes					
Ethene	0.0	52	220	44	
Propene	1.2	20	267	64	
1-Butene	1.1	−0.1	306	86	
Cyclopentene	0.7	33	290	75	29
1.3-Butadiene	0.0	110	279	80	25
Alkynes					
Ethyne	0.0	227	201	44	
Aromatics					
Benzene	0.0	83	269	82	34
Toluene	1.2	50	321	104	38
Ethylbenzene	2.0	30	361	128	42
Alkohols					
Methanol	5.7	−201	240	44	38
Ethanol	5.6	−235	283	65	43
1-Propanol	5.6	−258	325	87	47
1-Butanol	5.5	−274	363	110	51
Carboxylis acids					
Formic acid	4.7	−379	249		46
Acetic acid	5.8	−435	283	67	49
n-Butyric acid		−534	226		
Bencoic acid		−290	369		53

Continue

Compound	$\mu/$ 10^{-30} C m	$\Delta_f H^\varnothing/$ kJ mol^{-1}	$S^\varnothing/$ J mol^{-1} K^{-1}	$c_p/$ J mol^{-1} K^{-1}	$\Delta_V H^\varnothing/$ kJ mol^{-1}
Esters					
Methyl formate	5.9	−350	301		28
Ethyl formate	6.4	−371			28
Methyl acetate	5.7	−410	301		32
Ethylacetat	5.9	−443	363		36
Aldehydes					
Formaldehyd	7.8	−116	219	35	
Acetaldehyd	9.0	−166	264	57	26
Propionaldehyd	8.4	−192	305		29
Butyraldehyd	9.1	−205	345		34
Ketones					
Acetone	9.6	−218	295	75	31
Cyclohexanone		−230	322		42

Appendix 8.10
Order of magnitude of the reaction enthalpy $\Delta_R H$ for selected industrial reactions [Weissermel 2003].

Reaction	Products	$\Delta_R H$/kJ mol^{-1}
Reactions with O$_2$		
H$_2$ + 0.5 O$_2$	H$_2$O	−285
C + O$_2$	CO$_2$	−393
C + 0.5 O$_2$	CO	−111
CH$_4$ + 2 O$_2$	CO$_2$ + 2 H$_2$O	−890
CH$_4$ + NH$_3$ + 1.5 O$_2$	HCN + 3 H$_2$O	−473
CH$_3$OH + 0.5 O$_2$	HCHO + H$_2$O	−159
C$_2$H$_4$ + 0.5 O$_2$	Ethylen oxide	−105
C$_2$H$_4$ + 3 O$_2$	2 CO$_2$ + 2 H$_2$O	−1327
C$_2$H$_4$ + 0.5 O$_2$	CH$_3$CHO	−243
C$_2$H$_5$OH + 0.5 O$_2$	CH$_3$CHO + H$_2$O	−180
CH$_3$-CH = CH$_2$ + O$_2$	Acrolein + H$_2$O	−368
CH$_3$-CH = CH$_2$ + 0.5 O$_2$	Acetone	−255
Isopropanol + 0.5 O$_2$	Aceton + H$_2$O	−180
Acrolein + 0.5 O$_2$	Acrylic acid	−266

Appendix 8.10 Order of magnitude of the reaction enthalpy Δ_R for selected industrial reactions

Continue

Reaction	Products	$\Delta_R H$/kJ mol^{-1}
$CH_3\text{-}CH = CH\text{-}CHO + 0.5\, O_2$	$CH_3\text{-}CH = CH\text{-}COOH$	-268
Methacrolein $+ 0.5\, O_2$	Methacrylic acid	-252
Isobutyraldehyd $+ 0.5\, O_2$	Isobutyric acid	-312
$CH_2 = CH\text{-}C_2H_5 + 3O_2$	Maleic anhydride $+ 3\, H_2O$	-1315
$C_6H_6 + 4.5\, O_2$	Maleic anhydride $+ 2\, CO_2 + 2\, H_2O$	-1875
o-Xylol $+ 3O_2$	Phthalic anhydride $+ 3\, H_2O$	-1110
Naphthalene $+ 4.5\, O_2$	Phthalic anhydride $+ 2\, CO_2 + 2\, H_2O$	-1792
Reactions with H_2		
$C + 2\, H_2$	CH_4	-75
$CO + 3\, H_2$	$CH_4 + H_2O$	-206
$CO + 2\, H_2$	CH_3OH	-92
$CO_2 + 3\, H_2$	$CH_3OH + H_2O$	-50
$R\text{-}NO_2 + 3\, H_2$	$R\text{-}NH_2 + 2\, H_2O$	< -500
Butynol $+ 2\, H_2$	Butan diol	-251
$NC\text{-}(CH_2)_4\text{-}CN + 4\, H_2$	$H_2N\text{-}(CH_2)_6\text{-}NH_2$	-314
$C_6H_6 + 3\, H_2$	C_6H_{12}	-214
$C_6H_5\text{-}CH_3 + H_2$	$C_6H_6 + CH_4$	-126
$C_6H_5\text{-}NO_2 + 3\, H_2$	$C_6H_5 - NH_2 + 2\, H_2O$	-443
Dehydrogenation		
CH_3OH	$HCHO + H_2$	$+84$
$-CH_2\text{-}CH_3$	$-CH = CH_2 + H_2$	$+121$
$CH_4 + NH_3$	$HCN + 3\, H_2$	$+251$
$2\, CH_4$	$C_2H_2 + 3\, H_2$	$+337$
$C_2H_5\text{-}OH$	$CH_3\text{-}CHO + H_2$	$+84$
Isopropanol	Acetone $+ H_2$	$+67$
2-Butanol	Methyl ethyl ketone $+ H_2$	$+51$
n-Butane	$CH_3\text{-}CH = CH\text{-}CH_3 + H_2$	$+126$
Cyclohexanol	Cyclohexanone $+ H_2$	$+65$
$C_6H_5\text{-}C_2H_5$	$C_6H_5\text{-}CH = CH_2 + H_2$	$+121$

Continue

Reaction	Products	$\Delta_R H$/kJ mol^{-1}
Reactions with water		
$C + H_2O$	$CO + H_2$	$+119$
$CO + H_2O$	$CO_2 + H_2$	-41.2
$CH_4 + H_2O$	$3\,H_2 + CO$	$+205$
$HCN + H_2O$	$HCONH_2$	-75
$C_2H_4 + H_2O$	$C_2H_5\text{-}OH$	-46
$C_2H_4 + CO + H_2O$	$CH_3CH_2\text{-}COOH$	-159
$C_2H_2 + ROH$	$CH=CH\text{-}OR$	-125
Ethylene oxide $+ H_2O$	$HO\text{-}CH_2CH_2\text{-}OH$	-80
$H_2C=C=O + H_2O$	$CH_3\text{-}COOH$	-147
$CH_3\text{-}CH=CH_2 + H_2O$	Isopropanol	-50
Tetrahydrofuran $+ H_2O$	$HO\text{-}(CH_2)_4\text{-}OH$	-13
Phthalic anhydride $+ 2\,ROH$	Diesters of phthalic acid $+ H_2O$	-84
Reactions with HCl und Cl$_2$		
$CH_3OH + HCl$	$CH_3Cl + H_2O$	-33
$CH_2=CH\text{-}C_2H + HCl$	2-Chlor butadiene	-184
$C_2H_2 + HCl$	$CH_2=CH\text{-}Cl$	-99
$CH_2=CH\text{-}Cl + HCl$	$Cl\text{-}CH_2CH_2\text{-}Cl$	-71
$C_2H_4 + Cl_2$	$Cl\text{-}CH_2CH_2\text{-}Cl$	-180
$CH_3\text{-}CH=CH_2 + Cl_2$	$CH_3-CHCl-CH_2Cl$	-113
Others		
Ethylenoxid $+ CO_2$	Ethyl carbonate	-96
$C_6H_6 + C_2H_4$	$C_6H_5\text{-}C_2H_5$	-113
$C_6H_6 + CH_3\text{-}CH=CH_2$	Cumene	-113
$C_6H_6 + HNO_3$	Nitrobenzene	-117

Appendix 8.11
Antoine parameters of selected organic compounds.

$\lg (P/\text{Torr}) = A - \dfrac{B}{C + T/°C}$, (1 Torr = 1.33322 10^{-3} bar).

Compound	A	B	C	b.p./°C
Inorganic Compounds				
Carbon dioxide	9.66983	1295.524	269.243	− 78.42
Oxygen	6.68748	318.692	266.683	− 182.96
Xenon	38.26364	54 624.5	1 651.838	− 108.02
Alkanes				
Methane	6.34159	342.217	260.221	− 161.4
Ethane	6.82477	663.484	256.893	− 88.66
Propane	6.84343	818.54	248.677	− 42.11
Butane	6.82485	943.453	239.711	− 0.5
Pentane	6.84471	1060.793	231.541	36.07
Hexane	6.88555	1175.817	224.867	68.67
Cyclohexane	6.84941	1206.001	223.148	80.74
Alkenes				
Ethene	6.74819	584.291	254.862	− 103.78
Propene	6.82359	786.532	247.243	− 47.76
1-Butene	6.53101	610.261	228.066	− 6.09
1.3-Butadiene	6.85364	933.586	239.511	− 4.52
Alkynes				
Acetylene	6.57935	536.808	229.819	− 84.68
Aromatics				
Benzene	6.89272	1203.331	219.888	80.1
Methylbenzene	6.95805	1346.773	219.693	110.62
Ethylbenzene	6.9565	1423.543	213.091	136.18
Alkoholes				
Methanol	8.08097	1582.271	239.726	64.55
Ethanol	8.16556	1624.08	228.993	78.32
1-Propanol	7.74416	1437.686	198.463	97.15
2-Propanol	7.74021	1359.517	197.527	82.24
1-Butanol	7.36366	1305.198	173.427	117.73
2-Butanol	7.20131	1157	168.279	99.51
2-Methyl-1-propanol	7.29491	1230.81	170.947	107.89
2-Methyl-2-propanol	7.2034	1092.971	170.503	82.35
1-Pentanol	7.18246	1287.625	161.33	138

Continue

Compound	A	B	C	b.p./ °C
Alkohols				
1-Hexanol	8.1117	1872.743	202.666	
Cyclohexanol	6.2553	912.866	109.126	161.39
1.2-Ethane diol	8.09083	2088.936	203.454	197.49
Phenols				
Phenol	7.13301	1516.79	174.954	181.75
Amines				
Metylamine	7.3369	1011.532	233.286	− 6.28
Dimethylamine	7.08212	960.242	221.667	6.89
Trimethylamine	6.85755	955.944	237.515	2.87
Carboxylic acid				
Formic acid	4.97536	541.738	137.051	100.5
Acetic acid	7.38732	1533.313	222.309	117.9
Butyric acid	7.7399	1764.68	199.892	163.28
Esters				
Methyl formate	3.02742	3.018	− 11.88	32.47
Ethyl formate	7.00902	1123.943	218.247	54.01
Methyl acetate	7.06524	1157.63	219.726	56.93
Ethyl acetate	7.10179	1244.951	217.881	77.06
Aldehydes				
Ethanal	8.00552	1600.017	291.809	20.41
Butanal	6.38544	1330.948	210.833	123.71
Ketones, Ethers				
Acetone	7.11714	1210.595	229.664	56.1
1.4-Dioxane	7.43155	1554.679	240.337	101.29

Appendix 8.12
Properties of water.

Appendix 8.12.1
Formulas for calculating the physicochemical properties of water between 0 or 150 °C
(T in °C, P in bar) [Popiel 1998].

Saturation vapor pressure

$$P_s/\text{bar} = P_c \cdot exp\left\{[T_c/(273.15 + T)] \cdot (a_1 + a_2 \cdot \tau^{1.5} + a_3 \cdot \tau^3 + a_4 \cdot \tau^{3.5} + a_5 \cdot \tau^4 + a_6 \cdot \tau^{7.5})\right\}$$

- $P_C = 220.64$ bar
- $T_C = 647.096$ K
- $\tau = 1 - (273.15 + T)/T_C$
- $a_1 = -7.85951783$
- $a_2 = 1.84408259$
- $a_3 = -11.7866497$
- $a_4 = 22.6807411$
- $a_5 = -15.9618719$
- $a_6 = 1.80122502$

Density of liquid water at saturation vapor pressure

$$\rho_S/\text{kg m}^{-3} = a + b \cdot T + c \cdot T^2 + d \cdot T^{2.5} + e \cdot T^3$$

- $a = 999.79684$
- $b = 0.068317355$
- $c = -0.010740248$
- $d = 0.00082140905$
- $e = -2.3030988 \cdot 10^{-5}$

Density of liquid water at pressure above saturation pressure

$$\rho(T, P)/\text{kg m}^{-3} = P_S \cdot [1 + \kappa_P \cdot (P - P_s)]$$

where

$$\kappa_P/\text{bar}^{-1} = \left[\frac{(a + c \cdot T)}{(1 + b \cdot T + d \cdot T^2)}\right]^2$$

- $a = 0.007131672$
- $b = 0.011230766$
- $c = 5.369263 \times 10^{-5}$

Thermal expansion coefficient of liquid water

$$\beta/\text{K}^{-1} = -\frac{1}{\rho'_S} \cdot \frac{\partial \rho_S}{\partial T} = a + b \cdot T + c \cdot T^{1.5} + d \cdot T^2$$

- $a = -6.8785895 \times 10^{-5}$
- $b = 2.1687942 \times 10^{-5}$
- $c = -2.1236686 \times 10^{-6}$
- $d = 7.7200882 \times 10^{-8}$

Specific heat at constant pressure

$$c_{P,S}/\text{kJ kg}^{-1}\text{ K}^{-1} = a + b \cdot T + c \cdot T^{1.5} + d \cdot T^2 + e \cdot T^{2.5}$$

- $a = 4.2174356$
- $b = -0.0056181625$
- $c = 0.0012992528$
- $d = -0.00011535353$
- $e = 4.14964 \times 10^{-6}$

Heat of evaporation

$$\Delta_V H/\text{kJ kg}^{-1} = a + b \cdot T + c \cdot T^{1.5} + d \cdot T^{2.5} + e \cdot T^3$$

- $a = 2500.304$
- $b = -2.2521025$
- $c = -0.021465847$
- $d = 3.1750136 \times 10^{-4}$
- $e = -2.8607959 \times 10^{-5}$

Thermal conductivity

$$\lambda_S/\text{W m}^{-1}\text{K}^{-1} = a + b \cdot T + c \cdot T^{1.5} + d \cdot T^2 + e \cdot T^{0.5}$$

- $a = 0.5650285$
- $b = 0.0026363895$
- $c = -0.00012516934$
- $d = -1.5154918 \cdot 10^{-6}$
- $e = -0.0009412945$

Dynamic viscosity

$$\eta_S/\text{kg m}^{-1}\text{ s}^{-1} = \frac{1}{a + b \cdot T + c \cdot T^2 + d \cdot T^3}$$

- $a = 557.82468$
- $b = 19.408782$
- $c = 0.1360459$
- $d = -3.1160832 \times 10^{-4}$

Dynamic viscosity at pressures above the saturation vapor pressure

$$\eta(T, P)/\text{kg m}^{-1}\,\text{s}^{-1} = \mu_s \cdot \left[1 + \kappa_\mu \cdot (P - P_s)\right]$$

with the conpressibility factor:

$$\kappa_\mu/\text{bar}^{-1} = a + b \cdot T + c \cdot T^2 + d \cdot T^3 + e \cdot T^4$$

- $a = 0.0001335$
- $b = 5.57128 \times 10^{-6}$
- $c = -0.0061077357$
- $d = 0.3633062 \times 10^{-4}$
- $e = -0.8179944 \times 10^{-7}$

Surface tension

$$\sigma/\text{N m}^{-1} = a + b \cdot T + c \cdot T^2 + d \cdot T^3$$

- $a = 0.075652711$
- $b = -0.00013936956$
- $c = -3.0842103 \times 10^{-7}$
- $d = 2.75884365 \times 10^{-10}$

Critical parameters

- $P_c = 220.64$ bar
- $T_c = 647.096$ K
- $\rho_c = 332$ kg m^{-3}

Triple point

- $P_{TR} = 611.657$ Pa
- $T_{TR} = 0.01\,°\text{C}$
- $\rho_{s,TR} = 999.789$ kg m^{-3}
- $\rho_{v,TR} = 0.00485426$ kg m^{-3}.

Appendix 8.12.2
Properties of water ρ = density, c_p = heat capacity, α = thermal expansion coefficient, λ = thermal conductivity, η = Viscosity coefficient.

T/°C	P/bar	$\Delta_V H$/kJ kg^{-1}	ρ(liq) /kg m^{-3}	ρ(gas) /kg m^{-3}	c_p(liq) /kJ kg^{-1} K^{-1}	c_p(gas) /kJ kg^{-1} K^{-1}	α(liq) /10^{-3} K^{-1}	α(gas) /10^{-3} K^{-1}	λ(liq) /10^{-3} J m^{-1} K^{-1} s^{-1}	λ(gas) /10^{-3} J m^{-1} K^{-1} s^{-1}	η(liq) /10^{-6} kg m^{-1} s^{-1}	η(gas) /10^{-6} kg m^{-1} s^{-1}
0.01	0.006112	2501.0	999.8	0.004850	4.217	1.864	−0.0853	3.669	562	16.5	1791.4	9.22
10	0.012271	2477.4	999.7	0.009397	4.193	1.868	0.0821	3.544	582	17.2	1307.7	9.46
20	0.023368	2453.8	998.3	0.01720	4.182	1.874	0.2066	3.431	600	18.0	1002.7	9.73
30	0.042217	2430.3	995.7	0.03037	4.179	1.883	0.3056	3.327	615	18.7	797.7	10.01
40	0.073749	2406.5	992.2	0.05116	4.179	1.894	0.3890	3.233	629	19.5	653.1	10.31
50	0.12335	2382.6	988.0	0.08300	4.181	1.907	0.4624	3.150	640	20.3	547.1	10.62
60	0.19919	2358.4	983.1	0.1302	4.185	1.924	0.5288	3.076	651	21.1	466.8	10.94
70	0.31151	2333.8	977.7	0.1981	4.190	1.944	0.5900	3.012	659	22.0	404.4	11.26
80	0.47359	2308.8	971.6	0.2932	4.197	1.969	0.6473	2.958	667	22.9	355.0	11.60
90	0.70108	2283.3	965.1	0.4233	4.205	1.999	0.7019	2.915	673	23.8	315.0	11.93
100	1.01325	2257.3	958.1	0.5974	4.216	2.034	0.7547	2.882	677	24.8	282.2	12.28
110	1.4326	2230.5	950.7	0.8260	4.229	2.075	0.8068	2.861	681	25.8	254.9	12.62
120	1.9854	2202.9	942.8	1.121	4.245	2.124	0.8590	2.851	683	27.0	232.1	12.97
130	2.7012	2174.4	934.6	1.496	4.263	2.180	0.9121	2.853	684	28.1	212.7	13.32
140	3.6136	2144.9	925.9	1.966	4.285	2.245	0.9667	2.868	685	29.4	196.1	13.67
150	4.7597	2114.2	916.8	2.547	4.310	2.320	1.0237	2.897	684	30.8	191.9	14.02
160	6.1804	2082.2	907.3	3.259	4.339	2.406	1.0837	2.941	682	32.2	169.5	14.37
170	7.9202	2048.8	897.3	4.122	4.371	2.504	1.1475	3.001	679	33.8	158.8	14.72
180	10.027	2014.0	886.9	5.160	4.408	2.615	1.2162	3.078	674	35.4	149.3	15.07
190	12.552	1977.4	876.0	6.395	4.449	2.741	1.2906	3.174	669	37.2	141.0	15.42

Appendix 8.12 Properties of water

T/°C	P/bar	$\Delta_V H$/kJ kg^{-1}	ρ(liq) /kg m^{-3}	ρ(gas)	c_P(liq) /kJ kg^{-1} K^{-1}	c_P(gas)	α(liq) /10^{-3} K	α(gas)	λ(liq) /10^{-3} J m^{-1} K^{-1} s^{-1}	λ(gas)	η(liq) /10^{-6} kg m^{-1} s^{-1}	η(gas)
200	15.551	1939.0	864.7	7.865	4.497	2.883	1.3721	3.291	663	v39.1	133.6	15.78
210	19.080	1898.7	852.8	9.595	4.551	3.043	1.4623	3.432	656	41.1	126.9	16.13
220	23.201	1856.2	840.4	11.625	4.614	3.223	1.5629	3.599	648	43.4	121.0	16.49
230	27.979	1811.4	827.3	13.999	4.686	3.426	1.6763	3.798	639	45.7	115.5	16.85
240	33.480	1764.0	813.6	16.767	4.770	3.656	1.8658	4.036	629	48.3	110.5	17.22
250	39.776	1713.7	799.2	19.990	4.869	3.918	v1.9552	4.321	618	51.2	105.8	17.59
260	46.940	1660.2	783.9	23.742	4.986	4.221	2.1301	4.665	606	54.3	101.5	17.98
270	55.051	1603.1	767.8	28.112	5.126	4.574	2.3379	5.086	593	57.9	97.4	18.38
280	64.191	1541.7	750.5	33.215	5.296	4.996	2.5893	5.608	578	61.8	93.4	18.80
290	74.448	1475.5	732.1	39.198	5.507	5.507	2.8998	6.267	562	66.4	89.6	19.25
300	85.917	1403.5	712.2	46.255	5.773	6.144	3.2932	7.117	545	71.8	85.8	19.74
310	98.697	1324.5	690.6	54.648	6.120	6.962	3.8079	8.242	526	78.4	82.1	20.28
320	112.900	1239.1	666.9	64.754	6.586	8.053	4.5104	9.785	506	86.5	78.3	20.89
330	128.646	1139.0	640.4	77.144	7.248	9.589	5.5306	12.017	485	97.1	74.4	21.62
340	146.079	1026.8	610.2	92.755	8.270	11.920	7.1672	15.502	461	111.8	70.2	22.52
350	165.367	893.1	574.5	113.352	10.078	15.951	10.3944	21.733	436	134.2	65.7	23.72
360	186.737	722.6	528.3	143.467	14.987	26.792	19.2762	38.993	412	175.8	60.2	25.53
370	210.528	439.4	448.3	201.685	53.920	112.928	98.1843	170.915	420	308.0	51.4	29.35
374	220.64	0	322	322					830		38.2	38.2

Appendix 8.12.3
Density ρ / kg m^{-3} of water at different temperatures and pressures.

T/°C → Pressure/bar ↓	0	25	50	75	100	150	200	250	300	350	400	450	500
1	999.8	997.2	988.1	974.7	0.5895	0.5164	0.4603	0.4156	0.3790	0.3483	0.3223	0.2999	0.2805
5	1000.0	997.3	988.2	974.9	958.3	916.8	2.353	2.108	1.914	1.754	1.620	1.506	1.407
10	1000.3	997.6	988.5	975.1	958.6	917.1	4.056	4.297	3.877	3.540	3.262	3.027	2.825
20	1000.8	998.0	988.7	975.6	959.0	917.7	865.0	8.972	7.969	7.217	6.615	6.117	5.694
30	1001.3	998.5	989.3	976.1	959.5	918.2	865.8	14.17	12.02	11.05	10.07	9.274	8.611
40	1001.8	998.9	989.8	976.5	960.0	918.8	866.6	799.2	16.99	15.05	13.63	12.50	11.58
50	1002.3	999.3	990.2	976.9	960.5	919.4	867.3	800.4	22.06	19.25	17.30	15.81	14.59
60	1002.8	999.8	990.6	977.4	960.9	920.0	868.1	801.5	27.65	23.68	21.11	19.19	17.66
70	1003.3	1000.2	991.1	977.8	961.4	920.5	868.8	802.7	33.94	28.38	25.05	22.66	20.79
80	1003.8	1000.7	991.5	978.3	961.9	921.1	869.6	803.8	41.24	33.39	29.15	26.21	23.97
90	1004.3	1001.1	991.9	978.7	962.4	921.7	870.3	804.9	713.1	38.77	33.41	29.87	27.21
100	1004.8	1001.6	992.4	979.2	962.8	922.2	871.1	806.1	715.4	44.60	37.87	33.62	30.52
150	1007.2	1003.7	994.5	981.3	965.1	925.0	874.7	811.4	725.7	87.07	63.88	54.20	48.00
200	1009.7	1005.9	996.6	983.5	967.4	927.7	878.2	816.5	735.0	600.3	100.5	78.71	67.70
250	1012.1	1008.0	998.7	985.6	969.7	930.4	881.6	821.3	743.3	625.0	166.4	109.1	89.86
300	1014.5	1010.2	1000.7	987.7	971.9	933.0	884.9	826.0	751.0	643.4	356.4	148.6	115.2
350	1016.9	1012.3	1002.7	989.8	974.1	935.6	888.1	830.4	758.1	658.5	474.6	201.9	144.5
400	1019.2	1014.3	1004.8	991.8	976.2	938.1	891.3	834.7	764.7	671.4	523.4	270.6	178.1
450	1021.6	1016.4	1006.7	993.9	978.3	940.5	894.3	838.8	770.9	682.7	554.3	343.0	216.0
500	1023.9	1018.5	1008.7	995.9	980.4	943.0	897.3	842.7	776.7	692.9	577.4	402.0	257.0

Appendix 8.12.4
Specific heat capacity c_p/kJ kg^{-1} K^{-1} of water at different temperatures and pressures.

T/°C → Pressure/bar ↓	0	25	50	75	100	150	200	250	300	350	400	450	500
1	4.217	4.180	4.181	4.193	2.032	1.979	1.974	1.988	2.011	2.037	2.068	2.099	2.132
5	4.215	4.178	4.180	4.192	4.215	4.310	2.143	2.079	2.065	2.073	2.093	2.118	2.146
10	4.212	4.177	4.179	4.191	4.214	4.308	2.431	2.215	2.141	2.121	2.126	2.141	2.164
50	4.191	4.165	4.170	4.182	4.205	4.296	4.477	4.855	3.199	2.669	2.451	2.360	2.324
100	4.165	4.151	4.158	4.172	4.194	4.281	4.450	4.791	5.703	4.041	3.078	2.726	2.569
150	4.141	4.138	4.148	4.162	4.183	4.266	4.425	4.735	5.495	8.863	4.155	3.235	2.875
200	4.117	4.125	4.137	4.152	4.173	4.252	4.402	4.685	5.332	8.103	6.327	3.959	3.257
250	4.095	4.113	4.127	4.142	4.163	4.239	4.379	4.639	5.201	7.017	13.018	5.020	3.731
300	4.073	4.101	4.117	4.133	4.153	4.226	4.358	4.598	5.091	6.451	25.708	6.624	4.317
350	4.052	4.090	4.107	4.123	4.144	4.214	4.338	4.560	4.999	6.084	11.794	8.875	5.019
400	4.032	4.079	4.098	4.114	4.135	4.202	4.319	4.525	4.918	5.820	8.784	10.887	5.807
450	4.013	4.069	4.089	4.106	4.126	4.190	4.301	4.493	4.848	5.616	7.517	10.827	6.584
500	3.994	4.059	4.081	4.097	4.117	4.179	4.284	4.463	4.786	5.451	6.814	9.483	7.200

Appendix 8.12.5
Dynamic viscosity $\eta/10^{-6}$ kg m^{-1} s^{-1} of water at different temperatures and pressures.

T/°C → Pressure/ bar ↓	0	25	50	75	100	150	200	250	300	350	400	450	500
1	1792	890.8	547.1	378.3	12.28	14.19	16.18	18.22	20.29	22.37	24.45	26.52	28.57
5	1791	890.7	547.2	378.4	282.3	181.9	16.07	18.15	20.25	22.35	24.44	26.52	28.58
10	1790	890.6	547.2	378.6	282.4	182.0	15.93	18.07	20.20	22.32	24.42	26.51	28.58
50	1780	889.8	547.9	379.6	283.5	183.1	134.5	106.1	19.86	22.16	24.38	26.55	28.67
100	1769	888.9	548.7	380.8	284.8	184.4	135.7	107.5	86.39	22.18	24.49	26.72	28.90
150	1759	888.1	549.6	382.1	286.2	185.7	137.0	108.9	88.30	22.91	24.91	27.10	29.27
200	1749	887.4	550.5	383.4	287.5	186.9	138.2	110.2	90.05	69.15	25.96	27.77	29.82
250	1740	886.8	551.4	384.6	288.9	188.2	139.4	111.4	91.67	72.60	28.99	28.89	30.61
300	1731	886.4	552.3	385.9	290.2	189.5	140.5	112.6	93.19	75.30	43.67	30.78	31.70
350	1722	886.0	553.3	387.2	291.6	190.7	141.7	113.8	94.63	77.58	55.74	33.90	33.19
400	1714	885.8	554.3	388.5	292.9	191.9	142.8	115.0	96.00	79.60	61.26	39.03	35.17
450	1707	555.3	555.3	389.8	294.2	193.2	144.0	116.1	97.31	81.43	64.95	45.18	37.69
500	1700	556.4	556.4	391.1	295.6	194.4	145.1	117.2	98.57	83.10	67.81	50.69	40.70

Appendix 8.12.6
Self-diffusion coefficient $D/m^2\ s^{-1}$ [Lamb 1981] of water at different temperatures and pressures.

$T/\,°C$	P/bar	$D/10^{-9}\ m^2\ s^{-1}$
25	1	2.15 [Vogel 1982]
225	200	30.6
250	100	32.8
300	100	39.6
350	200	50.4
400	265	160
400	314	80.2
400	378	64.2
500	256	357
500	314	285
500	359	238
500	403	178

Appendix 8.12.7
Thermal expansion coefficient $\beta/10^{-3}$ K of water at different temperatures and pressures.

T/°C →	0	25	50	75	100	150	200	250	300	400	450	500
Pressure/ bar ↓												
1	−0.08518	0.2586	0.4623	0.6190	2.879	2.451	2.159	1.937	1.761	1.493	1.388	1.218
5	−0.08376	0.2590	0.4622	0.6185	0.7539	1.024	2.372	2.051	1.829	1.523	1.409	1.313
10	0.08199	0.2595	0.4620	0.6179	0.7530	1.022	2.728	2.218	1.922	1.562	1.437	1.333
50	0.06777	0.2633	0.4605	0.6132	0.7455	1.007	1.347	1.936	3.211	1.947	1.690	1.510
100	0.04989	0.2682	0.4589	0.6076	0.7366	0.0002	1.312	1.848	3.189	2.703	2.118	1.782
150	0.03201	0.2733	0.4574	0.6022	0.7281	0.9740	1.281	1.772	2.883	4.062	2.724	2.126
200	0.01424	0.2786	0.4562	0.5971	0.7200	0.9587	1.251	1.704	2.648	7.005	3.613	2.559
250	0.00331	0.2840	0.4551	0.5923	0.7122	0.9442	1.224	1.643	2.460	17.08	4.972	3.109
300	0.02052	0.2894	0.4542	0.5876	0.7047	0.9303	1.198	1.589	2.306	37.71	7.112	3.799
350	0.03728	0.2950	0.4534	0.5832	0.6975	0.9172	1.175	1.539	2.176	13.05	10.18	4.635
400	0.05347	0.3006	0.4528	0.5790	0.6907	0.9046	1.152	1.494	2.065	7.989	12.79	5.563
450	0.06896	0.3062	0.4524	0.5750	0.6841	0.8926	1.131	1.453	1.968	5.955	12.16	6.438
500	0.08364	0.3119	0.4520	0.5712	0.6777	0.8811	1.111	1.415	1.884	4.863	9.668	7.053

Appendix 8.12.8
Thermal conductivity $\lambda/10^{-3}$ W m^{-1} K^{-1} of water at different temperatures and pressures.

$T/°C \rightarrow$ Pressure/bar $\downarrow$	0	25	50	75	100	150	200	250	300	350	400	450	500
1	562.0	607.6	640.5	663.3	24.78	28.80	33.37	38.38	43.49	48.97	54.71	60.69	66.89
5	562.2	607.8	640.7	663.5	677.7	683.6	34.24	38.81	43.89	49.32	55.02	60.97	67.16
10	562.5	608.1	641.0	663.8	678.0	683.9	36.06	39.69	44.49	49.79	55.43	61.35	67.50
50	564.8	610.2	643.0	665.9	680.2	686.7	666.5	619.5	53.04	55.22	59.52	64.81	70.57
100	567.7	612.8	645.6	668.6	683.1	690.0	670.8	625.8	548.2	68.57	67.26	70.57	75.35
150	570.6	615.4	648.2	671.2	685.8	693.3	675.0	631.9	558.9	104.0	79.97	78.54	81.40
200	573.5	618.0	650.7	673.8	688.5	696.5	679.2	637.7	568.6	453.9	103.3	89.80	89.13
250	576.4	620.5	653.2	676.3	691.2	699.7	683.1	643.2	577.5	473.7	159.9	106.3	99.07
300	579.2	623.1	655.6	678.8	693.9	702.8	687.0	648.5	585.8	490.1	327.6	131.7	111.9
350	582.0	625.6	658.0	681.3	696.5	705.9	690.8	653.6	593.6	504.2	372.9	171.0	128.6
400	584.8	628.1	660.5	683.8	699.1	708.9	694.5	658.5	600.9	516.8	398.5	224.9	149.9
450	587.6	630.6	662.8	686.2	701.6	711.8	698.1	663.3	607.8	528.3	419.8	277.9	175.9
500	590.4	633.0	665.2	688.6	704.1	714.7	701.7	667.9	614.3	538.7	437.9	315.8	205.4

Appendix 8.12.9
Negative base ten logarithm of the ionic product of water $pK_W/\text{mol}^2\ \text{kg}^{-2}$ [Marshall 1981] at different temperatures.

T/°C →	0	25	50	75	100	150	200	250	300	350	400	450	500
Pressure/bar ↓													
Sat'd vapor	14.938	13.995	13.275	12.712	12.265	11.638	11.289	11.191	11.406	12.30	–	–	–
250	14.83	13.90	13.19	12.63	12.18	11.54	11.16	11.01	11.14	11.77	19.43	21.59	22.40
500	14.72	13.82	13.11	12.55	12.10	11.45	11.05	10.85	10.86	11.14	11.88	13.74	16.13
750	14.62	13.73	13.04	12.48	12.03	11.36	10.95	10.72	10.66	10.79	11.17	11.89	13.01
1000	14.53	13.66	12.96	12.41	11.96	11.29	10.86	10.60	10.50	10.54	10.77	11.19	11.81

Appendix 8.12.10
Relative static dielectric constant ε_r of water as a function of pressure and temperature.

T/°C →	0	25	50	75	100	200	300	400	500
Pressure/bar ↓									
1	87.82	78.46	69.91	62.24	1.00	1.00	1.00	1.00	1.00
60	88.05	78.65	70.09	62.42	55.59	34.89	1.11	1.07	1.05
100	88.28	78.85	70.27	62.59	55.76	35.11	20.39	1.17	1.11
200	88.75	79.24	30.63	62.34	56.11	35.52	21.24	1.64	1.32
1000	92.04	82.08	73.22	65.42	58.55	38.17	25.17	16.05	9.29
2000	95.20	84.94	75.89	67.95	61.00	40.66	28.07	19.69	13.86

Appendix 8.13
Properties of dry air (molar mass: $M = 28.966$ g mol^{-1}).

Composition of dry atmosphers near the earths surface (as of 1992) [Bliefert 2002].

Compound	Formular	Volume fraction
Nitrogen	N_2	78.084 %
Oxygen	O_2	20.946 %
Argon	Ar	0.934 %
Carbon dioxide	CO_2	354 ppm
Neon	Ne	18.18 ppm
Helium	He	5.24 ppm
Methane	CH_4	1.7 ppm
Krypton	Kr	1.14 ppm
Hydrogen	H_2	0.5 ppm
Dinitrogen monoxide	N_2O	0.3 ppm
Xenon	Xe	87 ppb
Carbon monoxide	CO	30–250 ppb
Ozone	O_3	10–100 ppb
Nitrogen dioxide	NO_2	10–100 ppb
Nitrogen oxide	NO	5–100 ppb
Sulfur dioxide	SO_2	< 1...50 ppb
Ammonia	NH_3	0.1–1 ppb
Formaldehyde	HCHO	0.1–1 ppb

Gas kinetic data of air under standard conditions
Mean particle velocity: $\bar{u} = 459$ m s^{-1}
Mean free path: $\Lambda = 6.63 \cdot 10^{-8}$ m
Collision number: $Z = 6.92 \cdot 10^9$ s^{-1}
Density: $\rho = 1.205$ kg m^{-3}

Appendix 8.13.1
Real gas factor $r = pV/RT$ of dry air at different temperatures an pressures.

T/°C →	−50	0	25	50	100	200	300	400	500
Pressure/bar ↓									
1	0.998	0.999	1.000	1.000	1.000	1.000	1.000	1.000	1.000
5	0.992	0.997	0.998	0.999	1.000	1.001	1.002	1.002	1.002
10	0.985	0.994	0.997	0.999	1.001	1.003	1.004	1.004	1.004
50	0.931	0.978	0.990	0.998	1.009	1.017	1.019	1.019	1.019
100	0.888	0.970	0.992	1.006	1.024	1.038	1.041	1.040	1.039
150	0.884	0.980	1.006	1.023	1.045	1.061	1.063	1.062	1.059
200	0.917	1.005	1.031	1.049	1.070	1.086	1.087	1.084	1.080
250	0.973	1.042	1.065	1.081	1.101	1.112	1.111	1.106	1.239
300	1.042	1.090	1.107	1.120	1.135	1.141	1.137	1.129	1.121
350	1.118	1.144	1.155	1.163	1.172	1.172	1.163	1.152	1.142
400	1.198	1.202	1.206	1.209	1.211	1.204	1.190	1.176	1.163
450	1.279	1.263	1.260	1.258	1.253	1.236	1.218	1.200	1.184
500	1.362	1.326	1.316	1.308	1.295	1.270	1.246	1.224	1.206

Appendix 8.13.2
Specific heat capacity c_p in kJ kg^{-1} K^{-1} of dry air at different temperatures and pressures.

T/°C →	−50	0	25	50	100	200	300	400	500
Pressure/bar ↓									
1	1.007	1.006	1.007	1.008	1.012	1.026	1.046	1.069	1.093
5	1.023	1.015	1.014	1.013	1.015	1.028	1.047	1.070	1.094
10	1.044	1.026	1.022	1.020	1.020	1.030	1.049	1.071	1.094
50	1.212	1.112	1.089	1.072	1.055	1.049	1.061	1.080	1.101
100	1.430	1.216	1.169	1.133	1.096	1.072	1.075	1.090	1.108
150	1.575	1.302	1.237	1.187	1.132	1.092	1.088	1.099	1.115
200	1.623	1.361	1.287	1.229	1.161	1.108	1.099	1.107	1.121
250	1.622	1.394	1.320	1.260	1.186	1.123	1.109	1.114	1.127
300	1.604	1.409	1.339	1.282	1.204	1.135	1.117	1.120	1.132
350	1.580	1.412	1.348	1.295	1.220	1.145	1.125	1.125	1.136
400	1.557	1.411	1.353	1.304	1.230	1.154	1.130	1.130	1.140
450	1.534	1.406	1.353	1.308	1.239	1.162	1.136	1.134	1.143
500	1.513	1.400	1.351	1.309	1.244	1.169	1.141	1.138	1.146

Appendix 8.13.3
Dynamic viscosity $\eta/10^{-3}$ mPa s at dry air at different temperatures and pressures.

T/°C →	− 50	0	25	50	100	200	300	400	500
Pressure/bar ↓									
1	14.55	17.10	18.20	19.25	21.60	25.70	29.20	32.55	35.50
5	14.63	17.16	18.26	19.30	21.64	25.73	29.23	32.57	35.52
10	14.74	17.24	18.33	19.37	21.70	25.78	29.27	32.61	35.54
50	16.01	18.08	19.11	20.07	22.26	26.20	29.60	32.86	35.76
100	18.49	19.47	20.29	21.12	23.09	26.77	30.05	33.19	36.04
150	21.09	21.25	21.82	22.48	24.06	27.39	30.56	33.54	36.35
200	25.19	23.19	23.40	23.76	24.98	28.03	31.10	34.10	36.69
250	28.93	25.49	25.38	25.42	26.27	28.87	31.68	34.53	37.01
300	32.68	27.77	27.25	27.28	27.51	29.67	32.23	34.93	37.39
350	36.21	30.11	29.35	28.79	28.84	30.55	32.82	35.42	37.80
400	39.78	32.59	31.41	30.98	30.27	31.39	33.44	35.85	38.15
450	43.42	34.93	33.51	32.36	31.70	32.21	34.06	36.36	38.54
500	46.91	37.29	35.51	34.06	32.28	33.15	34.64	36.86	38.96

Appendix 8.13.4
Thermal conductivity $\lambda/\text{W m}^{-1}\text{ K}^{-1}$ of dry air at different temperatures and pressures.

T/°C →	− 50	0	25	50	100	200	300	400	500
Pressure/bar ↓									
1	20.65	24.54	26.39	28.22	31.81	38.91	45.91	52.57	58.48
5	20.86	24.68	26.53	28.32	31.89	38.91	45.92	52.56	58.42
10	21.13	24.88	26.71	28.47	32.00	39.94	45.96	52.57	58.36
50	24.11	27.15	28.78	30.26	33.53	40.34	46.86	53.41	58.98
100	28.81	30.28	31.53	32.75	35.60	42.00	48.30	54.56	60.07
150	34.95	33.88	34.53	35.32	37.68	43.59	49.56	55.76	61.09
200	41.96	38.00	37.90	38.21	39.91	45.18	50.69	56.62	61.96
250	48.72	42.39	41.57	41.32	42.29	46.92	51.95	57.78	63.05
300	54.84	46.84	45.38	44.56	44.81	48.54	53.06	58.70	63.74
350	60.34	51.19	49.14	47.88	47.35	50.40	54.68	59.95	64.86
400	65.15	55.30	52.83	51.29	49.97	52.59	55.91	60.95	65.56
450	69.71	59.25	56.01	54.08	52.97	54.16	57.18	61.71	66.50
500	73.91	62.92	59.80	57.40	54.70	55.66	58.60	62.86	67.24

Appendix 8.14
Dimensionless characteristic numbers.

Name	Abbreviation	Equation	Physical meaning
Archimedes	Ar	$\dfrac{l^3 \cdot g \cdot \Delta \rho}{v^2 \cdot \rho}$	$\dfrac{\text{voyancy force}}{\text{inertial force}}$
Biot	Bi	$\dfrac{a \cdot l}{\lambda}$	$\dfrac{\text{convective heat transport}}{\text{heat transport in body}}$
Bodenstein	Bo	$\dfrac{u \cdot l}{D}$	$\dfrac{\text{moles introduced convertively}}{\text{moles introduced by diffusion}}$
Damköhler	Dam I	$\dfrac{U \cdot l}{u \cdot x}$	$\dfrac{\text{chemically converted moles}}{\text{moles introduced convertively}}$
	Dam II	$\dfrac{U \cdot l^2}{D \cdot x}$	$\dfrac{\text{chemically converted moles}}{\text{moles introduced by diffusion}}$
	Dam III	$\dfrac{H \cdot U \cdot l^2}{cp \cdot \rho \cdot u \cdot T}$	$\dfrac{\text{chemical heat evolution}}{\text{convective heat transport}}$
	Dam IV	$\dfrac{H \cdot U \cdot l^2}{\lambda \cdot \tau}$	$\dfrac{\text{chemical heat evolution}}{\text{heat transport by conduction}}$
Euler	Eu	$\dfrac{\Delta P}{\rho \cdot u^2}$	$\dfrac{\text{pressure force}}{\text{inertial force}}$
Fourier (heat)	Fo_q	$\dfrac{\lambda \cdot \tau}{\rho \cdot cp \cdot l^2}$	$\dfrac{\text{instationary heat flux}}{\text{convective flux}}$
Fourier (mass)	Fo_s	$\dfrac{D \cdot \tau}{l^2}$	$\dfrac{\text{instationry mass transport}}{\text{substantial mass transfer}}$
Froude	Fr	$\dfrac{u^2}{g \cdot l}$	$\dfrac{\text{inertial force}}{\text{gravity}}$
Galilei		$\dfrac{g \cdot l^3}{u^2}$	$\dfrac{\text{gravity}}{\text{inertial force}}$
Grashof	Gr	$\dfrac{g \cdot \gamma \cdot l^3 \cdot \Delta T}{v^2}$	$\dfrac{\text{thermal buoyoncy}}{\text{inertial force}}$
Lewis		$\dfrac{\lambda}{cp \cdot \rho \cdot D}$	$\dfrac{\text{conductive heat flux}}{\text{diffusive flux}}$
Mach	M	$\dfrac{u}{c}$	$\dfrac{\text{flow velocity}}{\text{speed of sound}}$
Newton	Ne	$\dfrac{F \cdot l}{m \cdot u}$	$\dfrac{\text{buoyancy force}}{\text{inertial force}}$
Nusselt	Nu	$\dfrac{a \cdot l}{\lambda}$	$\dfrac{\text{total heat transport}}{\text{heat transport by conduction}}$
Peclet (heat)	Pe_q	$\dfrac{u \cdot l \cdot cp \cdot \rho}{\lambda}$	$\dfrac{\text{heat transport convection}}{\text{heat transport by conduction}}$

Appendix 8.14 Dimensionless characteristic numbers

Name	Abbreviation	Equation	Physical meaning
Peclet (mass)	Pe_s	$\dfrac{u \cdot l}{D}$	$\dfrac{\text{mass transfer by convection}}{\text{mass transfer by diffusion}}$
Prandl	Pr	$\dfrac{cp \cdot \rho \cdot v}{\lambda}$	$\dfrac{\text{inner friction}}{\text{heat flux}}$
Reynolds	Re	$\dfrac{u \cdot l}{v}$	$\dfrac{\text{intertial force}}{\text{viscous force}}$
Schmidt	Sc	$\dfrac{v}{D}$	$\dfrac{\text{viscosity}}{\text{diffusion}}$
Sherwood	Sh	$\dfrac{\beta \cdot l}{D}$	$\dfrac{\text{total mass transfer}}{\text{mass transfer by diffusion}}$
Stanton (heat)	St_q	$\dfrac{a}{cp \cdot \rho \cdot u}$	$\dfrac{\text{total heat transport}}{\text{heat transport by convection}}$
Stanton (mass)	St_s	$\dfrac{\beta}{u}$	$\dfrac{\text{total mass transfer}}{\text{mass transfer by convection}}$
Weber	We	$\dfrac{u^2 \cdot \rho \cdot l}{\sigma}$	$\dfrac{\text{inertial force}}{\text{surface tension}}$

c speed of sound
c_p specific heat capacity
D Didiffusion coefficient
F force
g acceleration due to gravity
H Enthalpy
l characteristic length
m mass
P pressure
T temperature
U molar rate of conversion
u flow velocity
x molar fraction
α heat transfer coefficient
β mass transfer coefficient
γ thermal expansion coefficient
Δ difference
λ thermal expansion coefficient
v kinematic viscosity
ρ density
σ surface tension
τ time

Appendix 8.15
Important German regulations for handling of substances.

Constitution					
EU Law Federal laws are passed by parliaments (Federation, Land)	BImSchG	WHG	ChemG	GeräteSiG	
Regulations (concrete statements)	Störfall- verordnung	VawS	GefStoffV ChemVerbotsV ChemAltstoffV ChemPrüfV	VbF	DruckbV
Administrative regulations, technical rules (concrete measures for practive, interpretation and concretization of regulations)	TA-Luft TA-Lärm	VVAwS	TRGS	TRbF	TRB TRG

Precision of details

More easily changed

Appendix 8.16
Hazard and safety warnings.

R (Risk) phrases

R 1   explosive in dry state
R 2   detonable by percussion, friction, fire, or other ignition source
R 3   readily detonable by percussion, friction, fire, or other ignition source
R 4   forms highly sensitive explosive metal compounds
R 5   explosive on heating
R 6   explosive in the presence or absence of air
R 7   may cause fires
R 8   contact with combustible material may cause fire
R 9   contact with combustible material may cause explosion
R 10   flammable
R 11   readily ignitible
R 12   highly flammable
R 14   reacts violently with water
R 15   contact with water liberates extremely flammable gases

R 16	explosive when mixed with substances that support fire
R 17	spontaneously cumbustible in air
R 18	formation of explosive/readily ignitible vapor/air mixtures possible during use
R 19	may form explosive peroxides
R 20	harmful by inhalation
R 21	harmful in contact with skin
R 22	harmful if swallowed
R 23	toxic by inhalation
R 24	toxic in contact with skin
R 25	toxic if swallowed
R 26	very toxic by inhalation
R 27	very toxic in contact with skin
R 28	very toxic if swallowed
R 29	forms toxic gases on contact with water
R 30	can become readily ignitible during use
R 31	toxic gases produced on contact with acids
R 32	highly toxic gases produced on contact with acids
R 33	risk of cumulative effects
R 34	causes irritation
R 35	causes severe burns
R 36	irritating to eyes
R 37	irritating to respiratory system
R 38	irritating to skin
R 39	high risk of irreversible damage
R 40	irreversible damage possible
R 41	risk of serious damage to eyes
R 43	may cause sensitization by skin contact
R 44	risk of explosion on heating in confined space
R 45	may cause cancer
R 46	may cause heritable genetic damage
R 48	risk of serious damage to health on prolonged exposure
R 49	may cause cancer by inhalation
R 50	very toxic to aquatic organisms
R 51	toxic to aquatic organisms
R 52	damaging to aquatic organisms
R 53	may cause long-term adverse effects in the aquatic environment
R 54	toxic to plants
R 55	toxic to animals

R 56 toxic to soil organisms
R 57 toxic to bees
R 58 may cause long-term damage to the environment
R 59 hazardous to the ozone layer
R 60 may adversely affect reproduction
R 61 may damage the foetus in the womb
R 62 can possibly adversely affect reproduction
R 63 can possibly damage the foetus in the womb
R 64 may damage infants via maternal milk
R 65 may cause lung damage on swallowing
R 66 repeated contact may lead to chapped skin
R 67 vapors may cause drowsiness and beffudlement

S (Safety) phrases
S 1 keep in sealed container
S 2 keep away from children
S 3 keep cool
S 4 keep away from residential areas
S 5 store under ... (suitable liquid to be specified by manufacturer)
S 6 store under ... (inert gas to be specified by manufacturer)
S 7 keep container tightly closed
S 8 keep container dry
S 9 keep container in a well-ventilated location
S 12 do not seal container gas tight
S 13 keep away from foods, drinks, and feeds
S 14 keep away from ... (incompatible substances to be specified by manufacturer)
S 15 protect from heat
S 16 keep away from ignition sources; no smoking
S 17 keep away from flammable materials
S 18 take care when opening container and handling
S 20 do not eat or drink while working
S 21 do not smoke while working
S 22 do not inhale dust
S 23 do not inhale gas/smoke/vapor/aerosol
S 24 avoid contact with the skin
S 25 avoid contact with the eyes
S 26 on contact with eyes, rinse immediately with plenty of water and seek medical help
S 27 remove contaminated clothing immediately

S 28	on contact with the skin immediately wash with copious ... (to be specified by manufacturer)
S 29	do not allow to enter the sewers
S 30	never add water
S 33	take measures against electrostatic charging
S 34	avoid percussion and friction
S 35	waste and containers must be disposed of in a safe manner
S 36	wear appropriate protective clothing
S 37	wear appropriate protective gloves
S 38	wear respiratory equipment in case of inadequate ventilation
S 39	wear protective glasses/face mask
S 40	clean floors and contaminated objects with ... (to be specified by manufacturer)
S 41	do not inhale gases from explosions and fires
S 42	on spraying wear appropriate respiratory equipment (type to be specified by manufacturer)
S 43	In case of fire use ... (to be specified by manufacturer); (if water increases hazard, add: never use water)
S 44	in case of accident or illness, seek medical help; show this label if possible
S 45	in case of accident or illness, seek medical help immediately; show this label if possible
S 47	do not store at temperatures above ... °C (to be specified by manufacturer)
S 48	keep moistened with ... (suitable agent to be specified by manufacturer)
S 49	store only in original container
S 50	do not mix with ... (to be specified by manufacturer)
S 51	use only in well-ventilated areas
S 52	do not apply to large areas in residential and recreational buildings
S 53	avoid exposure; obtain special instructions before use
S 56	this material and its container must be disposed of as problem waste
S 57	to avoid contamination of the environment use suitable container
S 59	obtain infomation for reuse/recycling from producer/supplier
S 60	this material and its container must be disposed of as hazardous waste
S 61	avoid release to the environment; refer to special instructions/safety data sheet
S 62	on swallowing do not induce vomiting; seek medical help immediately and show label
S 63	on inhalation take accident victim into the fresh air
S 64	on swallowing wash out mouth with water (only if victim is concious)

Appendix 8.17
The 25 largest companies of the world in 2000.

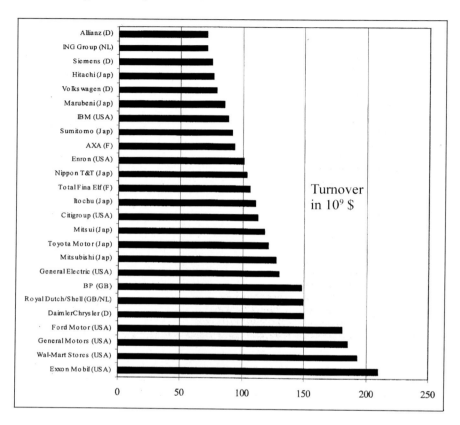

Appendix 8.18
The 25 largest companies in Germany in 2000.

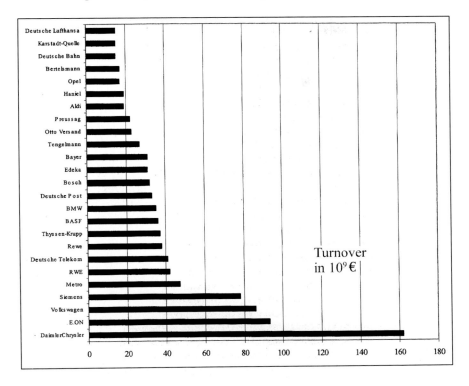

Appendix 8.19
Surface analysis methods.

Detection → Excitation ↓	Electron	Photons	Neutral particles	Ions	Phonons	Electric and magnet fields
Elektrons	AES, SAM, SEM, LEED, RHEED, EM, HEED, SES, EELS, TEM, IETS, EBIC	EMA, EDX	ESD	ESD		
Photons	XPS(ESCA), UPS	SES, OS, LM, FER, XRD, ESR, NMR, ELL, RFA	PD	LAMMA	PAS	PC, PDS, MPS, PVS
Neutral particles			AIM, NIS	MBT, FAB-MS	HAM	
Ions	INS	IMXA, IEX, PIXE	SNMS, IID	SIMS, IMP, IID, ISS, RBS		CDS
Phonons		TL	TDS		SAM	SDS
Electric and magnetic fields	FES, FEM, IETS, STM	FER		FIM, FIMS, FIAP, FD		MPS, SDS, PDS, PC, SPD, FEC, PVS, HE, CDS, CM

AES	Auger Elektron Spectroscopy
AIM	Adsorption Isotherm Measurements
CDS	Corona Discharge Spectroscopy
CM	Conductance
CPD	Contact Potential Difference
EBIC	Electron Beam Induced Current
EDX	Energy Dispersive X-Ray Spectroscopy
EELS	Electron Energy Loss Spectroscopy
EM	Electron Microscopy
EMA	Electron Microprobe Analysis
ESD	Electron Stimulated Desorption
ESR	Electron spin resonance
FAB-MS	Fast Atom Bombardement Mass Spectrometry
FEC	Field Effect of Conductance
FEM	Field-Emission Microscopy

FER	Field Effect of Reflectance
FES	Field-Emission Spectroscopy
FIAP	Field Ionization Atom Probe
FIM	Field Ion microscopy
FIMS	Field Ion Mass Spectroscopy
HAM	Heat of Adsorption Measurements
HE	Hall effect
HEED	High Energy Electron Diffraction
IETS	Inelastic Electron Tunneling Spectroscopy
IEX	Ion Excited X-Ray Fluorescence
IID	Ion Impact Desorption
IMXA	Ion Microprobe for X-Ray Analysis
INS	Ion Neutralisation Spectroscopy
ISS	Ion Scattering Spectroscopy ($\rightarrow$ LEIS)
LAMMA	Laser Microprobe Mass Analysis
LEED	Low Energy Electron Diffraction
LM	Light Microscope
MBT	Molecular Beam Techniques
MPS	Modulated Photoconductivity
NIS	Neutron Inelastic Scattering
NMR	Nuclear Magnetic Resonance
OS	Optical Spectroscopy
PAS	Photoacoustic Spectroscopy
PC	Photoconductivity
PD	Photodesorption
PDS	Photodischarge Spectroscopy
PIXE	Proton/Particle Induced X-ray Emission
PVS	Photovoltage Spectroscopy
RBS	Rutherford Backscattering ($\rightarrow$ HEIS)
RHEED	Reflection High Energy Electron Diffraction
SAM	Scanning-Auger Microscopy
SAM	Scanning Acoustic Microscopy
SDS	Surface Discharge Spectroscopy
SES	Secondary Electron Spectroscopy
SIMS	Secondary Ion Mass Spectrometry
SNMS	Sputtered Neutral Mass Spectrometry
SNMS	Secondary Neutral Mass Spectrometry
STM	Scanning Tunneling Microscopy

TDS	Thermo desorption Spectroscopy
TEM	Transmissions electron microscopy
TL	Thermoluminescence
UPS	Ultraviolett photoelectron Spectroscopy
X-AES	X-Ray Induced AES
XRD	X-Ray Diffraction
XRF	X-ray Fluorescence
XPS (ESCA)	X-Ray Photoelectron Spectroscopy

9 References

Achema 1991	R. Eckermann *Dechema-Informationssysteme für die Chemische Technik und Biotechnologie* Achema-Jahrbuch 1991, Band 1, Frankfurt 1990.
Adler 2000	R. Adler *Stand der Simulation von heterogen-gaskatalytischen Reaktionsabläufen in Festbettrohrreaktoren* Chem.-Ing.-Techn. 72(2000)555–564, Teil1. Chem.-Ing.-Techn. 72(2000)688–699, Teil2.
Albrecht 2000	C. Albrecht *Für hohe Temperaturen – Wärmeübertragung mit Salzschmelzen* CAV(10) (2000)50–52.
Amecke 1987	H.-B. Amecke *Chemiewirtschaft im Überblick* VCH, Weinheim, 1987.
Ameling 2000	D. Ameling *Die volkswirtschaftliche Bedeutung von Nickel und seinen Legierungen* Gefahrst. – Reinhaltung d. Luft, Berlin 60(2000)13–18.
Andrigo 1999	P. Andrigo, R. Bogatin, G. Pagani *Fixed bed reactor* Catalysis Today 52(1999)197–221.
Armor 1996	J.N. Armor *Global Overview of Catalysis* Applied Catalysis A: General 139(1996)217–228.
Arnzt 1993	D. Arntz *Trends in the chemical industry* Catalysis Today 18(1993)173–198.
Atkins 2002	P. W. Atkins *Physikalische Chemie* 3. Aufl., Wiley-VCH, Weinheim, 2002.
Auelmann I 1991	R. Auelmann *Wassereinsatz in der chemischen Industrie: Rationelle Nutzung* Chem. Ind. 9(1991)12–13.
Auelmann II 1991	R. Auelmann *Wassereinsatz in der chemischen Industrie: Trinkwasser wird geschont* Chem. Ind. 10(1991)10–12.
Auelmann III 1991	R. Auelmann *Wassereinsatz in der chemischen Industrie: Kühlwasser in Kreisläufen* Chem. Ind. 11(1991)10–12.

Aust 1999   E. Aust, St. Scholl
Wärmeintegration als Element der Verfahrensbearbeitung
Chem.-Ing.-Tech. 71(1999)674–679.

Baerns 1999   M. Baerns, H. Hofmann, A. Renken
Chemische Reaktionstechnik
3. Aufl., Wiley-VCH Weinheim, 1999.

Baerns 2000   M. Baerns, O. V. Buyevskaya
Catalytic Oxidative Conversion of Alkanes to Olefines and Oxygenates
Erdöl Erdgas Kohle 116(2000)25–30.

Bakay 1989   T. Bakay, K. Domnick
Produktionsintegrierter Umweltschutz – Verminderung von Reststoffen, gezeigt anhand ausgeführter Beispiele
Chem.-Ing.-Tech. 61(1989)867–870.

Bartels 1990   W. Bartels, H. Hoffmann, L. Rossinelli
PAAG-Verfahren (HAZOP)
Internationale Sektion der IVSS für die Verhütung von Arbeitsunfällen und Berufskrankheiten in der chemischen Industrie.
Gaisbergstr. 11, 6900 Heidelberg.

Bartholomew 1994   C. H. Bartholomew, W. C. Hecker
Catalytic reactor design
Chemical Engineering (1994)70–75.

BASF 1987   BASF
Acrylsäure rein
Techn. Inform. 1987.

BASF 1999   *BASF still tops global Top 50*
Chemical & Engineering News, July 26, (1999)23–25.

BASF 2001   *Vor dem Bleichen vergleichen*
BASF information, Zeitung für die Mitarbeiter der BASF, 2. Aug. 2001, S. 2.

Bauer 1996   M. H. Bauer, J. Stichlmair
Struktursynthese und Optimierung nichtidealer Rektifizierprozesse
Chem.-Ing.-Tech. 68(1996)911–912.

Baumann 1998   U. Baumann
Ab in den Ausguß?
Nachr. Chem. Tech. Lab. 46(1998)1049–1055.

Baur 2001   S. Baur, V. Casal, H. Schmidt, A. Krämer
SUWOX – ein Verfahren zur Zersetzung organischer Schadstoffe in überkritischem Wasser
NACHRICHTEN – Forschungszentrum Karlsruhe 33(2001)71–80.

Behr 2000   A. Behr, W. Ebbers, N. Wiese
Miniplants – Ein Beitrag zur inhärenten Sicherheit?
Chem.-Ing.-Tech. 72(2000)1157–1166.

Belevich 1996	P. Belevich *Selecting and applying flowmeters* Hydrocarbon Processing May(1996)67–75.
Bender 1995	H. Bender *Sicherer Umgang mit Gefahrstoffen* VCH, Weinheim, 1995.
Berenz 1991	R. Berenz *Jahrbuch der Chemiewirtschaft* *Das Rad nicht zweimal erfinden* VDI-Verlag, Düsseldorf, 1996.
Bermingham 2000	S. K. Bermingham et al. *A Design Procedure and Predictive Models for Solution Crystallisation Processes* AIChE Symp. Ser. 96(2000)250–264.
Berty 1979	B. J. M. Berty *The Changing Role of the Pilot Plant* CEP 9(1979)48–50.
Beßling 1995	B. Beßling, J. Ciprian, A. Polt, R. Welker *Kritische Anmerkungen zu den Werkzeugen der Verfahrensüberarbeitung* Chem.-Ing.-Tech. 67(1995)160–165.
Billet 2001	Billet *Separation Tray without Downcomers* Chem. Eng. Technol. 24(2001)1103–1113.
Bimschg 2001	*Bundesgesetzblatt* Jahrgang 2001, Teil I, Nr. 40.
Bio World 1997	Bio World *Bio World Patentrecherchen* Bio World (3)(1997)49–52.
Bisio 1997	A. Bisio *Catalytic process development:* *A process designers's point of view* Catalysis Today 36(1997)367–374.
Blaß 1984	E. Blaß *Aufgabengerechte Informationssammlung für die Verfahrensentwicklung* Chem.-Ing.-Tech. 56(1984)272–278.
Blaß 1985	E. Blaß *Methodische Entwicklung verfahrenstechnischer Prozesse* Chem.-Ing.-Tech. 57(1985)201–210.

Blaß 1989
E. Blaß
Entwicklung verfahrenstechnischer Prozesse
Otto Salle Verlag, Frankfurt, 1989.

Blaß 1996
E. Blaß
Jahrbuch der Chemiewirtschaft
Verfahrenstechnik im Wandel unserer Zeit
VDI-Verlag, Düsseldorf, 1996.

Bliefert 2002
C. Bliefert
Umweltchemie
2. Aufl., Wiley-VCH, Weinheim, 2002.

Blumenberg 1994
B. Blumenberg
Verfahrensentwicklung heute
Nachr. Chem. Tech. Lab. 42(1994)480–482.

Boeters 1989
H. D. Boeters, C. F. Müller
Handbuch Chemiepatent
2. Aufl., Juristischer Verlag, Heidelberg, 1989.

Bogenstätter 1985
G. Bogenstätter
Die chemische Fabrik der Zukunft: Verfahrenstechnische Weiterentwicklungen sind unerläßlich für wirtschaftliche und umweltfreundliche Produktion
Chem. Ind. 37(1985)275–277.

Böhling 1997
R. Böhling
Einsatz der zeit- und produktaufgelösten Temperatur- und Konzentrations-Programmierten Reaktionsspektroskopie zur Aufklärung der Wechselwirkung von Mo-V-(W)-(Cu)-Mischoxiden mit organischen Sonden
Doktorarbeit, TU Darmstadt, 1997.

Borho 1991
K. Borho, R. Polke, K. Wintermantel, H. Schubert, K. Sommer
Produkteigenschaften und Verfahrenstechnik
Chem.-Ing.-Techn. 63(1991)792–808.

Börnecke 2000
D. Börnecke
Basiswissen für Führungskräfte
Publicis MCD Verlag, Erlangen und München, 2000.

Brauer 1985
H. Brauer
Die Adsorptionstechnik – ein Gebiet mit Zukunft
Chem.-Ing.-Tech. 57(1985)650–663.

Brauer 1991
H. Brauer
Abwasserreinigung bis zur Rezyklierfähigkeit des Wassers
Chem.-Ing.-Tech. 63(1991)415–427.

Breyer 2000
N. Breyer
Nicht zu schnell und nicht zu langsam
PROCESS (2000)14–16.

Bringmann 1977	G. Bringmann, R. Kühn *Grenzwerte der Schadwirkung wassergefährdender Stoffe gegen Bakterien (Pseudomonas putida) und Grünalgen (Scenedesmus quadricauda) im Zellvermehrungshemmtest* Z. f. Wasser- und Abwasser-Forschung 10(1977)87–98.
Brunan 1994	C. R. Branan *Rules of Thumb for Chemical Engineers* Gulf Publishing Company, ouston, 1994.
Brunauer 1938	S. Brunauer, P.H. Emmett, E. Teller *Adsorption of Gases in Multimolecular Layers* J. Am. Chem. Soc. 60(1938)309.
Brunauer 1940	S. Brunauer, L. S. Deming, W. E. Deming, E. Teller *On a Theory of the van der Waals Adsorption of Gases* J. Am. Chem. Soc. 62(1940)1723.
Burst 1991	W. Burst, F. Heimann, G. Kaibel, S. Maier *Jahrbuch der Chemiewirtschaft* *Klein, aber fein* VDI-Verlag, Düsseldorf, 1991.
Buschulte 1995	Th. K. Buschulte, F. Heimann *Verfahrensentwicklung durch Kombination von Prozeßsimulation und Miniplant-Technik* Chem.-Ing.-Tech. 67(1995)718–723.
Busfield 1969	W. K. Busfield, D. Merigold *The Gas-phase Equilibrium between Trioxan and Formaldehyde: The Standard Enthalpy and Entropy of the Trimerisation of Formaldehyde* J. Chem. Soc. (A) (1969)2975–2977.
Buzzi-Ferraris 1999	G. Buzzi-Ferraris *Planning of experiments and kinetic analysis* Catalysis Today 52(1999)125–132.
Casper 1970	C. Casper *Strömungs- und Wärmeübergangsuntersuchungen an Gas/Flüssigkeits-Gemischen zur Auslegung von Filmverdampfern für viskose Medien* Chem.-Ing.-Tech. 42(1970)349–354.
Casper 1986	C. Casper *Die Wärmeübertragung an Zweiphasenfilmströmungen bei hoher Flüssigkeitsviskosität im gewendelten Strömungsrohr* Chem.-Ing.-Tech. 58(1986)58–59.
Cavalli 1997	L. Cavalli, A.Di Mario *Laboratory scale plants for catalytic tests* Catalysis Today 34(1997)369–377.

Cavani 1997	F. Cavani; F. Trifirò *Classification of industrial catalysts and catalysis for the petrochemical industry* Cataysis Today 34(1997)269–279.
Chemcad 1996	Chemcad III *Process Simulation Software* Chemstations, Inc., 1996.
Chemicalweek 1991	Chemicalweek *markets&economics* Chemicalweek 3(2001)38.
Chemie Manager 1998	Märkte, Unternehmen, Produkte *Der BASF-Chemistrytree* Chemie Manager 7(1998).
Chemiker Kalender 1984	H. U. v. Vogel *Chemiker Kalender* 3. Aufl., Springer-Verlag, Berlin, 1984.
Chemische Rundschau 1996	*Rohstoffpreise in Europa* Toluol: Spot-Markt ist tot Chem. Rundschau 26(1996)4.
Christ 1999	C. Christ *Integrated Enviromental Protection reduces Water Pollution* Chem. Eng. Technol. 22(1991)642–651.
Christ 2000	C. Christ *Umweltschonende Technologien aus industrieller Sicht* Chem.-Ing.-Tech. 72(2000)42–57.
Christmann 1985	A. Christmann et al. *Integrierte Verfahrensentwicklung* Chem. Ind. 37(1985)533–537.
CITplus	*Durchflussmessung* CITplus (1999)36–51.
Class 1982	I. Class, D. Janke *Stähle* Ullmann, 4. Auflage, Bd.22, S. 42–159, 1982.
Claus 1996	P. Claus, T. Berndt *Kinetische Modellierung der Hydrierung von Ethylacetat an einem Rh-Sn/SiO2-Katalysator* Chem.-Ing.-Techn. 68(1996)826–831.
Cohausz 1993	H. B. Cohausz *Patente & Muster* 2. Aufl., Wila Verlag, München, 1993.

Cohausz 1996	H. B. Cohausz *Info & Recherche* Wila Verlag, München, 1996.
Collin 1986	G. Collin *Prognosen in der chemischen Technik* Chem.-Ing.-Tech. 58(1986)364–372.
Contractor 1999	R. M. Contractor *Dupont's CFB technology for maleic anhydride* Chem. Eng. Sci. 54(1999)5627–5632)
Corbett 1995	R.A. Corbett *Effectively Use Corrosion Testing* Chem. Eng. Progress, April(1996)42–47.
Coulson 1990	J.M. Coulson, J.F. Richardson, J.R. Backhurst, J.H. Harker *Chemical Engineering* 4th Ed., Vol. 1, Pergamon Press, Oxford, 1990, pp. 286–292.
Crum 2000	J. Crum *High-nickel alloys in the chemical process industries* Stainless Steel World 12(2000)55–59.
DAdda 1997	M. DÀdda *Fixed-capital cost estimating* Catalysis Today 34(1997)457–467.
Damköhler 1936	G. Damköhler *Einflüsse der Strömung, Diffusion und des Wärmeüberganges auf die Leitung von Reaktionsöfen* Ztschr. Elektrochemie 42(1936)846–862.
Daun 1999	G. Daun, B. Bartenbach, M. Battrum *Praxisrelevante Simulationswerkzeuge für mehrstufige Batch-Prozesse* Chem.-Ing.-Tech. 71(1999)1253–1261.
DECHEMA 1953	*DECHEMA-Werkstoff-Tabelle* Herausgegeben von der DECHEMA, Frankfurt, in fortlaufenden Lieferungen, erste Lieferung 1953.
Dechema 1990	*Produktionsintegrierter Umweltschutz in der chemischen Industrie* DECHEMA, Frankfurt, 1990.
Denk 1996	O. Denk *Zur Aufstellung von Stoffmengenbilanzgleichungen auf der Basis komplexer Reaktionsgleichungssysteme* Chem.-Ing.-Tech. 68(1996)113–116.
Despeyroux 1993	B. Despeyroux, K. Deller, H. Krause *Auf den Träger kommt es an.* Chem.Ind. 116(1993)48–49.

DFG 1983	Toxikologische Untersuchungen an Fischen Verlag Chemie, Weinheim, 1983.
Dichtl 1987	E. Dichtl, O. Issing Vahlens großes Wirtschaftslexikon Verlag C. H. Beck, München, Verlag Franz Vahlen, München, 1987.
Dietz 2000	P. Dietz, U. Neumann Verfahrenstechnische Maschinen – Chancen der gleichzeitigen Prozeß- und Maschinenentwicklung Chem.-Ing.-Tech. 72(2000)9–16.
Dietzsch 1996	C. R. Dietzsch Patente – effizientes Entwicklungswerkzeug Transfer 1/2(1996)10–13.
DIN 38409	DIN 38409 Teil 41 Bestimmung des chemischen Sauerstoffbedarfs (CSB) DIN 38409 Teil 51 Bestimmung des Biochemischen Sauerstoffbedarfs (BSB)
DIN 38412	DIN 38412 Teil 8 Testverfahren mit Wasserorganismen (Gruppe L) Bestimmung der Wirkung von Wasserinhaltsstoffen auf Bakterien DIN 38412 Teil 15 Testverfahren mit Wasserorganismen (Gruppe L) Bestimmung der Hemmwirkung von Wasserinhaltsstoffen auf Fische
Dolder 1991	F. Dolder Patentmanagement im Betrieb: Geheimhalten oder patentieren? Management Zeitschrift 60(1991)64–68.
Donat 1989	H. Donat Apparate aus Sonderwerkstoffen Chem.-Ing.-Tech. 61(1989)585–590.
Donati 1997	G. Donati, R. Paludetto Scale up of chemical reactors Catalysis Today 34(1997)483–533.
Dreyer 1982	D. Dreyer, G. Luft Treibstrahlreaktor zur Untersuchung von hetzerogenkatalysierten Reaktionen Chemie-Technik 11(1982)1061–1056.
Drochner 1999	A. Drochner, R. Böhling, M. Fehlings, D. König, H. Vogel Konzentrationsprogrammierte Reaktionstechnik – Eine Methode zur Beurteilung des Anwendungspotentials instationärer Prozeßführungen bei Partialoxidationen Chem.-Ing.-Tech. 71(1999)226–230.

| Drochner 1999a | A. Drochner, K. Krauß, H. Vogel
Eine neue DRIFTS-Zelle zur In-situ-Untersuchung heterogen katalysierter Reaktionen
Chem.-Ing.-Techn. 71(1999)861–864. |

| Eggersdorfer 1994 | M. Eggersdorfer, L. Laupichler
Nachwachsende Rohstoffe – Perspektiven für die Chemie?
Nachr. Chem. Tech. Lab. 42(1994)996–1002. |

| Eggersdorfer 2000 | M. Eggersdorfer
Perspektiven nachwachsender Rohstoffe in Energiewirtschaft und Chemie
Spektrum der Wissenschaft Digest: Moderne Chemie II
1(2000)94–97. |

| Ehrfeld 2000 | W. Ehrfeld, V. Hessel, H. Löwe
Microreactors
Wiley-VCH, Weinheim, 2000. |

| Eichendorf 2001 | K. Eichendorf, E. Guntrum, C. Jochum, K.-J. Niemitz
Analyse zur Bewertung des Gefahrenpotenzials von prozessbezogenen Anlagen
Chem.-Ing.-Tech. 73(2001)809–812. |

| Eidt 1997 | C. M. Eidt, R. W. Cohen
Basic research at Exxon
Chemtech April (1997)6–10. |

| Eigenberger 1991 | G. Eigenberger
Aufgaben und Entwicklung in der Chemiereaktortechnik
World Congr. of Chem. Engng., 4; Strat. 2000, Proc. 16.–21.6.91, Karlsruhe-Frankf. a.M.:DECHEMA(1992)(1991)Juni, 1082/1102. |

| Eissen 2002 | M. Eissen, J. O. Metzger, E. Schmidt, U. Schneidewind
10 Jahre nach „Rio" – Konzepte zum Beitrag der Chemie zu einer nachhaltigen Entwicklung
Angew. Chem. 114(2002)402–425. |

| Emig 1993 | G. Emig
Neue reaktionstechnische Konzepte für die Verbesserung chemischer Verfahren
DECHEMA-Jahrestagung 1993 – Übersichtsvortrag. |

| Emig 1997 | G. Emig, R. Dittmayer
Simultaneous Heat and Mass Transfer and Chemical Reaction
aus G. Ertl, H. Knötzinger, J. Weitkamp
Heterogeneous Catalysis, Vol. 3
VCH, Weinheim, 1997, S. 1209–1252. |

| Engelbach 1979 | H. Engelbach; R. Krabetz; G. Dümbgen
Verfahren zur Herstellung von 3 bis 4-Atome enthaltenden β-olefinisch ungesättigten Aldehyden
Deutsche Offenlegungsschrift 2909597(1979). |

Engelmann 2001	F. Engelmann, W. Steiner *Dynamische Shift-Techniken zur schnellen Charakterisierung der Desaktivierungskinetik heterogener Katalysatoren* Chem.-Ing.-Techn. 73(2001)536–541.
Englund 1992	S. M. Englund, J. L. Mallory, D. L. Grinwis *Prevent Backflow* Chem. Eng. Progress, Feb. (1992)47–53.
Erdmann 1984	H. H. Erdmann, J. Kussi, K. H. Simmrock *Möglichkeiten und Probleme der Prozeß-Synthese* Chem.-Ing.-Tech. 56(1984)32–41.
Erdmann 1986	H. H. Erdmann, M. Lauer, M. Passmann, E. Schrank, K. H. Simmrock *Expertensysteme - ein Hilfsmittel der Prozeß-Synthese* Chem.-Ing.-Tech. 58(1986)296–307.
Erdmann 1986a	H.-H. Erdmann, H.-D. Engelmann, M. Lauer, K.H. Simmrock *Problemlösungen per Computer* Chemische Industrie 8(1986)682–686.
Erlwein 1998	E. Müller-Erlwein *Chemische Reaktionstechnik* B.G. Teubner, Stuttgart, 1998.
Ertl 1990	G. Ertl 1990 *Elementary Steps in Heterogeneous Catalysis* Angew. Chem. Int. Ed. Engl. 29(1990)1219–1227.
Ertl 1994	G. Ertl *Reaktionen an Festkörper-Oberflächen* Ber. Bunsenges. Phys. Chem. 98(1994)1413–1420.
Ertl 1997	G. Ertl, H. Knötzinger, J. Weitkamp *Handbook of Heterogeneous Catalysis* VCH Verlag, Weinheim, 1997.
Everett 1976	D. H. Everett, J. C. Powl *Adsorption in Slit-like and Cylindrical Micropores in the Henry's Law Region* JCS, Faraday I 72(1976)619–636.
Färber 1991	G. Färber *Jahrbuch der Chemiewirtschaft 1991* *Standardtechnik löst Probleme* VCH, Weinheim, S. 158–162.
Fattore 1997	V. Fattore *Importance and significance of patents* Cat. Today 34(1997)379–392.

Faz 1999	Chemische Forschung rasch genutzt Miniplants erleichtern die Erprobung neuer Verfahren / Die BASF als Beispiel Frankfurter Allgemeine Zeitung, 10.02.1999.
Fedtke 1996	M. Fedtke, W. Pritzkow, G. Zimmermann Lehrbuch der Technischen Chemie 6. Aufl., Deutscher Verlag für Grundstoffindustrie, Stuttgart, 1996.
Fehlings 1998	M. Fehlings, D. König, H. Vogel Explosionsgrenzen von Propen/Sauerstoff/Alkan-Mischungen und ihre Reaktivität an Bi/Mo-Mischoxiden Chemische Technik 50(1998)241–245.
Fehlings 1999	M. Fehlings, A. Drochner, H. Vogel Katalysatorforschung Heute Technische Universität Darmstadt thema Forschung 1(1999)136–142.
Fei 1998	W. Fei; H.-J. Bart Prediction of Diffusivities in Liquids Chem. Eng. Technol. 21(1998)659–665.
Fei 1998	W. Fei, H.J. Bart Voraussage von Diffusionskoeffizienten Chem.-Ing.-Techn. 70(1998)1309–1311.
Felcht 2000	U.-H. Felcht Chemie, Eine reife Industrie oder weiterhin Innovationsmotor? Universitätsbuchhandlung Blazek und Bergamann, Frankfurt 2000.
Felcht 2001	U.-H. Felcht, G. Kreysa Katalyse – Ein Schlüssel zum Erfolg in der Technischen Chemie. Festschrift anlässlich des 75. DECHEMA-Jubiläums.
Ferino 1999	I. Ferrino, E. Rombi Oscillating reactions Catalysis Today 52(1999)291–305.
Ferrada 1990	J. J. Ferrada, J.M. Holmes Developing Expert Systems Chem. Eng. Prog. 4(1990)34–41.
Fick 1855	A. Fick Ueber Diffusion Ann. Phys. und Chemie 94(1855)
Fild 2001	C. Fild, U. Jung Was der Chemie übrig bleibt Nachrichten aus der Chemie 49(2001)1080–1084.

Fitzer 1975	E. Fitzer, W. Fritz *Technische Chemie* Springer Verlag, Berlin, 1975, S. 173–175.
Fitzer 1989	E. Fitzer, W. Fritz *Technische Chemie* Springer-Verlag, Berlin, 3. Auflage, 1989.
Fluent 1998	FLUENT Deutschland GmbH Computational Fluid Dynamics Darmstadt, http://www.fluent.de, 1998
Forni 1997	L. Forni *Laboratory reactors* Catalysis Today 34(1997)353–367.
Forni 1997	L. Forni *Laboratory reactors* Catalysis Letters 34(1997)353–367.
Forni 1999	L. Forni *Mass and heat transfer in catalytic reactions* Catalysis Today 52(1999)147–152.
Forzatti 1997	P. Forzatti *Kinetic study of a simple chemical reaction* Catalysis Today 34(1997)401–409.
Forzatti 1999	P. Forzatti, L. Lietti *Catalyst deactivation* Catalysis Today 52(1999)165–181.
Franck 1999	E. U. Franck *Supercritical Water* Intenational Conference on the Properties of Water and Steam, Toro, Ottawa, p. 22–34, 1999. NRC Research Press 0-0660-17778-1
Franke 1999	D. Franke, W. Gösele *Hydrodynamischer Ansatz zur Modellierung von Fällungen* Chem.-Ing.-Tech. 71(1999)1245–1251.
Fratzscher 1993	W. Fratzscher, H.-P. Picht *Stoffdaten und Kennwerte der Verfahrenstechnik* 4. Aufl., Deutscher Verlag für Grundstoffindustrie, Leipzig, 1993.
Frey 1990	W. Frey, F. Heimann, S. Maier *Wirtschaftliche und technologische Bewertung von Verfahren* Chem.-Ing.-Tech. 62(1990)1–8.
Frey 1995	W. Frey *Wandel in der Verfahrenstechnik – Anforderungen an die Ausbildung* Chem.-Ing.-Tech. 67(1995)155–159.

Frey 1998	W. Frey, B. Lohe *Verfahrenstechnik im Wandel* Chem.-Ing.-Tech. 70(1998)51–63.
Frey 1998a	T. Frey, J. Stichlmair *Thermodynamische Grundlagen der Reaktivdestillation* Chem.-Ing.-Tech. 70(1998)1373–1381.
Frey 2000	Th. Frey, D. Brusis, J. Stichlmair, H. Bauer, S. Glanz *Systematische Prozesssynthese mit Hilfe mathematischer Methoden* Chem.-Ing.-Tech. 72(2000)813–821.
Fürer 1996	St. Fürer, J. Rauch, F. J. Sanden *Konzepte und Technologien für Mehrproduktanlagen* Chem.-Ing.-Tech. 68(1996)375–381.
Gärtner 2000	E. Gärtner *Computer und Indigo – Alles öko?* Nachrichten aus der Chemie 48(2000)1357–1361.
Geiger 2000	T. Geiger, C. Weichert *Thermische Abfallbehandlung* Chem.-Ing.-Tech. 72(2000)1512–1522.
Ghosh 1999	P. Ghosh *Rediction of Vapor-Liquid Equilibria using Peng-Robinson and Soave-Redlich-Kwong Equation of State* Chem. Eng. Technol. 22(1999)5.
Gilles 1986	E. D. Gilles, P. Holl, W. Marquardt *Dynamische Simulation komplexer chemischer Prozesse* Chem.-Ing.-Tech. 58(1986)268–278.
Gilles 1998	E. D. Gilles *Network Theory for Chemical Processes* Chem. Eng. Technol. 21(1998)121–132.
Gillett 2001	J.E. Gillett *Chemical Engineering Education in the Next Century* Chem. Eng. Technol. 24(2001)561–570.
GIT 1999	V. Hessel, W. Erfeld, K. Globig, O. Wörz *Mikroreaktionssysteme für die Hochtemperatursynthese* GIT Labor-Fachzeitschrift 10(1999)1100–1103.
Glanz 1999	S. Glanz, E. Blaß, J. Stickmair *Systematische vergleichende Kostenanalyse von destillativen und extraktiven Stofftrennverfahren* Chem.-Ing.-Tech. 71(1999)669–673.
Glasscock 1994	D. A. Glasscock; J. C. Hale *Process Simuation: the art and science of modeling* Chem. Engng. 11(1994)82–89

9 References

Gmehling 1977
J. Gmehling, U. Onken
Chemistry Data Series
Vapor-Liquid Equilibrium Data Collection
DECHEMA, Vol. I, Part 1, 1977.

Gmehling 1981
J. Gmehling, U. Onken, W. Arlt
Chemistry Data Series
Vapor-Liquid Equilibrium Data Collection
DECHEMA, Vol. I, Part 1a, 1981.

Gmehling 1992
J. Gmehling, B. Kolbe
Thermodynamik
2. Aufl., VCH, Weinheim, 1992, S. 183.

Göttert 1991
W. Göttert, E. Blaß
Zur rechnerischen Auswahl von Apparaten für die flüssig/flüssig-Extraktion
Chem.-Ing.-Tech. 63(1991)238–239.

Graesser 1995
U. Graeser, W. Keim, W. J. Petzny, J. Weitkamp
Perspektiven der Petrochemie
Erdöl Erdgas Kohle 111(1995)208–218.

Greß 1979
D. Greß, H. Hartmann, G. Kaibel, B. Seid
Einsatz von mathematischer Simulation und Miniplant-Technik in der Verfahrensentwicklung
Chem.-Ing.-Tech. 51(1979)601–611.

GVC.VDI 1997
GVC.VDI-Gesellschaft
Verfahrenstechnik/Chemieingenieurwesen
Herausgeber: Fachausschuß Aus- und Weiterbildung in Verfahrenstechnik, Graf-Recke-Str. 84, 40239 Düsseldorf, 3. vollständig überarbeitete Auflage 1997.

Haarer 1999
D. Haarer, H.J. Rosenkranz
Bayer's Interdisciplinary Research Philosophy
Adv. Mater. 11(1999)515–517.

Hacker 1980
I. Hacker, K. Hartmann
Probleme der modernen chemischen Technologie:
Heuristische Verfahren zum Entwurf Verfahrenstrechnischer Systeme
Akademie-Verlag, Berlin, 1980, S. 193–244.

Hagen 1996
J. Hagen
Technische Katalyse
VCH, Weinheim, 1996.

Hampe 1996
M. Hampe
Mikroplants zur Kreislaufführung in verfahrenstechnischen Prozessen
thema Forschung (TH-Darmstadt) 2(1996)98–108.

Handbook 1978
Handbook of Chemistry and Physics, C730
59nd Ed., 1978–1979.

Handbook 1991	R. C. Weast *Handbook of Chemistry and Physics* *Enthalpy of Combustion of Hydrocarbons* 72nd Ed., 1991–1992.
Hänny 1984	J. Hänny *Forschung und Entwicklung in der industriellen Unternehmung* Techn. Rundschau Sulzer 1(1984)4–10.
Hansen 1997	B. Hansen, F. Hirsch *Protecting Inventions in Chemistry* Wiley-VCH, Weinheim, 1997.
Hänßle 1984	P. Hänßle *Auswahlkriterien für organische Wärmeträger* Chemie-Technik 13(1984)92–101.
Harnisch 1984	H. Harnisch, R. Steiner, K. Winnacker *Chemische Technologie, Band 1* 4. Aufl., Carl Hanser Verlag, München, Wien, 1984.
Hartmann 1980	K. Hartmann, W. Schirmer, M. G. Slinko *Probleme der modernen chemischen Technologie* *Modellierung und Simulation* Akademie-Verlag, Berlin, 1980, S. 43–77.
Hauthal 1998	H. G. Hauthal *Verfahrenstechnik in Ludwigshafen* Nachr. Chem. Tech. Lab. 46(1998)41–42.
Hayes 1994	D. L. Hayes, A. C. Smith *What is that patent, trademark, or copyright worth?* CHEMTECH (1994)16–20.
Heimann 1998	F. Heimann *Labor- und Miniplanttechnik: Was bietet der Markt?* Chem.-Ing.-Tech. 70(1998)1192–1195.
Heinke 1997	G. Heinke, J. Korkhaus *Einsatz von korrosionsbeständigem Stahl in der chemischen Industrie* Chem.-Ing.-Techn. 69(1997)283–290.
Hellmund 1991	W. Hellmund; R. Auelmann; J. Wasel – Nielen; A. Rothert *Wassereinsatz in der chemischen Industrie (1): Rationelle Nutzung* Chem. Ind. 114(1991)12–13.
Hendershot 2000	D.C. Hendershot *Was Murphy Wrong?* Process Safety Progress 19(2000)65–68.
Henglein 1963	F. A. Henglein *Grundriß der chemischen Technik* Verlag Chemie, Weinheim, 1963.

Henker 1999	R. Henker, G. H. Wagner *Korrosionsbeständige Werkstoffe für Chemie-, Energie- und Umwelttechnik* *Erfahrungen beim Einsatz von Sonderedelstählen und Nickellegierungen in der chemischen Industrie* Korrosionsschutzseminar Dresden, 1999. TAW-Verlag, Wuppertal, ISBN 3-930526-16-6.
Henne 1994	H. J. Henne *Konzepte für noch mehr Sicherheit* Chemische Industrie 5(1994)23–25.
Henzel 1994	M. Henzel, W. Göpel *Oberflächenphysik des Feststoffes* Teubner Studienbücher, Stuttgart, 1994.
Herden 2001	A. Herden, C. Mayer, S. Kuch, R. Lacmann *Über die metastabile Grenze der Primär- und Sekundärkeimbildung* Chem.-Ing.-Tech. 73(2001)823–830.
Hezel 1985	C. Hezel *Aufstellung von Sicherheitsanalysen durch die Betreiber chemischer Anlagen* Chem.-Ing.-Tech. 58(1986)15–18.
Hirsch 1995	F. Hirsch, B. Hansen *Der Schutz von Chemie-Erfindungen* VCH, Weinheim, 1995.
Hirschberg 1999	H.G. Hirschberg *Handbuch Verfahrenstechnik und Anlagenbau* Springer-Verlag, Berlin, 1999.
Hochmüller 1973	K. Hochmüller VGB-Mitteilungen Sonderheft VGB, Speisewassertagung 1953.
Hofe 1998	Hofe *Ein neuer Strömungsrohrreaktor für die Kinetik schneller chemischer Reaktionen* GIT-Labor-Fachzeitschrift, Nr. 11(1998)1127–1131.
Hofen 1990	W. Hofen, M. Körfer, K. Zetzmann *Scale-up-Probleme bei der experimentellen Verfahrensentwicklung* Chem.-Ing.-Tech. 62(1990)805–812.
Hofmann 1979	H. Hofmann *Fortschritte bei der Modellierung von Festbettreaktoren* Chem.-Ing.-Techn. 51(1979)257–265.
Hofmann 1983	H. Hofmann, G. Emig *Systematik und Prinzipien der Auslegung chemischer Reaktoren* Dechema-Monographien, Band 94, Verlag Chemie, 1983.

Hopp 1990	V. Hopp *Der Rhein – größter Standort der chemischen Industrie in der Welt* Chemiker-Zeitung 114(1990)229–243.
Hopp 2000	V. Hopp *Die Zukunft hat schon längst begonnen* CIT plus 3(2000)6–8.
Hornbogen 1994	E. Hornbogen *Werkstoffe* Springer Verlag, Berlin, 6. Auflage, 1994.
Hörskens 1991	M. Hörskens *Jahrbuch der Chemiewirtschaft 1991* *BASF spielt einen Grand mit Vieren* VCH, Weinheim, S. 145–151.
Hou 1999	K. Hou, M. Fowles, R. Hughes *Effective Diffusivity Measurements on Porous Catalyst Pellets at Elevated Temperture and Pressures* TransIChemE 77(A)(1999)55–61.
Hou 1999	K. Hou, M. Fowles, R. Hughes *Effective Diffusivity Measurements on Porous Catalyst Pellets at Elevated Temperatures and Pressures* TransIChemE 77(A)(1999)55–61.
Hüls 1989	Hüls *Katalytische Abgasreinigung* Der Lichtbogen, Nr. 208, 38(1989)23.
Hüning 1984	W. Hüning, H. Nentwig, C. Gockel *Die verbrennungstechnische Entsorgung eines Chemiebetriebes* Chem.-Ing.-Tech. 56(1984)521–526.
Hüning 1986	W. Hüning *Aufgabenstellung und Konzeptfindung bei der thermischen Abluftreinigung* Chem.-Ing.-Tech. 58(1986)856–866.
Hüning 1989	W. Hüning, P. Reher *Verbrennungstechnik für den Umweltschutz* Chem.-Ing.-Tech. 61(1989)26–36.
Hyland 1998	M. A. Hyland *Projekte unter der Lupe* Chemie Technik 27(1998)64–65.
Ingham 1994	J. Ingham, I. J. Dunn *Desktop Simulation of Dynamic Chemical Engineering Processes* Chemical Technology Europe, Nov./Dec. (1994)14–20.

IUPAC 1985

IUPAC
Reporting physisorption data for gas/solid system
Pure Appl. Chem. 57(1985)603.

J. Stichlmair 1998

J. Stichlmair, T. Frey
Prozesse der Reaktivdestillation
Chem.-Ing. Tech. 70(1998)1507–1516.

Jäckel 1995

K.-P. Jäckel, M. Molzahn
Mit Miniplant die Kosten minimieren
Standort spezial 22(1995)24–26.

Jahrbuch 1991a

M. Kersten
Jahrbuch der Chemiewirtschaft 1991
Bruttostundenverdienste
VCH, Weinheim, S. 286.

Jahrbuch 1991b

M. Kersten
Jahrbuch der Chemiewirtschaft 1991
Forschungs- und Entwicklungsaufwendungen der chemischen Industrie
VCH, Weinheim, S. 302.

Jahrbuch 1991c

M. Kersten
Jahrbuch der Chemiewirtschaft 1991
Aufwendungen für den Umweltschutz in der chemischen Industrie
VCH, Weinheim, S. 303.

Jahrbuch 1991d

M. Kersten
Jahrbuch der Chemiewirtschaft 1991
Bruttostundenverdienste
VCH, Weinheim, S. 286.

Jahrbuch 1991e

M. Kersten
Jahrbuch der Chemiewirtschaft 1991
Gesamtumsatz der chemischen Industrie
VCH, Weinheim, S. 282.

Jahrbuch 1991f

M. Kersten
Jahrbuch der Chemiewirtschaft 1991
Preisindizes chemischer Anlagen
VCH, Weinheim, S. 280.

Jakubith 1998

M. Jakubith
Grundoperationen und chemische Reaktionstechnik
Wiley-VCH, Weinheim, 1998.

Jentzsch 1990

W. Jentzsch
Was erwartet die Chemische Industrie von der Physikalischen und Technischen Chemie?
Angew. Chem. 102(1990)1267–1273.

Johnstone 1957

R. E. Johnstone, M. W. Thring
Methods in Chemical Engineering
Pilot Plants, Models and Scale-up
McGrawHill, New York, 1957.

Jorisch 1998	W. Jorisch *Vakuumtechnik in der chemischen Industrie* Wiley-VCH, Weinheim, 1998.
Juhnke 1978	I. Juhnke, D. Lüdemann *Ergebnisse der Untersuchung von 200 chemischen Verbindungen auf akute Fischtoxizität mit dem Goldorfentest* Z. f. Wasser- und Abwasserforschung 11(1978)161.
Jung 1983	J. Jung *Aspekte der Vorausberechnung von Anlagenkosten bei der Projektierung verfahrenstechnischer Anlagen* Chemie-Technik 12(1983)9–18.
Kaibel 1987	G. Kaibel *Distillation Columns with Vertical Partitions* Chem. Eng. Technol. 10(1987)92–98.
Kaibel 1989	G. Kaibel, E. Blaß, J. Köhler *Gestaltung destillativer Trennungen unter Einbeziehung thermodynamischer Gesichtspunkte* Chem.-Ing.-Tech. 61(1989)16–25.
Kaibel 1989a	G. Kaibel, E. Blaß, J. Köhler *Gestaltung destillativer Trennungen unter Einbeziehung thermodynamischer Gesichtspunkte* Chem.-Ing.-Tech. 61(1989)16–25)
Kaibel 1989b	G. Kaibel, E. Blaß *Möglichkeiten zur Prozeßintegration bei destillativen Trennverfahren* Chem.-Ing.-Tech. 61(1989)104–112.
Kaibel 1990	G. Kaibel *Energieintegration in der thermischen Verfahrenstechnik* Chem.-Ing.-Tech. 62(1990)99–106.
Kaibel 1990a	G. Kaibel, E. Blass, J. Köhler *Thermodynamics – guideline for the development of distillation column arrangements* Gas Separation & Purification 4(1990)109–114.
Kaibel 1998	G. Kaibel *Abwärtsfahrweise und Zwischenspeicherung bei der diskontinuierlichen Destillation* Chem.-Ing.-Tech. 70(1998)711–713.
Kämereit 2001	W. Kämereit *Behälter durchleuchten* CAV Aug.(2001)18–19.
Kast 1988	W. Kast *Absorption aus der Gasphase* VCH, Weinheim, 1988.

Kaul 1999	C. Kaul, H. Exner, H. Vogel *Verhalten von anorganischen Katalysatormaterialien gegenüber überkritischen wäßrigen Lösungen* Mat.-wiss., u. Werkstofftech.. 30(1999)326–331.
Keil 1999	F. Keil *Diffusion und Chemische Reaktionen in der Gas/Feststoff Katalyse* Springer-Verlag, Berlin, 1999.
Kind 1990	J. Kind *Grundlagen der praktischen Information und Dokumentation Online-Dienste* 3. Aufl., Band 1, K. G. Saur, München, 1990, S. 366.
Kircherer 2001	A. Kircherer *Ökoeffizienz-Analyse als betrieblicher Nachhaltigkeitsindikator* Chem.-Ing.-Tech. 73(2001)404–406.
Klar 1991	W. Klar *Expertensysteme: gestern, heute, morgen* Elektronik 8(1991)108–123.
Klos 2000	T. Klos *Chemische Verfahren* Chem.-Ing.-Tech. 72(2000)1174–1181.
Knapp 1987	H. Knapp *Methoden zur Beschaffung von Information über Stoffeigenschaften* Chem.-Ing.-Tech. 59(1987)367–376.
Knauf 1998	R. Knauf, U. Meyer-Blumenroth, J. Semel *Einsatz von Membrantrennverfahren in der chemischen Industrie* Chem.-Ing.-Tech. 70(1998)1265–1270.
Kölbel 1960	H. Kölbel; J. Schulze *Projektierung und Vorkalkulation in der chemischen Industrie* Springer-Verlag, Berlin, 1960.
Kölbel 1967	H. Kölbel, J. Schulze *Neuentwicklungen zur Berechnung von Preisindices von chemische Anlagen* Chem. Ind. 19(1967)340 und 701.
König 1998	D. König *Partialoxidation von Propen zu Acrolein an einem Bi/Mo-Mischoxid-Katalysator -Prozeßentwicklung und Katalyse-* Doktorarbeit, TU Darmstadt, 1998 (D17).
Körner 1988	H. Körner *Optimaler Energieeinsatz in der Chemischen Industrie* Chem.-Ing.-Tech. 60(1988)511–518.

Krätz 1990	O. Krätz *7000 Jahre Chemie* Nikol Verlag, Hamburg, 1990.
Krammer 1999	P. Krammer, H. Vogel *Hydrolysis of Esters in Subcritical and Supercritical Water* J. Supercritical Fluids 16(2000)70–74.
Kraus 2001	O. E. Kraus *Managementwissen für Naturwissenschaftler* Springer-Verlag, Berlin, 2001.
Krauß 1999	K. Kraus, A. Dochner, M. Fehlings, H. Vogel *DRIFT-Spektroskopie zur In-situ-Untersuchung heterogen katalysierter Reaktionen* GIT Labor – Fachzeitschrift 43(1999)476–479.
Krekel 1985	J. Krekel, G. Siekmann *Die Rolle des Experiments in der Verfahrensentwicklung* Chem.-Ing.-Tech. 57(1985)511–519.
Krekel 1992	J. Krekel, R. Polke *Jahrbuch der Chemiewirtschaft 1992* *Qualitätssicherung – Unternehmensziel integriert in Verfahren* VCH, Weinheim, S. 206.
Kreul 1997	L. U. Kreul, G. Fernholz, A. Gorak, S. Engell *Erfahrungen mit den dynamischen Simulatoren DIVA, gPROMS und ABACUSS* Chem.-Ing.-Tech. 69(1997)650–653.
Krötz 1999	P. Krötz *Excellente Manager lernen gewinnen mit Teams* VDI Berichte Nr. 1519, 1999, S. 201–218.
Krubasik 1984	G. Krubasik *Management – Angreifer im Vorteil* Wirtschaftswoche Nr. 23(1984)48–56.
Kühn 1996	Kühn, Birett *Band 5, VI-4, Wassergefährdende Stoffe* Merkblätter Gefährliche Arbeitsstoffe 92. Erg. Lfg. 10(1996).
Kussi 2000	J. S. Kussi, H.-J. Leimkühler, R. Perne *Ganzheitliche Verfahrensentwicklung und -optimierung aus industrieller Sicht* Chem.-Ing.-Tech. 72(2000)1285–1293.
Kussi 2000a	J. S. Kussi, H. J. Leimküler, R. Perne *Overall Process Design and Optimization* AIChE Symp. Ser. 96(2000)315–319.
Laidler 1987	K.J. Laidler *Chemical Kinetics* 3. Ed., 1987.

Lamb 1981	W. J. Lamb, G. A. Hoffmann, J. Jonas *Self-diffusion in compressed supercritical water* J. Chem. Phys. 74(1981)6875–6880.
Landolt	Landolt-Börnstein *Kalorische Zustandsgrößen* 6. Aufl., II. Band, 4. Teil.
Last 2000	W. Last, J. Stichlmair *Bestimmung der Stoffübergangskoeffizienten mit Hilfe chemischer Absorptionen* Chem.-Ing.-Tech. 72(2000)1362–1366.
Lax 1967	D'Àns Lax *Taschenbuch für Chemiker und Physiker* *Zustandsgrößen von Wasser, Sättigungsdruck, Dichte, Verdampfungsenthalpie und relative Werte der thermodynamischen Funktionen* 3. Aufl., 1. Band, Springer-Verlag, Berlin, 1967.
Lenz 1989	H. Lenz, M. Molzahn, D.W. Schmitt *Produktionsintegrierter Umweltschutz – Verwertung von Reststoffen* Chem.-Ing.-Tech. 61(1989)860–866.
Leofanti 1997a	G. Leofanti et al. *Catalyst characterization: characterization techniques* Catalysis Letters 34(1997)307–327.
Leofanti 1997b	G. Leofanti et al. *Catalyst characterization: applications* Catalysis Today 34(1997)329–352.
Levenspiel 1980	O. Levenspiel *The Coming-Of-Age of Chemical Reaction Engineering* Chem. Eng. Sci. 35(1980)1821–1839.
Lieberam 1986	A. Lieberam *Expertensysteme für die Verfahrenstechnik* Chem.-Ing.-Tech. 58(1986)9–14.
Linnhoff 1981	B. Linnhoff, J.A. Turner *Heat-recovery networks: new insights yield big savings* Chemical Engineering, November 2(1981)56–70.
Linnhoff 1983	B. Linnhoff *New Concepts in Thermodynamics for Better Chemical Process Design* Chem. Eng. Res. Des. 61(1983)207–223.
Lintz 1999	H.G. Lintz, A. Quast *Partielle Oxidation im Integralreaktor: Möglichkeiten der mathematischen Beschreibung* Chem.-Ing.-Techn. 71(1999)126–131.

Lipphardt 1989 G. Lipphardt
Produktionsintegrierter Umweltschutz – Verpflichtung der Chemischen Industrie
Chem.-Ing.-Tech. 61(1989)855–859.

Loo 2000 L. Loo, O. Ebbeke
Restlos vernichten – Hightech-Verbrennungsanlagen in einem Chemie-Betrieb
Chemie Technik 29(2000)12–14.

Lowenstein 1985 J. G. Lowenstein
The Pilot Plant
Chemical Engineering 9(1985)62–76.

Lück 1983 G. Lück
Verfahrensentwicklung
Fortschritte der Verfahrenstechnik 21(1983)473–486.

Luft 1969 G. Luft
Die Berücksichtigung des Druckeinflusses auf die Geschwindigkeitskonstante bei der Berechnung Chemischer Hochdruckreaktoren
Chem.-Ing.-Techn. 12(1969)712–721.

Luft 1973 G. Luft
Kreislaufapparaturen für reaktionskinetische Messungen
Chem.-Ing.-Techn. 45(1973)596–602.

Luft 1978a G. Luft, O. Schermuly
Kreislaufreaktor mit Treibstrahlantrieb für feststoffkatalysierte chemische Reaktionen
Offenlegungsschrift 28 08 366, Anmeldetag: 27.2.1978.

Luft 1978b G. Luft, R. Römer, F. Häusser
Performance of Tubular and Loop Reactor in Kinetic Measurements
ACS Symposium Series, No. 65, Chemical Reaction Engineering, 1978.

Luft 1989 G. Luft, Y. Ogo
Activa Volumes of Polymerization Reactions
Polymer Handbook, 3th
John Wiley&Sons, 1989.

Lunde 1985 K. E. Lunde
Transfer pricing
Chemical Engineering, Feb. 4(1985)85–87.

Maier 1986 S. Maier, F.-J. Müller
Reaktionstechnik bei industriellen Hochdruckverfahren
Chem.-Ing.-Techn. 58(1986)287–295.

Maier 1990 S. Maier, G. Kaibel
Verkleinerung verfahrenstechnischer Versuchsanlagen – was ist erreichbar?
Chem.-Ing.-Tech. 62(1990)169–174.

Maier 1999	W.F. Maier *Kombinatorische Chemie – Herausforderung und Chance f ür die Entwicklung neuer Katalysatoren und Materialien* Angew. Chemie 111(1999)1294–1296.
Maier 2000	W. F. Maier, G. Kirsten, M. Orschel, P. A. Weiss *Combinatorial approaches to catalysts and catalysis* Chimica Oggi/chemistry today (2000)15–20.
Marchetti 1982	C. Marchetti *Die magische Entwicklungskurve* Bild der Wissenschaft 10(1982)115–128.
Marquardt 1992	W. Marquardt *Rechnergestützte Erstellung verfahrenstechnischer Prozeßmodelle* Chem.-Ing.-Tech. 64(1992)25–40.
Marquardt 1999	W. Marquardt *Von der Prozeßsimulation zur Lebenszyklusmodellierung* Chem.-Ing.-Techn. 71(1999)1119–1137.
Mars 1954	P. Mars, D.W. van Krevelen *Oxidations carried out by means of vanadium oxide catalysts* Special Supp. Chem. Eng. Sci. 3(1954)41.
Marshall 1981	W. L. Marshall, E. U. Frank *Ion Product of Water Substance* J. Phys. Chem. Ref. Data 10(1981)295–304.
Martino 2000	G. Martino *Catalysis for oil refining and petrochemistry, recent developments and future trends* Studies in Surface Science and Catalysis Vol. 130A, 12th International Congress on Catalysis, p. 83–103, Elsevier, 2000.
Masso 1969	A. H. Masso, D. F. Rudd *The Synthesis of System Designs* AIChE Journal 15(1969)10–17.
McCabe 1925	W. L. McCabe, E. W. Thiele *Graphical Design of Fractionating Columns* Ind. Eng. Chem. 17(1925)605–611.
Melzer 1980	W. Melzer; D. Jaenicke *Ullmanns Enzyklopädie der technischen Chemie* Verlag Chemie, Weinheim 1980, Band 5, S. 891f.
Mendoza 1998	V. A. Mendoza, V. G. Smolensky, J. F. Straitz *Do your flame arrestors provide adequate protection?* Hydrocarbon Processing Oct.(1998)63–69.
Menzinger 1969	M. Mezinger, R. L. Wolfgang *Bedeutung und Anwendung der Arrhenius-Aktivierungsenergie* Angew. Chem. 81(1969)446–452.

Menzler 1994	M. Menzler, W. Göpel *Oberflächenphysik des Festkörpers* Teubner, 2. Auflage, Stuttgart, 1994.
Mersmann 2000	A. Mersmann, K. Bartosch, B. Braun, A. Eble, C. Heyer *Möglichkeiten einer vorhersagenden Abschätzung der Kristallisationskinetik* Chem.-Ing.-Tech. 72(2000)17–30.
Mersmann 2000a	A. Mersmann, M. Löffelmann *Crystallization and Precipitation: The Optimal Supersaturation* Chem. Eng. Technol. 23(2000)11–15.
Messer 1998	A, Messer, V. Münch *Hochwertiges Werkzeug für die Verfahrenssimulation* Chem.-Ing.-Tech. 70(1998)29.
Messtechnik 1998	*Kleine Ströme kein Problem* Chemie Technik 27(1998)22–24.
Meßumformer 1999	*Marktübersicht Meßumformer* CITplus, Nr. 3, 2(1999)34.
Metivier 2000	P. Metivier *Catalysis for fine chemicals: An industrial perspective* Studies in Surface Science and Catalysis Vol. 130A, 12th International Congress on Catalysis, p. 167–176, Elsevier, 2000.
Meyer-Galow 2000	E. Meyer-Galow *Wo liegt die Zukunft der chemischen Industrie?* Nachrichten aus der Chemie 48(2000)1234–1238.
Miller 1965	C. A. Miller *New Cost Factors – Give Quick, Accurate Estimates* Chemical Engineering 13(1965)226–239.
Miller 1997	J. A. Miller *Basic research at DuPont* CHEMTECH, April (1997)12–16.
Misono 1999	Misono *New catalytic technologies in Japan* Catalysis Today 51(1999)369–375.
Mittelstraß 1994	J. Mittelstraß *Grundlagen und Anwendung – Über das schwierige Verhältnis zwischen Forschung, Entwicklung und Politik* Chem.-Ing.-Tech. 66(1994)309–315.
Molzahn 2001	M. Molzahn, K. Wittstock *Verfahrensingenieure für das 21. Jahrhundert* Chem.-Ing.-Tech. 73(2001)797–801.

Moro 1993

Y. Moro-oka
New Aspects of Spillover-Effects in Catalysis
Elsevier Science Publishers, 1993.

Mosberger 1992

E. Mosberger
Jahrbuch der Chemiewirtschaft 1992/93
Führungskonzept mit zahlreichen Varianten
Verlagsgruppe Handelsblatt GmbH, Düsseldorf, S. 200–205.

Müller – Erlwein 1998

E. Müller – Erlwein
Chemische Reaktionstechnik
B.G. Teubner, Stuttgart, 1998.

Münch 1992

V. Münch
Patentbegriffe von A bis Z
VCH, Weinheim, 1992.

Muthmann 1984

E. Muthmann
Ermittlung der Investitionskosten von Chemieanlagen in verschiedenen Projektphasen
Chem.-Ing.-Tech. 56(1984)940–941.

Narin 1993

F. Narin, V. M. Smith, Jr., M. B. Albert
What patents tell you about your competition
CHEMTECH, Feb. (1993)52–59.

Niemantsverdriet 2000

J. W. Niemantsverdriet
Spectroscopy in Catalysis
2nd ed., Wiley-VCH, Weinheim, 2000.

Notari 1991

B. Notari
C-1 Chemistry, a critical review
Catalytic Sci. and Technology 1(1991)55–76.

Ochi 1982

K. Ochi, S. Hiraba, K. Kojima
Prediction of solid-liquid equilibria using ASOG
J. Chem. Eng. Japan, 12(1982)59–61.

OECD 1981

Modifizierter Zahn-Wellens Test
OECD Guidelines for Testing of Chemicals, Paris 1981.

Oh 1998

M. Oh, I. Moon
Framework of Dynamik Simulation for Complex Chemical Processes
Korean J. Chem. Eng. 15(3)(1998)231–242.

Onken 1997

U. Onken
The Development of Chemical Engineering in German Industry and Universities
Chem. Eng. Technol. 20(1997)71–75.

Ostrovskii 1997

N. M. Ostrovskii
Probleme in Erprobung und Vorhersage der Einsatzdauer von Katalysatoren
Khim. prom 6(1997)61–72.

Ozero 1984	B. J. Ozero, R. Landau Encyclopedia of Chemical Processing and Design, 1st ed, Vol. 20, Marcel Dekker, New York, 1984, S. 274–318.
Ozonia	*Die großtechnische Erzeugung und Anwendung von Ozon* Firmenschrift der Ozonia Switzerland
Palluzi 1991	R. P. Palluzi *Understand Your Pilot-Plant Options* Chemical Engineering Progress, Jan. (1991)21–26.
Pasquon 1994	I. Pasquon *What are the important trends in catalysis for the future?* Applied Catalysis A: General 113(1994)193–198.
Perego 1997	C. Perego, P. Villa *Catalyst preparation methods* Catalysis Today 34(1997)281–305.
Perego 1999	C. Perego, S. Peratello *Experimental methode in catalytic kinetics* Catalysis Today 52(1999)133–145.
Perlitz 1985	M. Perlitz, H. Löbler *Brauchen Unternehmen zum Innovieren Krisen?* Zeitschrift für Betriebswirtschaft 55(1985)424–450.
Perlitz 2000	M. Perlitz *Ein europäisches 21. Jahrhundert?* Elektronik (2000)70–75.
Pernicone 1997	N. Pernicone *Scale-up of catalyst production* Catalysis Today 34(1997)535–547.
Peterssen 1965	E. E. Peterssen *Chemical Reactor Analysis* Prentice – Hall, Englewood Chiffs, 1965.
Petrochemie 1990	Die Petrochemie Erdöl und Kohle-Erdgas-Petrochemie 43(1990)375–376.
Petzny 1997	W. Petzny, G. Rossmanith *Petrochemie und nachwachsende Rohstoffe* Erdöl Erdgas Kohle 113(1997)294–296.
Petzny 1999	W. J. Petzny, K. J. Mainusch *Neue Wege zu Petrochemikalien* Erdöl Erdgas Kohle 115(1999)597–602.
Pilz 1985	V. Pilz *Sicherheitsanalysen zur systematischen Überprüfung von Verfahren und Anlagen – Methoden, Nutzen und Grenzen* Chem.-Ing.-Tech. 57(1985)289–307.

Pinna 1998
F. Pinna
Supported Metal Catalysts Preparation
Catalysis Today 41(1998)129–137.

Platz 1986
J. Platz,. H. J. Schmelzer
Projektmanagement in der industriellen Forschung und Entwicklung
Springer-Verlag, Berlin, 1986.

Platzer 1996a
B. Platzer
Massen- und ortsfeste Betrachtungsweisen bei der Reaktormodellierung
Chem.-Ing.-Tech. 68(1996)769–781.

Platzer 1996b
B. Platzer
Basismodelle für Reaktoren – eine kritische Bilanz
Chem.-Ing.-Tech. 68(1996)1395–1403.

Platzer 1999
B. Platzer, K. Steffani, S. Grobe
Möglichkeiten zur Vorausberechnung von Verweilzeitverteilungen
Chem.-Ing.-Techn. 71(1999)795–807.

Plotkin 1999
J. S. Plotkin, A. B. Swanson
New technologies key to revamping petrochemicals
Oil&gas Journal 13(1999)108–116.

Poe 1999
C. Poe, K. Bonnell
Selecting Low-flow Pumps
Chemical Engineering, Feb. (1999)78–82.

Pohle
H. Pohle
Chemische Industrie – Umweltschutz, Arbeitsschutz, Anlagensicherheit
VCH, Weinheim, 1991.

Poling 2001
B. E. Poling, J. M. Prausnitz, J. P. O' Connell
The properties of gases and liquids
McGraw-Hill, New York, 5ed, 2001.

Popiel 1998
C.O. Popiel, J. Wojtkowiak
Simple Formulas for thermophysical Properties of Liquid Water for Heat Transfer Calculations
heat transfer engineering 19(1998)87–101.

Porter 1980
M. E. Porter
Competitive Strategy
The Free Press, New York, 1980.

Porter 1985
M. E. Porter
Technology and Competitive Advantage
J. Business Strategy (1985)60–89.

Prausnitz 1969
J. M. Prausnitz
Molecular Thermodynamics of Fluid-Phase Equilibria
Prentice-Hall International Series in the Physical and Chemical Engineering Sciences, 1969.

Prinzing 1985	P. Prinzing, R. Rödl, D. Aichert *Investitionskosten-Schätzung für Chemieanlagen* Chem.-Ing.-Tech. 57(1985)8–14.
Puigjaner 1991	L. Puigjaner, A. Espuna *Computer-Oriented Process Engineering* Elsevier, Amsterdam, 1991.
Qi 2001	M. Qi, M. Lorenz, A. Vogelpohl *Mathematische Lösung des Zweidimensionalen Dispersionsmodells* Chem.-Ing.-Tech. 73(2001)1435–1439.
Quadbeck 1990	H. J. Quadbeck-Seeger *Chemie für die Zukunft – Standortbestimmung und Persepektiven* Angew. Chem. 102(1990)1213–1224.
Quadbeck 1997	H.-J. Quadbeck-Seeger *Chemie Rekorde* Wiley-VCH, Weinheim,1997.
Raichle 2001	Raichle, Y. Traa, J. Weitkamp *Aromaten: Von wertvollen Basischemikalien zu Überschusskomponenten* Chem.-Ing.-Tech. 73(2001)947–955.
Rauch 2003	J. Rauch *Multipurpose Plants* Wiley-VCH, Weinheim, 2003.
Reher 1991	P. Reher *Jahrbuch der Chemiewirtschaft 1991* *Produktion mit umweltgerechter Entsorgung* VCH, Weinheim, S. 213–217.
Reichel 1995	H.-R. Reichel *Gebrauchsmuster- und Patentrecht – praxisnah* Kontakt & Studium, 3.Aufl., Band 278, expert verlag, Renningen – Malmsheim, 1995.
Reisener 2000	G. Reisener, M. Schreiber, R. Adler *Korrektur reaktionskinetischer Daten des realen Differential-Kreislaufreaktors* Chem.-Ing.-Tech. 72(2000)1192–1195.
Robbins 1979	L. A. Robbins *The Miniplant Concept* CEP, Sep. (1979)45–47.
Romanow 1999	S. Romanow *Catalysts – They just keep things reacting* Hydrocarbon Processing, July (1999)15.
Roth 1991	L. Roth, G. Rupp *Sicherheitsdatenblätter* VCH, Weinheim, 1991.

Rothert 1992	A. Rothert *Jahrbuch der Chemiewirtschaft 1992* *Die Grenzen des Machbaren* VCH, Weinheim, S. 28.
Sandler 1999	S. I. Sandler *Chemical and Engineering Thermodynamics* John Wiley&Sons, Inc., New York, 1999.
Santacesaria 1997	E. Santacesario *Kinetics and transport phenomena* Catalysis Today 34(1997)393–400.
Santacesaria 1997a	E. Santacesario *Kinetics and transport phenomena in heterogeneous gas-solid and gas-liquid systems* Catalysis Today 34(1997)411–420.
Santacesaria 1999	E. Santacesario *Fundamental chemical kinetics: the first step to reaction modelling and reaction enigneering* Catalysis Today 52(1999)113–123.
Sapre 1995	A. V. Sapre, J. R. Katzer *Core of Chemical Reaction Engineering: One Industrial View* Ind. Eng. Chem. Res. 34(1995)2202–2225.
Sattler 2001	K. Sattler *Thermische Trennverfahren* 3. Aufl., Wiley-VCH, Weinheim, 2001.
Sattler 2000	K. Sattler, W. Kasper *Verfahrenstechnische Anlagen* *Planung, Bau und Betrieb* Wiley-VCH, Weinheim, Band 1 und 2, 2000.
Scharfe 1991	G. Scharfe, B. Sewekow *Produktionsintegrierter Umweltschutz: Das Beispiel Bayer* Chemische Industrie (1991)17–20.
Scheiding 1989	W. Scheiding, H. Hartmann, P. Tönnishoff *SATU86 – Simulationsrahmen zur dynamischen Prozeßsimulation für den industriellen Einsatz* Chem.-Ing.-Tech. 61(1989)292–299.
Schembecker 1996	G. Schembecker, K. H. Simmrock *Alternativen schnell und zuverlässig* Standort Spezial 20(1996)18.
Schembra 1993	M. Schembra; J. Schulze *Schätzung der Investitionskosten bei der Prozeßentwicklung* Chem.-Ing.-Tech. 65(1993)41–47.

Schierbaum 1997	B. Schierbaum, D. Rosskopp, D. Bröll, H. Vogel *Integriertes Verfahren zur Reinigung von carbonsäurehaltigen Prozeßabwässern* Chem.-Ing.-Tech. 69(1997)519–523.
Schlicksupp 1977	H. Schlicksupp *Kreative Ideenfindung in der Unternehmung* Walter de Gruyter, Berlin, 1977.
Schlögl 1998	R. Schlögl *Des Kaisers neue Kleider* Angew. Chem. 110(1998)2467–2470.
Schneider 1965	K. W. Schneider *Großanlagen in der Mineralölindustrie und Petrochemie* Chem.-Ing.-Tech. 37(1965)875–879.
Schneider 1998	D. F. Schneider *Build a Better Process Model* Chemical Engineering Progress, April (1998)75–85.
Schönbucher 2002	A. Schönbucher *Thermische Verfahrenstechnik* Springer Verlag, Berlin, 2002.
Scholl 1995	S. Scholl, A. Polt, S. Krüger *Ausgewählte Problempunkte bei Simulation und Gestaltung rektifikativer Trennprozesse* Chem.-Ing.-Tech. 67(1995)166–170.
Scholl 1996	S. Scholl *Zur Bestimmung von Verschmutzungswiderständen industrieller Wärmeträger* Chem.-Ing. Tech. 68(1996)124–127.
Scholl 1997	A. Scholl *Eine Maschine, sieben Funktionen* *Dauerbrenner Flüssigkeitsringverdichter* Chemie Technik 26(1997)48–62.
Schuch 1992	G. Schuch, J. König *Jahrbuch der Chemiewirtschaft 1992* *Mehrproduktanlagen – Stetiger Wandel aus Erfahrung* VCH, Weinheim, 1993, S. 195–206.
Schuler 1995	H. Schuler *Prozeßsimulation* VHC, Weinheim 1995.
Schulz 1990	A. Schulz – Walz *Meß- und Dosiertechnik beim Betrieb von Miniplants* Chem.-Ing.-Tech. 62(1990)453–457.

9 References

Schwenk 2000	E. F. Schwenk *Sternstunden der frühen Chemie* Verlag C. H. Beck, München, 2. Auflage, 2000.
Seider 1999	W. D. Seider, J. D. Seader, D. R. Lewin *Process Design Principles* John Wiley&Sons, Inc., New York, 1999.
Semel 1997	R. Semel *Verfahrensbearbeitung heute* Nachr. Chem. Tech. Lab 45(1997)601–607.
Senkan 1998	S. M. Senkan *Hight-throughput screening of solid-state catalyst libraries* Nature 394(1998)350–353.
Senkan 1999	S. M. Senkan, S. Ozturk *Discovery and Optimization of Heterogeneous Caalysts by Using Combinatorial Chemistry* Angew. Chem. Int. Ed. 38(1999)791–795.
Senkan 2001	S. Senkan *Kombinatorische heterogene Katalyse* Angew. Chemie 113(2001)322–341.
Shinnar 1991	R. Shinnar *The future of the chemical industries* CHEMTECH, Jan. (1991)58–64.
Siegert 1999	M. Siegert, J. Stichlmair, J.-U. Repke, G. Wozny *Dreiphasenrektifikation in Packungskolonnen* Chem.-Ing.-Tech. 71(1999)819–823.
Solinas 1997	M.A. Soloinas *Methodologies for economic and financial analysis* Catalysis Today 34(1997)469–481.
Sowa 1997	Ch. J. Sowa *Process Development: A Better Way* Chemical Engineering Progress, Feb. (1997)109–112.
Sowell 1998	R. Sowell *Why a simulation system doesn't match the plant* Hydrocarbon Processing, March (1998)102–107.
Specht 1988	G. Specht, K. Michel *Integrierte Technologie- und Marktplanung mit Innovationsportfolios* Zeitschrift für Betriebswirtschaft 58(1988)502–520.
Splanemann 2001	R. Splanemann *Production Simulation – A Strategic Tool to Enable Efficient Production Processes* Chem. Eng. Technol. 24(2001)571–573.

Srinivasan 1997	R. Srinivasan et al. *Micromachined Reactors for Catalytic Partial Oxidation Reactions* AIChE Journal 43(1997)3059.
Standort 1995 Felcht	U.-H. Felcht *Millionenbeträge einsparen* Standort spezial 22(1995)17.
Steele 1984	G. L. Steele jr. *Common LISP: The Language* Digital Press, 1984.
Steinbach 1999	A. Steinbach *Verfahrensentwicklung und Produktion* CITplus, Heft 3, 2(1999)28–29.
Steiner 1967	R. Steiner, G. Luft *Die simultane Abhängigkeit der Reaktionsgeschwindigkeit von Druck und Temperatur* Chem. Engng. Sci. 22(1967)119–126.
Stephanopoulos 1976	G. Stephanopolous, A.W. Westerberg *Evolutionary Synthesis of Optimal Process Flowsheets* Chemical Engineering Science 31(1976)195–204.
Steude 1997	H. E. Steude, L. Deibele, J. Schröter *MINIPLANT-Technik – ausgewählte Aspekte der apparativen Gestaltung* Chem.-Ing.-Tech. 69(1997)623–631.
Stewart 1993	I. Stewart *Spielt Gott Roulette? Uhrwerk oder Chaos?* Insel Verlag, Frankfurt, Leipzig, 1993.
Streit 1991	B. Streit *Lexikon Ökotoxikologie* VCH, Weinheim, 1991.
Strube 1998	J. Strube, H. Schmidt-Traube, M. Schulte *Auslegung; Betrieb und ökonomische Betrachtung chromatographischer Trennprozesse* Chem.-Ing.-Tech. 70(1998)1271–1279.
Strube 1998	J. Strube, H. Schmidt-Traub, M. Schulte *Auslegung, Betrieb und ökonomische Betrachtung chromatographischer Trennprozesse* Chem.-Ing.-Tech. 70(1998)1271–1270.
Sulzer	Sulzer Chemtech AG *Fraktionierte Kristallisation* Industriestr. 8 CH-9471 Buchs/ Switzerland

9 References

Sulzer 1987
: Sulzer
Kreiselpumpen Handbuch
Gebrüder Sulzer AG, Winterthur, Schweiz, 2. Auflage, 1987.

Swodenk 1984
: W. Swodenk
Umweltfreundlichere Produktionsverfahren in der chemischen Industrie
Chem.-Ing.-Tech. 56(1984)1–8.

Szöllosi 1998
: M. Szöllözi-Janze
Fritz Haber 1868–1934
Verlag C.H. Beck, München, 1998, S. 287.

Teifke 1997
: J. Teifke
Die Flüssigkeitsring-Vakuumpumpe als verfahrenstechnische Maschine
Vakuum in Forschung und Praxis 2(1997)85–92.

Thiele 1939
: E. W. Thiele
Relation between Catalytic Activity and Size of Particle
Ind. Eng. Chem. 31(1939)916–920.

Thomas 1994
: J. M. Thomas
Wendepunkt der Katalyse
Angew. Chem. 106(1994)963–989.

Tröster 1985
: E. Tröster
Sicherheitsbetrachtungen bei der Planung von Chemieanlagen
Chem.-Ing.-Tech. 57(1985)15–19.

Tostmann 2001
: K.-H. Tostmann
Korrosion – Ursachen und Vermeidung
Wiley-VCH, Weinheim, 2001.

Trotta 1997
: R. Trotta, I. Miracca
Approach to the industrial process
Catalysis Today 34(1997)429–446.

Trum 1986
: P. Trum, N. R. Iudica
Expertensysteme – Aufbau und Anwendungsgebiete
Chem.-Ing.-Tech. 58(1986)6–9.

Uhlemann 1996
: J. Uhlemann, V. Garcia, M. Cabassud, G. Casamatta
Optimierungsstrategien für Batchreaktoren
Chem.-Ing.-Tech. 68(1996)917–926.

Ullmann's Encyclopedia
: Ullmann's Encyclopedia of Industrial Chemistry
6th ed., Wiley-VCH, Weinheim, 2002.

Ullmann's 1 Ohara 1985
: T. Ohara
Acrylic Acid and Derivatives
Ullmann's, 5th ed., Vol. A1, 1985, S. 161ff.

Ullmann's 2 Henkel 1992
: K.-D. Henkel
Reatortypes and their Industrial Applications
Ullmann's, 5th ed., Vol. B4; 1992; S. 87ff.

Van Heek 1999	K.H. Van Heek *Entwicklung der Kohlenwissenschaft im 20. Jahrhundert (1)* Erdöl Erdgas Kohle 115(1999)546–551.
VCI 1995	Verband der Chemischen Industrie *Jahresbericht 1994/95* Frankfurt, 1995.
VCI 2000	*Jahresbericht* Verband der Chemischen Industrie e. V., 30. Juni 2001, 60329 Frankfurt.
VCI 2001	*Chemiewirtschaft in Zahlen 2001* Verband der Chemischen Industrie e. V., Karlstr. 21, 60329 Frankfurt. www.chemische-industrie.de
VDI-Wärmeatlas	*VDI-Wärmeatlas- Berechnungsblätter für den Wärmeübergang* Herausgeber: Verein Deutscher Ingenieure, VDI-Gesellschaft Verfahrenstechnik und Chemieingenieurwesen (GVC) VDI Verlag, 6te Auflage, Düsseldorf, 1991.
Villermaux 1995	J. Villermaux *Future Challenges in Chemical Engineering Research* Tans IChemE 73(1995)105–109.
Vogel 1981a	H. Vogel, A. Weiss *Transport Properties of Liquids.* *I. Self-Diffusion, Viscosity and Density of Nearly Spherical and Disk Like Molecules in the Pure Liquid Phase* Ber. Bunsenges. Phys. Chem. 85(1981)539–548.
Vogel 1981b	H. Vogel, A. Weiss *Transport Properties of Liquids.* *II. Self-Diffusion of Almost Spherical Molecules in Athermal Liquid Mixtures* Ber. Bunsenges. Phys. Chem. 85(1981)1022–1026.
Vogel 1982	H. Vogel *Transporteigenschaften reiner Flüssigkeiten und binärer Mischungen* Doktorarbeit Darmstadt 1982, D17.
Vogel 1997	H. Vogel *Entwicklung von chemischen Prozessen: Stand der Technik und Trends in der Forschung* Technische Hochschule Darmstadt thema Forschung 1(1997104–106).
Vogel 1999	H. Vogel et al. *Chemie in überkritischem Wasser* Angew. Chem. 111(1999)3180–3196.

Vogel 2001
H. Vogel
rational catalyst design für Acrylsäure
CHEManager, Polymer Forschung Darmstadt,
Deutsches Kunststoff-Institut
GIT Verlag, Darmstadt, 2001, S. 34–35.

Volmer 1931
M. Volmer, W. Schultz
Kondensation an Kristallen
Zeitschrift für Phys. Chem. 156(1931)1–22.

Vvedensky 1994
D. Vvedensky
Partial Differential Equations with Mathematica
Addison-Wesley Publishing Company, Wohingham, England, 1994.

Wagemann 1997
K. Wagemann
Wie neue Technologien entstehen
Nachr. Chem. Tech. Lab. 45(1997)273–275.

Wagener 1972
D. Wagener
Kostenermittlung im Anlagenbau für die chemische Industrie
Technische Mitteilungen 65(1972)542–547.

Walter 1997
D. W. Walter
The future of patent searching
CHEMTECH 27(1997)16–20.

Wang 1999
S. Wang, H. Hofmann
Strategies and methods for the investigation of chemical reaction kinetics
Chem. Eng. Sci. 54(1999)1639–1647.

Watzenberger 1999
O. Watzenberger, E. Ströfer, A. Anderlohr
Instationär-oxidative Dehydrierung von Ethylbenzol zu Styrol
Chem.-Ing.-Tech. 71(1999)150–152.

Weber 1996
K. H. Weber
Inbetriebnahme verfahrenstechnischer Anlagen
VDI Verlag, Düsseldorf, 1996.

Wedler 1997
G. Wedler
Lehrbuch der Physikalischen Chemie
4. Aufl.; VCH; Weinheim; 1997.

Weierstraß 1999
Weierstraß
Dynamische Prozeßsimulation
Chem.-Ing.-Techn. 71(1999)1118.

Weissermel 2003
K. Weissermel, H.-J. Arpe
Industrial Organic chemistry
4th Aufl., Wiley-VCH, Weinheim, 2003.

Weisz 1954	P. B. Weisz, D. Prater *Interpretation of Measurements in Experimental Catalysis* Adv. Catal. 6(1954)143–196.
Wengerowski 2001	J. Wengerowski *Anlagenbau – von der Idee bis zur Inbetriebnahme* Bioworld 1(2001)12–13.
Westerterp 1992	K. R. Westerterp *Multifunctional Reaktors* Chemical Engineering Science 47(1992)2195–2206.
Wiesner 1990	J. Wiesner *Produktionsintegrierter Umweltschutz in der chemischen Industrie* DECHEMA, Frankfurt, 1990.
Wijngaarden 1998	R.J. Wijngaarden, A. Kronberg, K.R. Westerterp *Industrial Catalysis* Wiley-VCH, Weinheim, 1998.
Willers 2000	Y. Willers, U. Jung *Gibt es das überhaupt: „Spezialchemikalien"?* Nachrichten aus der Chemie 48(2000)1374–1377.
Wilson 1971	G. T. Wilson *Capital investment for chemical plant* Brit. Chem. Eng. Proc. Tech. 16(1971)931.
Wirth 1994	W.-D. Wirth *Patente lesen, nicht nur zählen* Nachr. Chem. Tech. Lab. 42(1994)884.
Wörsdörfer 1991	U. Wörsdörfer, U. Müller-Nehler *Jahrbuch der Chemiewirtschaft 1991* *Ein Gewinn für die Forschung* VCH, Weinheim, S. 183–186.
Wörz 1995	O. Wörz *Process development via a miniplant* Chemical Engineering and Processing 34(1995)261–268.
Wörz 2000	O. Wörz, K.-P. Jäckel, T. Richter, A. Wolf *Mikroreaktoren – Ein neues Werkzeug für die Reaktorentwicklung* Chem.-Ing.-Techn. 72(2000)460–463.
Wörz 2000a	O. Wörz *Wozu Mikroreaktoren?* Chemie in unserer Zeit 34(2000)24–29.
Wozny 1991	G. Wozny *Jahrbuch der Chemiewirtschaft 1991* *Software macht den Fortschritt* VCH, Weinheim, 1992, S. 153–158.

Wozny 1991a   G. Wozny, L. Jeromin
Dynamische Prozeßsimulation in der industriellen Praxis
Chem.-Ing.-Techn. 63(1991)313–326.

Yau 1995   Te-Lin Yau, K. W. Bird
Manage Corrosion with Zirconium
Chem. Eng. Progress (1995)42–46.

Yaws 1988   C. Yaws, P. Chiang
Enthalpy of formation for 700 major organic compounds
Chemical Engineering (1988)81–88.

Yaws 1994   C. Yaws, X. Lin, L. Bu
Calculate Viscosities for 355 Liquids
Chemical Engineering April(1994)119–128.

Zeitz 1987   M. Zeitz
Simulationstechnik
Chem.-Ing.-Tech. 59(1987)464–469.

Zevnik 1963   F. C. Zevnik, R. L. Buchanan
Generalized correlation of proces investment
Chem. Eng. Progress 59(1963)70–77.

Zlokarnik 1989   M. Zlokarnik
Umweltschutz – eine ständige Herausforderung
Chem.-Ing.-Tech. 61(1989)378–385.

Zlokarnik 2000   M. Zlokarnik
Scale-up in Chemical Engineering
Wiley-VCH, Weinheim, 2000.

Subject Index

Subject Index

A

ABC analysis 11
absorption 136–143, 194
abstracts, chemical 268
acetone-water mixture 126
acid
– acrylic 7, 246
– corrosion 231
acrolein 254
acrylic acid 7, 246
acrylonitrile 77
active / activation / activity 16, 20
– coefficients, activity 98
– energy, activation 249
– substance, active 8
adiabatic reaction control 70
adsorption 38, 165–166, 196
– isotherm 40, 42
aging 23
air
– compressed air 189, 346
– properties of dry air 411
– waist-air purification 194
ammoxidation 77
amortization time 363
amplitude 152
analysis
– ABC analysis 11
– benefit, cost / benefit analysis 11
– eco-efficiency analysis 351
– energetic analysis 121
– organization 326
– sensitivity analysis 361–362
– surface analysis methods 422–424
– vector analysis 379
– wastewater (see there) 198–199
Antoine
– equation 96, 266
– parameters 264, 397–398
arbitrary functions 381
area 386
aromatics, prices 275
Arrhenius equation 248, 373
ASPEN 108, 289
athermal mixtures 99
atomic masses 382
autocatalytic reaction 251
automation of a miniplant 301
azeotropic distillation 134
azetropes 132

B

backmixing 58
balance
– line 103
– mass balance 261–263, 334
– of materials 299
basic products 5–6
batch distillation 111–115
belt filters 171
benefit, cost / benefit analysis 11
Bernoullis equation 173
Berzelius, JJ 16
BET model 40
binary mixtures 96, 390
binding energy 39
binodal curve 149
biocatalyst 17
biofilters 196
biowashers 196
BlmSchG 315
blower 185
BOD_5 (biological oxygen demand wastewater analysis) 198
Bodenstein number 64, 414
bond strengths 39
Bosch, C 18
brainstorming 12
Brunauer-Emmet-Teller isotherm 41
bubble-cap plate 127
bubble-through measurement 205
Buckingham Π theory 67

C

calculation / calculus
– additional costs, calculation method 337
– differential 376
– integral 376
– production, calculation of 339–350
caloric 388
capital-dependent costs (depreciation) 349
carbon
– DOC (dissolved organic carbon wastewater analysis) 199
– steel 234
– TOC (total organic carbon wastewater analysis) 199
Cartesian coordinate 54
cascade 60
– control 217–218
cash
– cow 272
– flow 362–363
catalyst 16–51
– biocatalyst 17
– combustion, catalytic 193
– external catalyst efficiency 34
– heterogenous catalysis 17

– – heterogeneously catalyzed reaction 248
– homogenous catalysis 17
– mixed-oxide 26
– performance 20–28
– rational catalyst design 19
– sphere 47
catcracking 274
Celanese 3
celcius scale 202
cell model 66
centrifuges 171
– pumps, centrifugal 179
characteristic numbers, dimensionless 414–415
checking the individual steps 294–295
chemical(s)
– abstracts 268
– basic 275, 340
– – prices 275, 340
– companies, market capital 3
– composition 28
– data 239–261
– equilibrium 240
– fine chemicals 7
– industry 3
– plant 15
– potential 144
– reaction technology 51–79
– specialty 8
chemisorption 165–166
chemistry
– combinatorial 19
– petrochemistry 5, 273
– *Reppe* chemistry 6
chromatographic separation 169
Ciba SC 3
Clariant 3
clarification plant 197–199
Clausius-Clapeyron equation 96, 266
COD (chemical oxygen demand wastewater analysis) 198
coking 22, 274
– delayed 274
– flexicoking 274
– fluid 274
collision diameters 36
column diameter 106
combinatorial chemistry 19
combustion 192–195
– catalytic 193
– plants 192–195
– thermal 192
– – heat of 239
commissioning 326

companies 420–422
– market capital, chemical companies 3
compensation time 213
competitors 268
composite curve 342
– cold 342
– hot 342
compressed air 189, 346
compressors 172, 184–185
– isothermal compression 185
– polytropic compression 185
concentrations 390
condensers 92
conditions, supercritical 69
conduction, thermal 81
conductive term / conductivity 52, 208
conjunction 219
conservation
– of enthalpy 52
– of impetus 53
– of mass 52
construction
– of models 323–325
– pole point construction 147
– staircase construction 105
contact dryer 169
continuity equation 52
control technology 211–220
– cascade control 217–218
– circuit control 211, 213
– closed-loop control 211–218
– open-loop control 218–220
– process control system 203
– ratio control 218
– reaction control (*see there*) 70–71
controller
– *D* controller 214
– *I* controller 213
– output 211
– *P* controller 213
– parameters 216
– *PI* controller 214
– *PID* controller 214
– position controller 211
convection 82
– dryer 169
convective term 52
conversion 257
– of various units to SI units 385
cooling water 188–189, 345
copper 226
Coriolis flow meter 209
corrosion
– acid 231

Subject Index | 469

– behavior 229
– chemical corrosion test 231
– oxygen 231
– protection 232
costs / prices 3–4, 270–272, 275, 336–350
– additional costs 337
– aromatics 275
– basic chemicals 275, 340
– capital-dependent costs (depreciation) 349
– cost / benefit analysis 11
– development 270–271
– energy sources 275, 341–346
– feedstock cost 340–341
– fixed costs 4
– infrastructure costs 339
– intermediates, prices 340
– ISBL investment costs 336–339
– maintenance costs 349
– market price 272, 275
– olefines 275
– OSBL investment costs 339
– overheads 349
– price indices 336
– for production 3–4, 11, 25–27, 339–350
– raw materials 275, 340
– staff costs 348
– variable costs 4
– waste-disposal 347–348
countercurrent extraction 150
CPU 220
cracking 274
– catcracking 274
– hydrocracking 274
creative thinking 12
critical
– data 242
– values 391
crystallization 155–165
– film 162
– – falling-film crystallization 162
– melt 161–162
– mixed-crystal formation 158
– solution 159–161
– suspension 162
cubic state equation 95
cumulative residence-time curve 61
cylindrical coordinates 377
cylindrical
– coordinates 54
– single pore 46

D

Damköhler number 57, 414
data

– banks 287
– chemical 239–261
– critical 247
– ecotoxicological 264
– kinetic 248
– of mixtures 265–266
– physiochemical 263–266
– safety 264
– study reports 331–357
– technical data sheets 323
– thermodynamic 393
deactivation functions 24
decomposition temperature 223
degression
– coefficient 336
– method 336
delay
– coking, delayed 274
– time 213
demand 272
density 264, 386
depreciation (capital-dependent costs) 349
design
– basic design phase 308, 318
– of reactors 74–79
desorption 38, 136, 142–143
development costs 270–271
dielectric constant 410
differential
– calculus 376
– partial differential equations 380
– reaction differential circulation 247
– selectivity 257, 259
diffusion
– coefficient 37
– – self-diffusion coefficient 407
– equation 381
– film diffusion 32–34, 50
– in free space 35–37
– *Knudson* diffusion 35, 37
– *Poiseuille* diffusion 35
– pore diffusion 35, 37, 45
– surface diffusion 35, 38
dimensionless characteristic numbers
      414–415
diphyl 92
disjunction 219
dispersion model 63
distillation 111–120, 132–134
– azeotropic 134
– batch 111–115, 117
– column 119–120
– continuous 115
– extractive 133

470 | Subject Index

– heavy vacuum distillate 274
– steam 142
– two-pressure 132
– vapor-entrainment 136, 142–143
distribution function 61
divergence 379
dividing walls 119
Döbereiner, JW 16
DOC (dissolved organic carbon wastewater analysis) 199
Dow / UCC 3
drag coefficient 177
dry air, properties of 189, 346
drying 167–169
– contact dryer 169
– convection dryer 169
dual-flow plates 128

E

earnings 350
eco-efficiency analysis 351
ecological fingerprint 351
economic risk 361–363
ecotoxicological data 264
Eddy current flow meter 209
ethyl acetate 251
elasticity, modulus of 229
elastomers 227
electrical energy 188, 343
– dielectric constant 410
electrodialysis 169
elements, list of 382
Eley-Rideal kinetics 43
enamel 227
end product 278
energy 388
– activation energy 249
– analysis, energetic 121
– binding energy 39
– electrical 188
– prices for energy sources 275, 341–346
– refrigeration energy 346
– supply 186–189
engineering phase
– basic 308, 311, 318
– detail 312, 318
enthalpy
– conservation of 52
– of evaporation 186
– of formation 239
– reaction of 241–242
– – free enthalpy of 242
– – *Gibbs* standard enthalpy of formation 242
– – *Gibbs* standard reaction enthalpy 242

– temperature-enthalpy diagram 342
entropy, standard 243
EOX (extractable organic halogens wastewater analysis) 199
equilibrium
– chemicals 240
– constant K 242
– phase 55, 265
– thermodynamic 240–245
– vapor-liquid 266
errors / error function 65, 284
ethylene oxide 7, 246
EU law 416
Eulers equation 173
eutetic system 158
evaluation
– preliminary study 283
– technology evaluation 350–352
evaporator / evaporation 90–91, 99, 186, 264–266
– enthalpy of 186
– falling-film 90
– flash 99
– heat of 264, 266, 344
– helical-tube 91
– thin-layer 91
experimental work, assessment of 357
expert systems 292
explosion
– combinatorial 286
– diagram 221
– limits 265
extraction 143–155
– column 156
– countercurrent 150
– one-stage 150

F

factor
– F factor 176
– labyrinth factor 37
– loading factor 176
Fahrenheit scale 202
falling-film
– crystallization 162
– evaporator 90
feed ratio 103
feedback 211
feedstock cost 340–341
Fenskes plate number 108
ferrous metals 226
Ficks law 35
– first law 35
– second law 35

field level 203
film
– crystallization 162
– – falling-film 162
– diffusion 32–34, 50
filters
– belt filters 171
– biofilters 196
filtration 170–172
fine chemicals 7
fingerprint, ecological 351
firms
Aventis 3
fixed costs 4
flammable liquids 222
flares 192
flash
– evaporation 99
– point 222
floating-body flow meter 208
flow
– cash flow 362–363
– diagram, basic 332
– dual-flow plate 128
– measurement 206–210
– – *Coriolis* flow meter 209
– – *Eddy* current flow meter 209
– – floating-body flow meter 208
– – magnetic-induction flow meter 208
– – thermal flow meter 209
– – ultrasonic flow meter 210
– P & I flow studies 321–322
– *Poiseuille* flow 38
– process flow diagram 333
– profile 175
– – laminar 175
– – plug 175–176
– – turbulent 175
– velocity 177
– waste-disposal flow diagram 344
fluid
– coking 274
– dynamics 172
– phases, special processes for 169
fluidized-bed reactor 77
force 387
formaldehyde 241
formation
– enthalpie of 239
– *Gibbs* standard enthalpy of formation 242
– mixed-crystal 158
– seed 158
fouling 22, 90
– factor 356

Fouriers first law 81
free
– path 36
– space 35
– – diffusion in free space (*see also* diffusion) 35–37
frequency 152
fuels 345
fugacity 242
– coefficients 98
function plans 323–324
future 370

G

gas
– critical values for some gases 391
– kinetic gas theory 35
– natural gas 273, 369
– off-gas purification 194–196
– permeation 169
– phase 94
– real gas factor 412
– vacuum gas oil 274
gas-liquid equilibria 94
GHSV 21
Gibbs standard
– enthalpy of formation 242
– reaction enthalpy 242, 394
Gibbs-Thomson equation 156
Gilliland diagram 109
gradient 379
graphite 227
group-contribution methods 243

H

Haber-Bosch process 19
Hagen-Poiseuille law 175
hastelloy C 234
Hatta number 50
hazard potential 220
heat
– capacity 228, 243–244, 264, 389, 392
– of combustion 239
– of evaporation 264, 266, 344
– exchangers 84
– – indirect 93
– – plate 87
– – spiral 87
– – tube-bundle 87
– of melting 264
– of reaction 239–240
heat-transfer 80
– coefficient 82, 84, 389
helical-tube evaporator 91

Henrys
– constant 138
– law 137
Hess equation 239
heterogenous catalysis 17
heuristic rules 121
high-throughput screening 19
Hoechst 3
vant *Hoff* equation 244
Hofmann-Schoenemann method 67
homogenous catalysis 17
HTU-NTU method 141
hydrocracking 274
hydrodynamics 172
hyperbolic functions 375

I

ICI 3
ideas 12
ignition
– energy 222
– temperature 222
impregnation 25
incineration 277
– plants 348
indexing word / subject index 467–469
industrial research and development 3–4
infinite plate number 104
infrastructure 335
– costs 339
integral
– calculus 376
– reactor 247–248
– selectivity 257
intermediates 5–6
– prices 340
investment costs 336–339
– ISBL 335
– OSBL 339
– return on investment (*see there*) 358–361
ion / ionic
– exchange 167
– product, ionic 410
iron, soft 226
ISBL 335
– investment costs 336–339
isobars 96
isochore 96
isotherm 38–41, 96
– *Brunauer-Emmet-Teller* isotherm 41
– compression, isothermal 185
– *Langmuir* isotherm 39–40
– sorption isotherm 38
– – adsorption 40, 42

K

key technologies 18
Kickhoff meeting 313
kinetics 31–50, 245–256
– data 248
– *Eley-Rideal* kinetics 43
– gas theory 35
– *Langmuir-Hinshelwood* kinetics 42
– macrokinetics 32, 55
– *Mars-van Krevelen* kinetics 44
– microkinetics 31, 55
Knudson diffusion 35, 37
Krupp 234

L

labyrinth factor 37
Langmuir isotherm 39–40
Langmuir-Hinshelwood kinetics 42
Laplacian equation 381
law 416
LD_{50} 265
learning curve 9–10
length 385
Lever rule 149
LHSV 21
licensing (*see also* patenting situation) 267–270, 314–317
lifetime 22–24
logical elements 219
Linnhoff analysis 342
liquid
– distribution 130
– flammable 222
– gas (*see there*) 35, 94, 169, 194–195, 273
– gas-liquid equilibria 94
– oil (*see there*) 273–274
– pumps 179, 184
– residues 192
literature 427–464
loading 38, 146–147
– factor 176
location 271

M

machines 356
macrokinetics 32, 55
macropores 31
magnetic-induction flow meter 208
maintenance costs 349
malotherm 92
market
– capital, chemical companies 3
– growth 272
– price 272

– situation   272–273
mass transport   50
Mars-van Krevelen kinetics   44
mass   386
– atomic masses   382
– balance   261–263, 334
– conservation of mass   52
materials
– balance   299
– for chemical apparatus   226
– metallic   233
– raw (*see* raw materials)   5–6, 273–275
– selection   224–236
– support material (Tragermaterialien)   27
McCabe-Thiele
– diagram   98
– method   102, 146
measurement   202–210
– bubble-through measurement   205
– flow measurement (*see there*)   206–210
– measuring level   205–206
– point   216
– pressure measurement   205
– temperature measurement   202
mechanical processes   170–172
– stress, mechanical   387
– and thermal stability   228–229
melt crystallization   161–162
melting, heat of   264
membrane separation   169, 196
mesopores   31
methanol-water mixture   97
Michaelis-Menten mechanism   249
microkinetics   31, 55
microplant   296–297
micropores   31
microreactors   79
miniaturization, limits of   300–302
miniplant   281, 296–298
– automation of   301
– integrated   281
– technology   297–298
mistakes   284
Mitasch, A   18
mixed-crystal formation   158
mixed-oxide catalyst   26
mixer-settler stage   145
mixer-settlers   154
mixture
– acetone-water   126
– athermal   99
– binary   96, 390
– data of   265–266
– methanol-water   97

– regular   99
model
– BET model   40
– cell model   66
– construction of models   323–325
– dispersion model   63
– NRTL model   266
molar fraction   146
molecular velocity   36
Montz A3   128

N

Nabla operator   379
Naviers equation   172
negation   219
Nernst distribution law   145
nonisothermal reactors   68–74
Novartis   3
NPSH (net positive suction head)   181
NRTL
– equation   99
– model   266
Nusselt number   34, 85, 414

O

oil
– consumption   273
– vacuum gas oil   274
olefines, prices   275
one-stage extraction   150
on-stream time   24
orientational study   283
orifice plate   207
OSBL   335
– investment costs   339
osmosis, reverse   169
Ostwald, W   16
overall
– efficiencya   344
– factor   337
overhead costs   349
oxygen
– BOD$_5$ (biological oxygen demand wastewater analysis)   198
– COD (chemical oxygen demand wastewater analysis)   198
– corrosion   231
– propene-oxygen-nitrogen   221

P

P & I flow studies   321–322
packings   127
parallel reaction   255
particle size   30

Subject Index

patent / patenting and licensing situation 267–270, 314–317
– claims 269
– World patents index 268
Peer, M 18
pervaporation 169
petrochemistry 5, 273
petroleum 273–274
phase
– composition 29
– design phase, basic 308, 318
– diagram 95, 158
– engineering phase (*see there*) 308, 311, 318
– equilibrium 55, 265
phthalic acid 246
physiochemical data 263–266
Pictet-Troutman rule 266
pilot plant 281, 296, 302–303
pinch 342
pipelines 174–179
– pipe diameter 177
plant
– capacity 276–277
– chemical plant 15
– – clarification plant 197–199
– – combustion plant 192–195
– incineration plants 348
– microplant 296–297
– miniplant (*see there*) 281, 296–298, 301
– pilot plant 281, 296, 302–303
– production plant 281
– safety 220–223
– sewage treatment plant 277, 348
– trial plants 296
plastics 235
– thermoplastics 227
plate 104–108, 127
– bubble-cap 127
– dual-flow 128
– efficiency 105
– heat exchanger 87
– number
– – *Fenskes* 108
– – infinite 104
– – *Underwoods* 108
– orifice 207
– sieve 128
– spring-plate manometer 206
– valve 127
plug flow 175–175
Poiseuille
– diffusion 35
– flow 38
poisoning 22

pole point construction 147
polytropic reaction control 71
pore
– cylindrical single pore 46
– diffusion 35, 37, 45
– structure 30
– type 31
porosity 37
portfolio matrix 272
position controller 211
potential
– chemicals 144
– equation 381
– hazard 381
power 388
– law 249
Prandtl number 85, 414
precipitation 25
pre-exponential factor 249
preproject 308
pressure 387
– drop 78, 131, 177–178
– measurement 205
– transducer 206
– two-pressure distillation 132
price (*see* costs) 3–4, 11, 25–27, 270–272, 275, 336–339
process
– control system 203
– description 322
– development 11–12, 281–284
– flow diagram 333
– study 283
product
– basic products 5–6
– end product 278
– ionic 410
– processing
– – thermal 80–170
– – mechanical 170–172
– question mark products 272
– star products 272
production
– costs (*see there*) 3–4, 11, 25–27, 336–351
– plant 281
– processes 7
– structure 4–11
project
– execution 313
– preproject 308
– study 283
– team 367
promotors 29
propene 254

propene-oxygen-nitrogen 221
properties of water 399–401
pulsation 152–153
pumping station 180
pumps 179–185, 344
– centrifugal 179
– liquid 179, 184
– ring 183–184
– rotary 180
– vacuum 182–183

Q
quench 93
question mark products 272

R
Raoults law 98
ratio control 218
rational catalyst design 19
raw materials 5–6, 273–275
– fossil 6
– prices 275, 340
– renewable 6, 369
reaction
– autocatalytic 251
– control 70–71
– – adiabatic 70
– – polytropic 71
– differential circulating 247
– enthalpies 241, 394
– – free enthalpy of 242
– heat of reaction 239–240
– heterogeneously catalyzed reaction 248
– parallel 255
– rate 246
– sequential 255
– surface 42
– technology, chemical 51–79
– term 52
reactor 51–79
– design of reactors 74–79
– fluidized-bed reactor 77
– ideal reactors 56–57
– integral reactor 247–248
– microreactors 79
– nonisothermal reactors 68–74
– real reactors 63
– salt-bath-bundle reactor 78
– stirred-tank reactor 75
– tubular reactor 76
rectification 102
– reactive 134
Redlich-Kwong equation 95
reflux ratio 103–105
– minimum 104, 108
– optimum 105
– total 104
refrigeration 189
– energy 346
regular mixtures 99
reliability, technical 352–357
renewable raw materials 6, 369
Reppe chemistry 6
research 3–4
– evaluation preliminary study 283
residence-time
– behavior 60–61
– cumulative residence-time curve 61
– distribution function 61
residues 192
– long 274
– short 274
resistance thermometer 204
return on investment 358–361
– dynamic return 360–361
– static return 358–360
reverse osmosis 169
Reynolds number 85, 414
Rhone-Poulenc 3
ring pump 183–184
– liquid 184
risk (R) phrases 416–418
river water 188
rotary pump 180

S
safety (S)
– concept 319
– data 264
– phrases 418–419
– plant 220–223
– studies 317–320
salt-bath-bundle reactor 78
scalar field 379
scale-up factor 282
scale-up law 153
Schmidt number 34, 414
screening, high-throughput 19
sedimentation 172–174
seed formation 158
selectivity 16, 20, 145, 257–259
– differential 257, 259
– integral 257, 259
self-diffusion coefficient 407
sensitivity analysis 361–362
sensors 202
separation processes
– chromatographic 169

476 | Subject Index

– thermal and mechanical 80–172
sequential reaction 255
setpoint 211
sewage treatment plant 277, 347
Sherwood number 34, 414
short-cut methods 108
sieve plates 128
simulation programs 288–291
sintering 23
sleeping dog 272
slop system 201
solution / solubility 156, 264
– crystallization 159–161
solvent, minimum amount of 151
sorption 38
– adsorption (*see there*) 38, 165–166, 196
– chemisorption 165–166
– isotherm 38
space, free space 35
space-time yield 21
specialists 367
specialty chemicals 8
specific units, method of 338
specification 278
sperical coordinates 377
spinning-basket principle 248
spiral heat exchanger 87
split range 218
spring-plate manometer 206
staff costs 348
staircase
– construction 105
– method 140
star products 272
start-up 327
steam 343
– distillation 142
– network 186
steel 226
– carbon 234
– V2A 234
– V4A 234
Stefan-Boltzmann law 80
STN International 268
Stokes equation 172
storage 190
stress, mechanical 387
stripping 136–143
studies
– evaluation preliminary study 283
– P & I flow studies 321–322
– process study 283
– project study 283, 317–320
– safety studies 317–320

– study reports 331–357
subject index 467–479
submicropores 31
Sulzer BX 128
Sulzer Chemtech 162
supercritical
– conditions 69
– media 169
supersaturated solution 156
supersolubility curves 158
support (Tragermaterialien) 27
– material 29
surface
– analysis methods 422–424
– diffusion 35, 38
– reaction 42
– structure 30
– tension 401
suspension crystallization 162

T

tank
– farms 190
– stirred tank
– – batch stirred tank 57
– – ideal continuous stirred tank 56
– – reactor, stirred-tank 75
tantalum 226
taylor series 373
team work 367
– project team 367
technology / technical
– chemical reaction 51–79
– control technology 211–220
– data sheets, technical 323
– evaluation 350–352
– miniplant 297–298
– reliability, technical 352–357
– *S* curve 9
temperature
– decomposition 223
– ignition 222
– measurement 202
– profile 21
temperature-enthalpy diagram 342
tetrahydrofuran 133
thermal
– combustion 192
– conduction / conductivity equation 81, 228, 381, 389
– expansion 228
– mechanical properties and thermal stability 228–229
– thermal flow meter 209

Subject Index

- nonisothermal reactors 68–74
- processes 80–170
- radiation 80
thermodynamic
- data 393–394
- equilibrium 240–245
thermoelement 204
thermometer, resistance of a metal 204
thermoplastics 227
Thiele modulus 45
thin-layer evaporator 91
throttle devices 207
tie line 3 149
time
- compensation time 213
- delay time 213
- lifetime 22–24
- on-stream 24
- space-time yield 21
titanium 226
TOC (total organic carbon wastewater analysis) 199
toxicity 264
training 326
transducer 211
- pressure transducer 206
transformation 378
trial plants 296
triangular diagram 149, 159
triangular scheme 258
trigonometry 374
triple point 401
tube-bundle heat exchanger 87
tubular reactor 76
- ideal 57
tungsten 226
two-pressure distillation 132

U

Ubbelohde viscosimeter 175
ultrasonic flow meter 210
Underwoods plate number 108
UNIQUAC equation 99
units to SI units, conversion of 385
URAS 210

V

vacuum
- gas oil 274
- heavy vacuum distillate 274
- pump 182–183
valve plates 127
vapor permeation 196
vapor-entrainment distillation 136, 142–143

vapor-liquid equilibria 266
variable costs 4
vector
- analysis 379
- field 379
velocity 387
- flow 177
- molecular 36
ventilators 184
virial equation 95
visbreaking 274
viscosity 264, 389
- dynamic 389
- kinematic 389
- *Ubbelohde* viscosimeter 175
volume 386
VOX (volatile organic halogens wastewater analysis) 199

W

van der Waals
- constants 391
- equation 95
waste-air purification 194
washer, biowashers 196
waste-disposal 192–201
- costs 347–348
- flow diagram 334
- site 277, 348
- situation 277
wastewater
- analysis 198–199
- - biological oxygen demand (BOD_5) 198
- - chemical oxygen demand (COD) 198
- - dissolved organic carbon (DOC) 199
- - extractable organic halogens (EOX) 199
- - total organic carbon (TOC) 199
- - volatile organic halogens (VOX) 199
- purification and disposal 197–201
- treatment 197
- - biological 197
- - chemical 197
- - physical 197
water
- acetone-water mixture 126
- cooling 188–189, 345
- methanol-water mixture 97
- properties of water 399–401
- river water 188
- wastewater purification and disposal 197–201
water-gas shift reaction 244
wave equation 381
WHSV 21

Wilson equation 99
World patents index 268

Z
Zahn-Wellens test 199
Ziegler-Nichols method 215
zirconium 226